AF868814

Smart Materials

Lenore Rasmussen
Editor

Smart Materials

Considerations on Earth and in Space

Springer

Editor
Lenore Rasmussen
Ras Labs, Inc.
Boston, MA, USA

ISBN 978-3-030-70512-1 ISBN 978-3-030-70514-5 (eBook)
https://doi.org/10.1007/978-3-030-70514-5

© Springer Nature Switzerland AG 2022
This work is subject to copyright. All rights are reserved by the Publisher, whether the whole or part of the material is concerned, specifically the rights of translation, reprinting, reuse of illustrations, recitation, broadcasting, reproduction on microfilms or in any other physical way, and transmission or information storage and retrieval, electronic adaptation, computer software, or by similar or dissimilar methodology now known or hereafter developed.
The use of general descriptive names, registered names, trademarks, service marks, etc. in this publication does not imply, even in the absence of a specific statement, that such names are exempt from the relevant protective laws and regulations and therefore free for general use.
The publisher, the authors, and the editors are safe to assume that the advice and information in this book are believed to be true and accurate at the date of publication. Neither the publisher nor the authors or the editors give a warranty, expressed or implied, with respect to the material contained herein or for any errors or omissions that may have been made. The publisher remains neutral with regard to jurisdictional claims in published maps and institutional affiliations.

This Springer imprint is published by the registered company Springer Nature Switzerland AG
The registered company address is: Gewerbestrasse 11, 6330 Cham, Switzerland

To my husband, Henrik, for his love, wisdom, wit, and keen insights, and to my sons Paul, Lars, and Carl.

Preface

The inception of this book came about from the interest of the CASIS-ISS-Ras Labs space experiment on the International Space Station. Much of this work was captured and presented to a general audience at the US Department of Energy's Princeton Plasma Physics Laboratory (PPPL) at Princeton University. Since the dawn of humanity, there has been a fascination with the sky over our heads, and with our greater understanding of the cosmos. Smart materials, also known as intelligent materials, are a relatively new and rapidly evolving field. Durable smart materials that respond to stimuli by a physical change will lead to unprecedented advances for humanity, both here on Earth and as we explore the universe.

Editor's Background, excerpted in part from the Ronald E. Hatcher Science on Saturdays Series, PPPL at Princeton University, on February 3, 2018:

Moderator Andrew Zwicker, PPPL: Tell us what teachers had an influence on you.

Many teachers influenced me. My physics teacher at Pennsboro High School, Mr. Gregory Dodd, also taught Chemistry II. This was a small high school. I graduated in a class of 45 from Pennsboro High School in West Virginia. There were only three of us who wanted to take that second year of chemistry, which the high school graciously offered. Mr. Dodd made it happen. He was also the vice principal and would set the clocks in the school by his watch every morning – we called it Dodd time. Mr. Timothy McCartney taught math, and even though calculus wasn't offered at the time, he let me borrow his books. At the end of the school year, he asked for written feedback on his trigonometry course. I critiqued his course and also thanked him for being like a father to us seniors in trig class. He was kind and would listen, taught us a lot of math, and prepared us for college and the world. As a freshman at Virginia Tech, the instruction style gravitated to of learning *how* things worked. My freshman chemistry professor, Prof. George Sanzone at Virginia Tech, would have office hours and he was there. One lecture he was talking so fast about semiconductors that I couldn't read my notes, so I went and asked him if he could explain semiconductors again. He said, "But that's not going to be on the test, that was just extra." I said, "That's fine, I just can't read my notes." He responded, "You want to know how it works?" "Yes!" "But it's not going to be on the test." "That's okay." He then said, "Dear child, sit down" and he pulled books off the shelf and we went into how, really how, semiconductors work. I was a biochemistry major because of the pre-vet track, but I tacked on a chemistry double major and he was my second course advisor throughout my undergraduate career. In graduate school, I learned the discipline and rigor of working with viruses with Prof. John Johnson and through Prof. Michael Rossmann's group at Purdue University. Then I was back at Virginia Tech with Prof. James McGrath and Prof. Thomas Ward, with their encyclopedic knowledge of polymer chemistry, Prof. Garth Wilkes, with his creativity and Prof. Kathyn Uhrich's creative insights at Rutgers University [multi-university research initiative between Virginia Tech and Rutgers University to complete my PhD], who relayed their knowledge and how to think critically.

I love animals. I grew up on a farm, so all through middle school and high school, I was going to be a vet. That was part of the attraction to Virginia Tech, which has a veterinary school. By my freshman year at Virginia Tech, I fell in love with the chemistry, so was on the fence. I thought about med school and explored all three options, including a dual MD-PhD. After interviewing people, most people, even in those dual tracks, tended to pick one. I still have a lot of pets. I still take care of birds that fall out of nests. But I gravitated to science. The veterinary interest is what got me into science and science is what kept me in science. I never looked back. If I see an animal in distress, I'm still very involved with animal care, but research has been my calling.

How did we get the name Synthetic Muscle™? We were in the 2013 business accelerator MassChallenge in Boston. We were calling our material electroactive polymers, specifically contractile electroactive polymers (EAPs). Meanwhile, the MassChallenge folks (thank you!) were calling me the Synthetic Muscle Lady. They suggested that we call our EAPs Synthetic Muscle. My argument was that muscle only contracts, while our EAPs contract and expand. They then strongly suggested we call it Synthetic Muscle, so we did. People remember Synthetic Muscle™!

Boston, MA, USA Lenore Rasmussen

Acknowledgments

I would like to thank and acknowledge Springer and their affiliates for all their help in bringing this book to fruition: Brinda Megasyamaian, Deepika Sam, Silembarasan Panneerselvam, Michael McCabe, Zoe Kennedy, Brian Halm, and many others. Many thanks to the robotics engineering chairs at Worcester Polytechnic Institute, past and present: Prof. Michel Gennert and Prof. Xiao Jing, and to so many others at WPI in the Robotics Engineering Department and across multiple departments. I would like to thank the program managers at the National Science Foundation for all their help and guidance. Many thanks to Tom Ryden, Joyce Sidopoulos, Fady Saad and others at MassRobotics. A big thank you to the many folks at the CASIS and NASA who helped get the Synthetic Muscle™ Experiment into space and back to Earth again. I would like to thank the US Army's Picatinny Arsenal and Natick Labs. And many thanks to the US Department of Energy, especially the folks at the Princeton Plasma Physics Laboratory, and to Princeton University. Finally, I would like to acknowledge the many investigators around the world who are diligently working in the fields of smart materials. We all are making science fiction a reality and we are all advancing humanity, here on Earth and in the cosmos.

Contents

1 **Synthetic Muscle™ for Deep Space Travel and Other Applications on Earth and in Space** 1
Lenore Rasmussen, Peter N. Vicars, Calum R. Briggs, Tianyu Cheng, Margot Meredith, Leila N. Albers, Simone Rodriguez, M. Damaris Smith, Matthew Bowers, Edward A. Clancy, Charles Gentile, Lewis Meixler, George Ascione, Nicole Allen, Robert Hitchner, James Taylor, Laurie Bagley, Daniel Hoffman, Ramona Gaza, Leon Moy, Patrick Mark, Dan Prillaman, Robert Nodarse, Michael Menegus, Jo Ann Ross-Ratto, Christopher Thellen, Danielle Froio, Matthew Maltese, Thomas Seacrist, Cosme Furlong, Payam Razavi, Greig Martino, Alex Zhong, Shannon Carey, Ben Secino, Logan Valenza, Catherine Poirier, Charles Sinkler, Dylan Corl, Surbhi Hablani, Tyler Fuerst, Sergio Gallucci, Whitney Blocher, and Stephanie Liffland

2 **Q&A with NASA Astronaut Yvonne Cagle, MD Moon and Mars Missions RBE 533 "Smart Materials, Actuation, and Biomimicry," Worcester Polytechnic Institute on October 8, 2019** 49
Yvonne Cagle

3 **Shape Control of Large Lightweight Mirrors with Dielectric Elastomer Actuation** 61
R. D. Kornbluh, D. S. Flamm, H. Prahlad, K. M. Nashold, S. Chhokar, R. Pelrine, D. L. Huestis, J. Simons, T. Cooper, and D. G. Watters

4 **The Hydro Muscle and CRFC Valve: An Efficient and Compact Fluidic Robotic System** 83
Marko Popovic, Shannon Moffat, Julia D'Agostino, Ellie Clarrissimeaux, Eric Carkin, and Matthew Bowers

5 Contractile Pneumatic Artificial Muscle Generates Extension by Actuating an Integrated Pushrod 101
Norman M. Wereley, Erica G. Hocking, Benjamin K. S. Woods, and Curt S. Kothera

6 Soft Pneumatic Actuators: Modeling, Control, and Application 129
Erik Howard Skorina and Cagdas D. Onal

7 Active Thermal Control of Satellites with Electroactive Materials 221
Frédéric Vidal, Gwendoline Petroffe, Layla Beouch, Sophie Cantin, Claude Chevrot, Pierre-Henri Aubert, and Jean-Paul Dudon

Index 255

Chapter 1
Synthetic Muscle™ for Deep Space Travel and Other Applications on Earth and in Space

Lenore Rasmussen, Peter N. Vicars, Calum R. Briggs, Tianyu Cheng, Margot Meredith, Leila N. Albers, Simone Rodriguez, M. Damaris Smith, Matthew Bowers, Edward A. Clancy, Charles Gentile, Lewis Meixler, George Ascione, Nicole Allen, Robert Hitchner, James Taylor, Laurie Bagley, Daniel Hoffman, Ramona Gaza, Leon Moy, Patrick Mark, Dan Prillaman, Robert Nodarse, Michael Menegus, Jo Ann Ross-Ratto, Christopher Thellen, Danielle Froio, Matthew Maltese, Thomas Seacrist, Cosme Furlong, Payam Razavi, Greig Martino, Alex Zhong, Shannon Carey, Ben Secino, Logan Valenza, Catherine Poirier, Charles Sinkler, Dylan Corl, Surbhi Hablani, Tyler Fuerst, Sergio Gallucci, Whitney Blocher, and Stephanie Liffland

---

L. Rasmussen (✉) · P. N. Vicars · C. R. Briggs · T. Cheng · M. Meredith · M. D. Smith
Ras Labs, Inc., Boston, MA, USA
e-mail: rasmussl@raslabs.com

L. N. Albers
Ras Labs Technical Advisory Board, Quincy, MA, USA

Vertex Pharmaceuticals, Boston, MA, USA

S. Rodriguez
Ras Labs, Inc., Boston, MA, USA

Cognizant Technology Solutions Corporation, Dallas, TX, USA

M. Bowers
Ras Labs, Inc., Boston, MA, USA

Robotics Engineering, Worcester Polytechnic Institute, Worcester, MA, USA

E. A. Clancy
Electrical Engineering and Computer Science, Worcester Polytechnic Institute, Worcester, MA, USA

C. Gentile · L. Meixler · G. Ascione · N. Allen · R. Hitchner · J. Taylor · L. Bagley
US Department of Energy's Princeton Plasma Physics Laboratory, Princeton University, Princeton, NJ, USA
e-mail: lbagley@pppl.gov

D. Hoffman
Gas and Fluid Dynamics Laboratory, Department of Mechanical and Aerospace Engineering, Princeton University, Princeton, NJ, USA

© Springer Nature Switzerland AG 2022
L. Rasmussen (ed.), *Smart Materials*,
https://doi.org/10.1007/978-3-030-70514-5_1

R. Gaza
Leidos, Exploration & Mission Support, NASA Johnson Space Center, Houston, TX, USA
e-mail: JSC-public-inquiries@mail.nasa.gov; http://www.nasa.gov/centers/johnson/home/index.html

L. Moy · P. Mark · D. Prillaman · R. Nodarse · M. Menegus
Organic Materials Branch, US Army Combat Capabilities Development Command – Armament Center (CCDC-AC), Picatinny Arsenal, NJ, USA
e-mail: usarmy.pica.acc.mbx.acc-nj-sbo@mail.mil; usarmy.pica.acc.mbx.acc-nj-sbo@mail.mil; usarmy.pica.acc.mbx.acc-nj-sbo@mail.mil; usarmy.pica.acc.mbx.acc-nj-sbo@mail.mil; usarmy.pica.acc.mbx.acc-nj-sbo@mail.mil; https://home.army.mil/picatinny/; https://home.army.mil/picatinny/index.php/doing-business-us

J. A. Ross-Ratto · C. Thellen · D. Froio
US Army Natick Soldier Research, Development, and Engineering Center, Natick Labs, Natick, MA, USA
e-mail: usarmy.natick.ccdc-sc.mbx.publicinquiry@mail.mil; usarmy.natick.ccdc-sc.mbx.publicinquiry@mail.mil; usarmy.natick.ccdc-sc.mbx.publicinquiry@mail.mil; https://www.army.mil/natick

M. Maltese · T. Seacrist
Center for Injury Research and Prevention, The Children's Hospital of Philadelphia, Philadelphia, PA, USA

C. Furlong · P. Razavi
Department of Mechanical Engineering, Worcester Polytechnic Institute, Worcester, MA, USA

G. Martino
United Prosthetic, Inc., Dorchester, MA, USA

A. Zhong
Purdue University, West Lafayette, IN, USA

S. Carey · B. Secino
Robotics Engineering, Worcester Polytechnic Institute, Worcester, MA, USA

L. Valenza
Florida Institute of Technology, Melbourne, FL, USA

C. Poirier
Trinitity College, Hartford, CT, USA

C. Sinkler
Electrical Engineering and Computer Science, Worcester Polytechnic Institute, Worcester, MA, USA

Bose Corporation, Framingham, MA, USA

D. Corl
RVCC, Branchburg, NJ, USA

S. Hablani
Skidmore College, Saratoga Springs, NY, USA

T. Fuerst · S. Gallucci · W. Blocher
Clarkson University, Potsdam, NY, USA

S. Liffland
University of North Carolina at Chapel Hill, Chapel Hill, NC, USA

Abstract For deep space travel, new materials are being explored to assist humans in dangerous environments, such as high radiation, extreme temperature, and extreme pressure. Synthetic Muscle™ is a class of electroactive polymer (EAP)-based materials and actuators that shape-morph at low voltage (1.5 V to 50 V), sense pressure (gentle touch to high impact), and attenuate force. These EAPs can survive and work in environments where humans cannot safely enter due to extreme environments or due to contagions that have no cure. From the Ras Labs-CASIS-ISS Experiment, the flown Synthetic Muscle™ samples compared well to the ground control samples, even after over a year on the International Space Station. Replicating human grasp has implications in robotics and prosthetics. EAP linkages can be actuated and EAP pressure sensors placed at the fingertip regions of robotic grippers for tactile feedback. With autonomy, artificial intelligence, machine learning, and EAP and other smart material technologies all coming together, there is an incredible fusion of mechanical and biological concepts to make truly innovative biomimetic motion. Smart materials will allow humanity to advance and survive on Earth and in space: on the ISS National Laboratory, the planned Moon base, the anticipated Mars settlements, and beyond.

Keywords Electroactive polymer · EAP · Smart material · Intelligent material · Sensor · Gripper · End effector · Robot · Collaborative robot · Humanoid

1.1 Excerpts from the Princeton Plasma Physics Lab Ronald E. Hatcher Science on Saturdays Series, Princeton University: https://www.pppl.gov/events/science-saturday-synthetic-muscle-deep-space-travel Plus Additional Material

1.1.1 Ras Labs-CASIS-ISS NL Synthetic Muscle™ Experiment: Resistance to Radiation UA-2014-119

The space work was a team effort between Ras Labs; the US Department of Energy's Princeton Plasma Physics Laboratory at Princeton University (PPPL) for preliminary radiation testing; the US Army CCDC-AC, which helped with the G-force testing to make sure our materials could withstand up to 10 Gs anticipated for the launch to get off the planet; the Center for the Advancement of Science in Space (CASIS); and the National Aeronautics and Space Administration (NASA), with their expertise, weekly meetings, and coordination for the Synthetic Muscle™ Experiment. A lot of interns from different universities and a variety of disciplines worked on this project. One team helped get the experiment ready to go up to the International Space Station (ISS), and another team helped with the analysis back on Earth. With our industry partners, we often must be very quiet about much of the work we're doing. With the Ras Labs-CASIS-ISS Synthetic Muscle™ Experiment, we were expected and encouraged to publish this work and were able to put a lot of students on the space project. This endeavor made all of us think universally.

What is Synthetic Muscle™? What are electroactive polymers (EAPs)? Synthetic Muscle™ is a contractile electroactive polymer (EAP). Electroactive polymers are smart materials. Smart materials, or intelligent materials, respond to stimuli. It could be electricity for EAPs. It could be pressure, such as for expandable stents, or it could be heat, such as for thermo-responsive materials (Fig. 1.1). It could be light, like for those who wear transition spectacle lens, which is a smart material where the light causes a change in the lens so that the eyes remain comfortable in bright light. Electroactive polymers respond to electricity. This is seen in ionic shape-morphing EAPs; in dielectric elastomeric actuators (DEAs), such as the actuators in the Merbot [1, 2]; in ionic polymer metal composites (IPMCs), such as the large biomimetic Venus flytrap [3]; and in piezoelectricity, such as in the small RoboBee [4, 5] and other robotic insects (Fig. 1.1). There are a multitude of other EAP actuations, such as carbon nanotube (CNT) yarns with nylon [6] and elastomeric alcohol-based actuators [7]. Ras Labs' EAPs respond to electricity by contraction and, when the electricity polarity is reversed (the positive charged internal electrode switched to a negative charge), expansion, which can be cycled back and forth. This movement can be captured as actuation.

Humanity is in the fourth revolution of the industrial era – the robotics era. Everything in the room around you, in your car, when you go shopping, has been touched by a robot and by automation at some point. When we think about motion in the industrial era, it is very metal based. It is very hard, very durable, and very strong. When we think about mechanical motion, we think about how a train moves, or when we open up an analog watch, the inner workings and how they mesh. Or when we think about motion, we think about how a bird flies, how a fish swims through the water, or how we walk, run, and manipulate tools, and never the twain shall meet. Even in Ben Franklin's time, Luigi Galvani took a frog leg, zapped it with electricity, and could get that leg to twitch and move (Fig. 1.2), but it would decompose in a few days [9]. Synthetic Muscle™, electroactive polymers, are man-made materials. They can withstand and work for long periods of time in environments where living organisms cannot survive. Now the twain can meet. This is leading to the combination of an incredible infusion of mechanical and biological concepts to make truly innovative biomimetic motion. There is a whole field

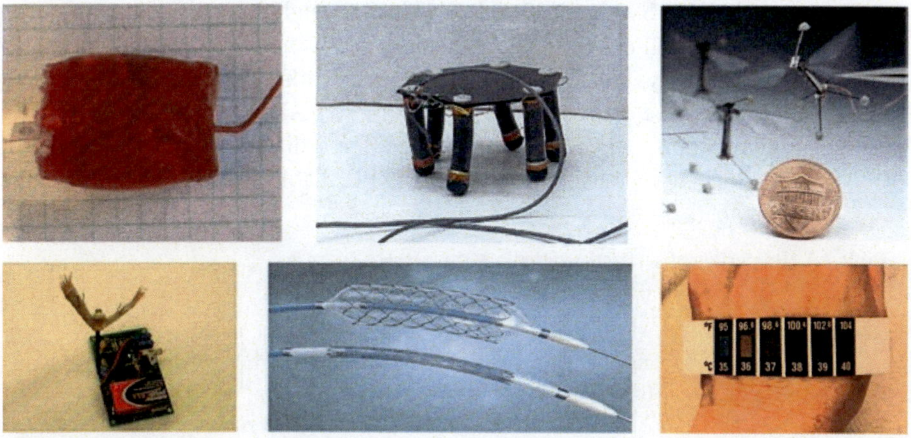

Fig. 1.1 L to R: Large 9-g EAP actuator. (Photo: Ras Labs), Merbot [2], swarm of robotic insects [8], Venus flytrap using IPMC [3], expandable heart stents (Shutterstock), and a thermo-responsive thermometer. (Photo: Ras Labs)

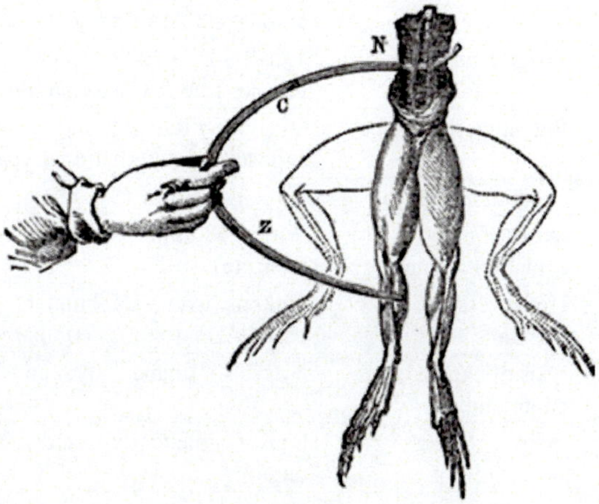

Fig. 1.2 Luigi Galvani's classic electric experiment with a frog leg [9]

blossoming out of this, including in pneumatics, known as soft robotics. Synthetic Muscle™ operates between 0 and 50 volts and can use off-the-shelf batteries. These EAPs also act as variable resistors, so can be used as sensors, from gentle touch to high impact. One thing that is missing in current prosthetic hand and robotic gripper design is a streamlined and intuitive way to get tactile feedback, what we know as touch. This can be tied in with actuation, i.e., integrated with life-like shape-morphing motion.

Our hands and how we use them three-dimensionally are a big part of what defines us as a species and are part of our beauty and human expression. When looking at prosthetics, I was disappointed in what was available. With upper limb prosthetics, you have the mannequin like hands that look great, but are not functional. Or you have hooks, which are still used today, and now the more sophisticated robot-like hands, and not much in between. The goal was to make a material that would work in a certain way to bridge that gap between form and function.

Today we do have Synthetic Muscle™. This is a highly reactive actuation using an EAP-based material at low voltage [10–13]. This is perhaps not as important for the prosthetic industry, but for the robotics industry, these materials are also very robust (Table 1.1). The class of materials that these EAPs are composed of tend to be radiation resistant. The EAPs were tested for radiation resistance at the US Department of Energy's Plasma Physics Lab [PPPL] at Princeton University and in

Table 1.1 Properties of selected Synthetic Muscle™ EAPs.

Property	Value
Strength	Abe to withstand~2000 N, 1200 impacts at 908 N (tested in collaboration with children's Hospital of Philadelphia) and over 10 G (tested by US Army)
	Abe to withstand 3,000,000 compression cycles (5–30 psi, ambulation mimicry, tested by O&P B2B partner)
Durometer	Shore O: 5 to 34
Power requirement	Linear range between 0 and 50 V
Force output	0.01 mN (with less than 2 g actuator, US Army project)
Workoutput ratio	Move objects 1:18 actuator weight: total load weight,~2× human work output (US Army project)
Temperature tolerance	4 to 408 K (−269 to 135°C, tested at US DOE's Princeton Plasma Physics Lab & Cava Lab, Princeton University)
Radiation resistance	Over 305 kRad gamma radiation (tested at PPPL); broad spectrum exposure on International Space Station
Vacuum tolerance	1.5 × 10–^{5} Torr
Onset of electro-actuation at nano-level	Within 48 milliseconds (tested at CHSLT Lab, Worcester Polytechnic Institute)
High pressure tolerance	2200 ps (over ~1500 meters of ocean depth, tested by New England based AUV robotics company)
Axis of expansion & contraction	3-axes simultaneously; can be controlled to desired direction
Fatigue testing	Sensors able to withstand 1,000,000+ cycles (1-kg load at ~1 Hz) ongoing

space on the International Space Station [ISS] [14, 15]. These EAPs can also withstand high and low pressures and high vacuum. With exposure to high vacuum, however, it will dry out over time. Like real muscle, Synthetic Muscle™ operates best moist and can be easily encapsulated or coated. These EAPs can withstand extreme temperatures, up to 135 °C, where they do not decompose, but undergo a phase change. Selected Synthetic Muscle™ EAPs were tested at various cold temperatures, including even down to between 2 and 4 K (−271 to −269 °C) at the Cava [Solid State Chemistry] Lab at Princeton University. The EAP took a while to thaw and then worked fine [10, 12, 16]. That is partly how we got into the space program, because we had already done extreme temperature testing. These EAP materials also attenuate force. What this means is that these materials could potentially be used, whether in space or on Earth, as protective equipment. We weren't sure why our EAPs were attenuating force. These materials are rubbery. Most rubbery things don't attenuate force very well [often amplify force], but these EAPs did attenuate force. This work was done in collaboration with Children's Hospital of Philadelphia (CHOP) and Rowan University [17]. High-speed photography was used to analyze impacts using a 6-foot drop tower. These materials are robust hydrogels. They are not a pure liquid, they are not a pure solid, but they are non-Newtonian materials with aspects of both liquids and solids. Force attenuation of impacts was observed from the load cells (Fig. 1.3) [17]. When the high-speed photography was observed, a rippling effect in the hydrogels occurred upon impact. Much like when you toss a

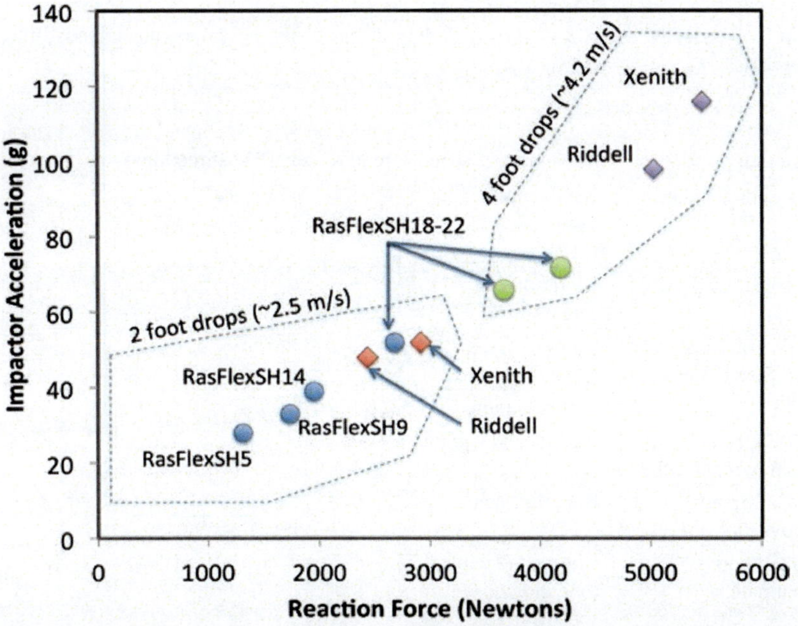

Fig. 1.3 Ras Labs' EAPs (RasFlex™ samples) vs. encapsulated foam inserts in Xenith and Riddell top of the market football helmets; high impact. RasFlex™ samples also survived the full 6-foot drop tower test [17]

stone into a pond and see ripples extending outward in the water, that is what we saw in these EAP hydrogels. The force attenuation in these elastomeric EAPs may be due to their non-Newtonian liquid-like behavior dispersing that impact force laterally through the semisolid matrix.

There are all sorts of electroactive smart materials, smart materials, and actuating-type systems (Table 1.2, short list). Industrial piezoelectric materials have been around since World War II and are still used today in radar, microbalances, and timing mechanisms. Electric energy can be inputted to get mechanical vibration motion out, and piezoelectric materials can be mechanically compressed to generate electricity out. They are a two-way electroactive material. Piezoelectric materials have been used to make some really clever devices such as the robotic flybot's insect-like motion [18]. Dielectric elastomeric [actuators] (DEAs) [1, 2, 19–22] are a little more developed than the ion-containing contractile EAPs. They have fast actuation and can be very strong. DEAs do require high energy requirements, operating in the kilovolt range. Ionic polymer metal composites (IPMCs) [23, 24] operate at low voltage, but tend to be limited by size. IPMCs bend back and forth with electric input and reversed electric polarity. Shape memory alloys (SMAs), like nitinol [alloys of titanium and nickel], are heat activated. The actuation is strong and fast. SMAs can also be considered electroactive because they have high electric resistance. They can be activated by electricity using even a 9-volt battery, where the heat from the electric resistance creates heat, which then actuates the SMA. Pneumatics were added to this table because they are very much in the news

Table 1.2 Selected smart materials including electroactive polymers, electroactive metals, and pneumatics

Materials	Movement & speed	Low energy requirements	Contract & expand (all directions)	Durability/ toughness
Ras Labs Synthetic Muscle™	Bend, contract & expand speed variable	✓	✓	✓
Piezoelectric EAPs	Vibration fast actuation	✓		✓
Dielectric Elastomes (SRI, StetchSense, Parker)	Contract (z-axis) & expand (xy-axes) Fast actuation			✓
IMPCs (Prof. Shahinpoor & Others)	Bend speed variable	✓		✓
Shape Memory Alloys (Nitinol™)	Bend fast actuation			✓
Soft Robotics Pneumatics	Contract & expand (one direction) speed variable			✓
Early EAPs (Profs. Tanaka, Osada, Gong, Shiga)	Bend, contract & expand slow actuation	✓	✓	

with soft robotics. McKibben artificial muscles [the precursor to pneumatic artificial muscles (PAMs)] [25, 26] were first developed as an actuated system in the 1950s by Joseph McKibben to assist polio patients by using air pressure in a rubber bladder sheathed with a strong braided mesh [27, 28]. Pneumatic actuators are very strong and tend to contract and expand in one direction, but the actuation direction can be directed and can be multidimensional. Pneumatic pressure is used extensively in industry. Soft robotic grippers with pneumatic actuation, typically silicone elastomers, can be used for gentle grip [29], though the energy requirements to provide pneumatic pressure, pressurizing air, are somewhat cumbersome and noisy. Early electroactive polymers, from the work of Prof. Yoshihito Osada [30–47], Tohru Shiga [48, 49], and Prof. Toyoichi Tanaka [50–59] [and others] [60], tended to be with materials in the polyacrylamide family, which were very electroactive, but not very strong nor load bearing. Ras Labs' EAPs were synthetically tailored to make them strong and durable enough for real-world applications. They were developed to be robust and were found to be able to withstand extreme environments. Synthetic Muscle™ can be contracted and expanded by simply reversing the electric polarity (switch the electric input of the electrode embedded in the EAP from positive to negative). The electric polarity can be cycled to create contraction-expansion cycles in these EAPs. One of the biggest advantages of these EAPs is that they operate at low voltages, typically less than 50 V. Off-the-shelf batteries can be used to operate these EAPs.

With Synthetic Muscle™-based actuator systems, both modes, contraction and expansion, can be used simultaneously. For a project involving rotational actuation, what really helped with the speed was to use both modes simultaneously to get the object to roll one way and then the other [by switching the electric polarity], and this also provided a sink for the other electrode. The work output using both modes of EAP actuation simultaneously was roughly twice that of human work output. Human work output is highly variable, but is about 1:10, which is the weight of our bodies to the weight of our bodies plus what we can push or move [61]. Human muscle only contracts. Human muscles are built with antagonistic pairing, partly for stability, but also so that when we move our limbs, we can then move them back. With the rotational actuation project using both modes simultaneously, contraction and expansion in adjacent zones, work outputs of 1:18 were observed. This was using a 2-gram EAP actuator system to rotate a 33.5-gram cylinder, which is a ratio of 1:18 for the EAP actuator system to the total weight of the device that was being rotated [62]. Upon reflection, this made sense. By using both contraction and expansion simultaneously, rather than just contraction (like biological muscles), we were able to almost double the work output. How would we be designed, or how would a robotic arm or hand be designed, if the "muscles" in it could do both modes – if the muscles (or actuators) could both contract and expand – arranged around a joint? This opens up conceptual designs well beyond biomimicry.

Biomimicry is especially important in prosthetics to replicate and restore form and function for the missing limb(s), but why is biomimicry [and beyond] important in robotics? Robots and automation are of course used to do repetitious tasks, but there is also a strong need for robots, in space or on Earth, where you cannot send a human and wouldn't even want to send in a dog. We do have the technology to send in robots rather than people or even animals for these extremely hazardous situations. Often when we get into these situations, what is needed is something that we consider fairly simple – turning knobs and pushing buttons – to mitigate that situation. Actually, turning knobs for robots is very difficult using traditional gripper technology. For us, it is very simple to turn a knob because of our anatomy, not only of the thumb but also of the wrist. Our strength is in that diagonal direction, that oblique movement – the way you swing a hammer and the way you hold a pen – it is all in the diagonal. It is exciting where we can go, in prosthetics and robotics, with human-like grasp. In industry, robots are often partitioned away from humans for safety reasons. Collaborative robots are also known as cobots: There is a push for safe robot-human interactions in industrial settings, for robots to work safely collaboratively and intuitively side by side with humans for pick-and-sort and other applications. Biomimetic design also applies toward humanoids, also known as hubots, for collaborative work and performance in hazardous environments [and in environments with contagions that have no cure].

Synthetic Muscle™-based actuation is a system. It is the EAP(s), the wiring, the coating(s), how they attach, and how they are put in into place; all of this comes into play. Eq. 1 describes the correlation between electroactivity and other attributes of the system. The electroactivity of the EAP is directly proportional to the voltage. If slower or less pronounced movement is needed, the voltage can be lowered. If faster or more pronounced movement is needed, the voltage can be raised [up to a point, typically voltage is toggled between 1.5 and 50 V]. The longer the time period and the longer the EAP is exposed to a voltage level, the more size change will be observed [in both contraction and expansion and up to a point. The EAP's size will plateau given enough time]. In Synthetic Muscle™, the ionic content or ionic concentration within the EAP itself is directly proportional to its electroactivity. These EAPs are tough hydrogels, and the electrolytic solution also comes into play and enhances electroactivity [up to a point, too much electrolyte can cause an undesirable phase change in these materials]. The electroactivity is inversely proportional to the volume and hardness of the EAP. The harder or firmer the EAP, the less flexible it is, and the less it can shape-morph (contract or expand). The softer the EAP, the more flexible it is, the more it can shape-morph. The volume or initial size of the EAP is a proportional consideration because the shape-morphing is an osmotic effect. The electroactivity of the EAP is inversely proportional to its volume. The percent size change is greater and faster, in terms of relative proportionality, the smaller the initial size of the EAP. Large EAPs shape-morph, proportionally, much slower than small EAPs. Rather than making a large bulk Synthetic Muscle™-based actuator, it worked better to make three smaller EAP actuators, for example, three EAP cylinders, and have them work together [10].

$$\mathrm{Ea}\,\alpha\,(\mathrm{V})(\mathrm{T})(\mathrm{IC})[\mathrm{X}]/(\mathrm{Vol})(\mathrm{H}) \tag{1.1}$$

where Ea is electroactivity, V voltage, T time, IC ionic content of EAP, [X] molar electrolyte concentration, Vol EAP volume (time dependence), and H EAP hardness.

The contraction that these EAPs undergo when electrically stimulated is a threefold effect:

1. The strong negative charged cathode, which is placed external to the EAP, attracts the positively charged electrolyte cations out of the EAP.
2. The localized pH effect at the positive charged anode, which is embedded in the EAP, protonates the weak acid groups along the polymer chains of the EAP, making the EAP neutrally charged, so that the positively charged electrolyte cations in the electrolyte solution are no longer attracted to the side groups of the EAP.
3. The osmotic effects following the ion gradient from the rapid cation migration from the electric input cause a large water loss out of the flexible EAP.

All three of these effects cause a rapid and pronounced contraction in the EAPs. The first two events occur in tandem, followed quickly by the third effect of osmosis, causing a rapid and pronounced contraction in these EAPs [10].

Reversing the electric polarity causes expansion by a similar threefold effect:

1. The strong negative charged cathode, which is now the internal embedded electrode in the EAP, attracts the positively charged electrolyte cations back into the EAP.
2. The localized pH effect at the negatively charged cathode, which is now embedded in the EAP, deprotonates the weak acid groups along the polymer chains of the EAP, making the EAP negatively charged again, so the positively charged electrolyte cations are attracted back to the side groups of the EAP.
3. The osmotic effects following the ion gradient from the rapid cation migration from the (reversed) electric input cause a large water gain back into the flexible EAP.

All three of these effects cause a rapid and pronounced expansion in the EAPs. By flipping the electric polarity, a huge salt gradient is created (step 3), this time inside the EAP, thus a large influx of water back into the EAP osmotically, resulting in expansion. Again, the first two events occur in tandem, followed quickly by the third effect of osmosis, causing a rapid and pronounced expansion these EAPs [10].

If we had tried to get our materials on the International Space Station as its own goal, that would have been very difficult because the odds are so low. There have been about 3,000 experiments flown on the ISS. This collaborative work was one of them. So how did we get there? How did we get the attention of CASIS? Ras Labs was in the business accelerator MassChallenge in 2013. My primary role at Ras Labs is CTO. I'm a tech head, learning business as I go, so treated MassChallenge as a crash MBA course. There were several groups that we presented to [pitches] to try to get funding, and the only group that had technical questions, which meant that I could fully answer their questions, was with the CASIS group. There were two astronauts on the panel asking about my technology. They were impressed with our previous extreme temperature research that demonstrated survivability of our EAPs down to 4 K, close to absolute zero. We discussed that if our EAPs were indeed proven to be radiation resistant, then these materials and actuators could be used for deep space exploration – send it out there and wherever it lands, it can do its thing. More importantly, our nearer-term interest is for that safe human-robot interaction: whether it is somewhere you cannot send an astronaut, or it is an astronaut in distress and is an emergency retrieval mission, to be able to do the job without causing more harm while mitigating the situation.

The ISS orbits the Earth at an approximate altitude of 407 km. At this altitude, the radiation profile differs significantly from that on the Earth's surface, and there is a significant increase in the presence and abundance of ionizing radiation. Space radiation is primarily comprised of atoms in which electrons have seen stripped away through interstellar acceleration. All three of the main kinds of space radiation, particles trapped in the Earth's magnetic field, particles ejected into space during solar flares, and galactic cosmic rays, which are high-energy and heavy ions from outside the solar system, represent forms of dangerous ionizing radiation. Ionizing radiation has the ability to knock electrons out of orbit within an atom upon contact, thus causing the formation of ions and radicals. As it passes through matter, including synthetic material and living tissue, ionizing radiation loses energy, which is absorbed by the conducting material. This poses a danger to living organisms, as the ionization of water and other cell components can damage DNA molecules in the path of the radiation, causing breaks and mutations in DNA strands, including large clusters of breaks, which are not easily repaired by cells. These mutations can cause varying forms of cancer and pose a serious health risk to those exposed. Other exposure risks include both acute and late nervous system complications, degenerative effects, circulatory and digestive diseases, and, in severe cases, death (Table 1.3) [63].

For the EAPs themselves, two prior runs of radiation testing on the EAP materials were done at PPPL even before setting up the matrix experiments. Testing of EAP matrices was then done at PPPL before the ISS experiment. After the initial radiation ground experiments, a matrix was designed of EAPs with different additives and coatings. Free radical inhibitors were of interest. Various coatings were also investigated and combinations of additives and coatings. The interest in free radical inhibitors was because of what happens in space. With biologics exposed to radiation, you get the long-term effects with the DNA, mutations, and problems such as cancer down the road, but you also get near-term effects with high radiation with the formation of free radicals and ions within the cell, and that causes a lot of immediate damage. Because our EAPs operate best moist, we were concerned about the same thing, that radicals could form and degrade our materials. The additives added were free radical inhibitors for this experiment. Various coatings were also explored. Gamma radiation testing was done on Earth at PPPL. The materials were characterized [electroactivity, pH, peroxide level, durometer, and material

Table 1.3 Organ specific exposure limits for astronauts [63]

Exposure interval	Blood forming organs	Eye	Skin
30 Days	25 rem	100 rem	150 rem
Annual	50 rem	200 rem	300 rem
Career	150–400 rem [200 + 7.5(age – 30) for men] 100–300 rem [200 + 7.5(age – 38) for women]	400 rem	600 rem

integrity]. An experiment was then designed and put in space. The matrix for the space experiment was derived from the on-Earth experimentation. When the space experiment was back on Earth, the EAPs were tested for electroactivity, pH, durometer, and material integrity to see if there was any changes or breakdown.

Starting with ground experimentation for testing radiation resistance of Synthetic Muscle™, at PPPL, we used their Cs-137 source. Cs-137 is a strong gamma emitter with a beta profile as part of the breakdown (Table 1.4) [64]. First, samples from our Generation 3 EAP materials and two of our Generation 4 EAP formulations were investigated (Fig. 1.4). For the space experiment, the space exposure was only planned to be for 90 days. The ISS is low orbital. The folks at PPPL suggested that if we were going to really test our EAPs for radiation resistance, why not do the equivalent of a trip to Mars and back? This was the first radiation experiment performed, and we were not able to determine much difference between the control samples (no radiation) and irradiated samples. So next we exposed the Generation 3 and the two Generation 4 EAP samples using the Cs-137 pellet at PPPL to the equivalent of a trip to Jupiter and beyond. For this experiment, we nested the samples into small Pyrex® petri dishes, where the glass acted as an indicator. Glass is interesting and is a good visual indicator for radiation. Glass when first made, with no pigments, is clear. Over time, glass turns blue or green over decades due to the ionizing effects from the cosmos and from our own star, the sun, particularly if the glass is outside, like the glass insulating caps in old electric lines. Over centuries, glass will turn brown and eventually even black. In the second exposure, with the experiment now nested in the small petri dishes, the glass turned from clear to brown in that radiation exposure, i.e., the equivalent of centuries worth of ionizing radiation over a long weekend (Fig. 1.4). One of the Generation 4 formulations, Gen 4B, did not survive that second exposure. Instead of a clear homogenous polymer, it looked like porridge [severe material degradation readily observable]. The Gen 3 and Gen 4A EAPs were selected for subsequent ground and space experimentation.

For the ground flight readiness matrix experimentation, there was only a 2-inch by 2-inch square area for uniform exposure from the Cs-137 pellet. Tiny EAP samples, 28 samples, had to fit within this small space. The matrices for this ground experimentation are shown in Fig. 1.5. A duplicate, with 28 samples, was the control, with no radiation exposure for comparison. The four number 1 positions were the control positions, with no additives and no coatings. Positions 2 through 4 had

Table 1.4 Radiation profile for Cs-137 [64]

Physical half life	Radiation type	Energy		Specific activity	Tenth value thickness [avg. gamma]	Max beta range
		Maximum	Average			
30.17 years	Beta	0.512 MeV	0.157 MeV	87 Ci/g	Concrete=13 cm	Air = 1300 cm
		94.6% abundance				
		1.173 MeV	0.415 MeV			Plastic = 0.53 cm
		5.4% abundance			Lead = 1.7 cm	
	Gama	0.662 MeV				Aluminum = 0.25 cm
		85% abundance				

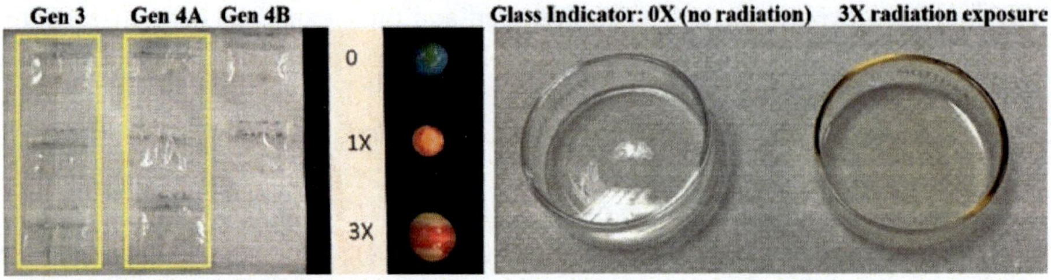

Fig. 1.4 Preliminary radiation experiments using Cs-137. EAP formulations in *yellow boxes* chosen for further ground and space experimentation. (Photo credits: Ras Labs)
Note: Preliminary radiation experiments on Gen 3 and Gen 4 EAPs, First exposure of Gen 3 and Gen 4 Synthetic Muscle™ (1X) May 20 – 21, 2014, Total run-time of 21.7 hours, Cs-137 source, 116.229 kRad total dose. This is more than 20X LD-human100 (lethal human dose) and much more gamma radiation exposure than would be encountered on a 90-day exposure on the ISS NL. This is more equivalent to a trip to Mars and back in terms of gamma radiation exposure. The EAP samples, both third and fourth generation, held up well. Second exposure of Gen 3 and Gen 4 Synthetic Muscle™ (3X) June 2-4, 2014, Total run-time of 45.33 hours, Cs-127, 326.455 kRad total dose. This is on top of the previous exposure, so the level of exposure is beyond the gamma radiation that would be encountered on the ISS NL, and more equivalent to a trip to Jupiter. The third and fourth generation EAP samples (Gen 3 and Gen 4A) using a certain synthetic strategy held up well, in terms of material integrity, pH, peroxide formation (0 ppm) and electroactivity, while another synthetic strategy in one of the fourth generation EAP samples (Gen 4B) did not maintain material integrity from the second round of radiation exposure and had approximately 7 ppm peroxide (indicative of free radicals, all other samples had 0 ppm peroxide), so for flight readiness, the Gen 4B synthetic strategy was ruled out and not used

the free radical inhibitor additives. Positions 5 through 7 had different coatings. All of the EAP samples, experimental and control, had fairly similar results in terms of durometer, durability, electroactivity, and pH and were also tested for free radicals [minimal] (Fig. 1.6). What we did notice, which was more apparent in the space experiment, was there a slight yellowing in the experimental samples that had been exposed to radiation. Yellowing can be indicative of accelerated aging in polymers [in plastics]. Electroactivity was analyzed for all the samples (Fig. 1.6), and then we picked the winners to go into the space experiment (Fig. 1.7).

It is very difficult to get of the planet. The amount of shaking and G-forces that the space experiment would encounter during liftoff had to be taken into account. G-force testing to over 10 Gs was performed on the EAPs themselves at US Army's CCDC-AC's Labs, with excellent results, in preparation for space flight to the ISS National Laboratory [15]. For the space experiment, four aluminum cages were milled out and then oxygen plasma-treated titanium backing put into place [below the EAP samples]. Aluminum screening was positioned above the EAP samples, which were bolted into place for each cage, to serve as windows throughout the duration of the space experiment. This was so that we would be able to observe what was happening with the Synthetic Muscle™ samples by NASA downlinks back to Earth [from the ISS]. In addition to the specialized coatings as part of the space experiment, the cages had clear plastic coatings below and above the aluminum screening windows and a final encapsulation for each cage in Teflon® perfluoroalkoxy alkane (PFA) bags heat sealed and taped with Kapton® tape [as a fire prevention measure] (Fig. 1.8).

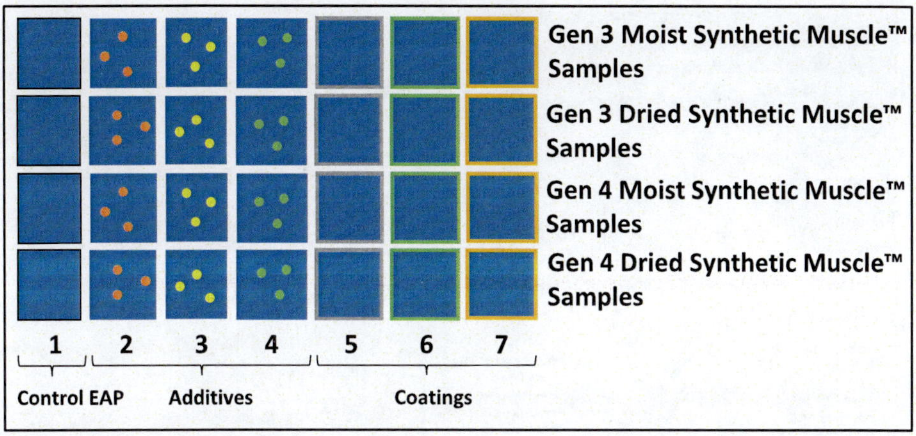

Fig. 1.5 Radiation resistance ground experimentation of Generation 3 and Generation 4 Synthetic Muscle™ with various additives and coatings.
Note: Flight readiness I, radiation experiment at PPPL on Earth, multiple Synthetic Muscle™ samples, 2X: Total run-time 47.8 hours, Cs-137, for a total dose of 305.931 kRad to a matrix of to a matrix of Gen 3 EAPs and a matrix of Gen 4 (Gen 4A) EAPs, with sample midpoint 5.25 cm away from the Cs-137 source, June 25-27, 2014: Four Synthetic Muscle™ sample matrices were made, each on an oxygen plasma treated titanium metal coupon. The control (position 1) had no additive and no coatings. The free radical inhibitors butylated hydroxytoluene (BHT) (position 2), butylated hydroxyanisole (BHA) (position3), and γ-tocopherol (E) (position 4) were included in each matrix. Mylar® coating (position 5), Natick Army Lab multi-ply coating (position 6), and Danfloss® goldsilicone layered coating (position 7) were included in each matrix. For the Gen 3 and Gen 4 matrices, both moist and dry microclimates were investigated. Two Gen 3 matrices, with one matrix moist and one matrix dry, were exposed to gamma radiation and two were not (held back as controls). Two Gen 4 matrices, with one matrix moist and one matrix dry, were exposed to gamma radiation and two were not (held back as controls). In terms of gamma radiation, this exposure was more than what was anticipated on the ISS for 90+ days and more equivalent to a trip to Mars and back. All samples had 0 ppm peroxide and similar pH values after the Cs-137 exposure. The significant change from the previous radiation exposure experimentation was that color change was observed in the Synthetic Muscle™ samples adhered to an oxygen plasma treated titanium coupon and then exposed to gamma radiation, with visible yellowing, compared to the matrices that were not exposed to gamma radiation, which were much more clear and colorless. Yellowing can be indicative of accelerated aging in polymers

The Ras Labs-CASIS-ISS NL Synthetic Muscle™ Experiment UA-2014-119 went up on the SpaceX CRS-6, which was launched into orbit on April 14, 2015, at 4:10 PM EST [20:10:14 UTC]. It was then supposed to come back on the SpaceX CRS-7 90 days later. The SpaceX Dragon CRS-7 exploded soon after liftoff. So, this experiment on the ISS that was only supposed to be up for 90 days was actually on the ISS for over a year [393 days]. As nervous as we were, "How are our EAPs holding up and withstanding this?" – it led to a better experiment with that long space exposure.

Just getting ready for the launch took about twice as long as we anticipated, the safety requirements, double-checking everything, working with NASA and CASIS, getting SDS data sheets in place, and working with Army to test the Gen 3 and Gen 4 EAPs in their ballistics range [to well above 10 Gs]. For any space experiment on the ISS NL, a lot of planning is required and thinking through every single worst-case scenario. We had to think about the direction of the experiment for the launch. Jim

	Generation 3 (no exposure)	Generation 3 (2x exposure)	Generation 4 (no exposure)	Generation 4 (2x exposure)
Peroxide test	0 ppm	0 ppm	0 ppm	0 ppm
pH test	2 – 3	2 – 3	2 – 3	2 – 3
Shore O Hardness	2.5 – 4.5	2.5 – 4.5	2.5 – 4.5	3.5 – 4.5
Electroactivity	Electroactive	Electroactive	Electroactive	Electroactive

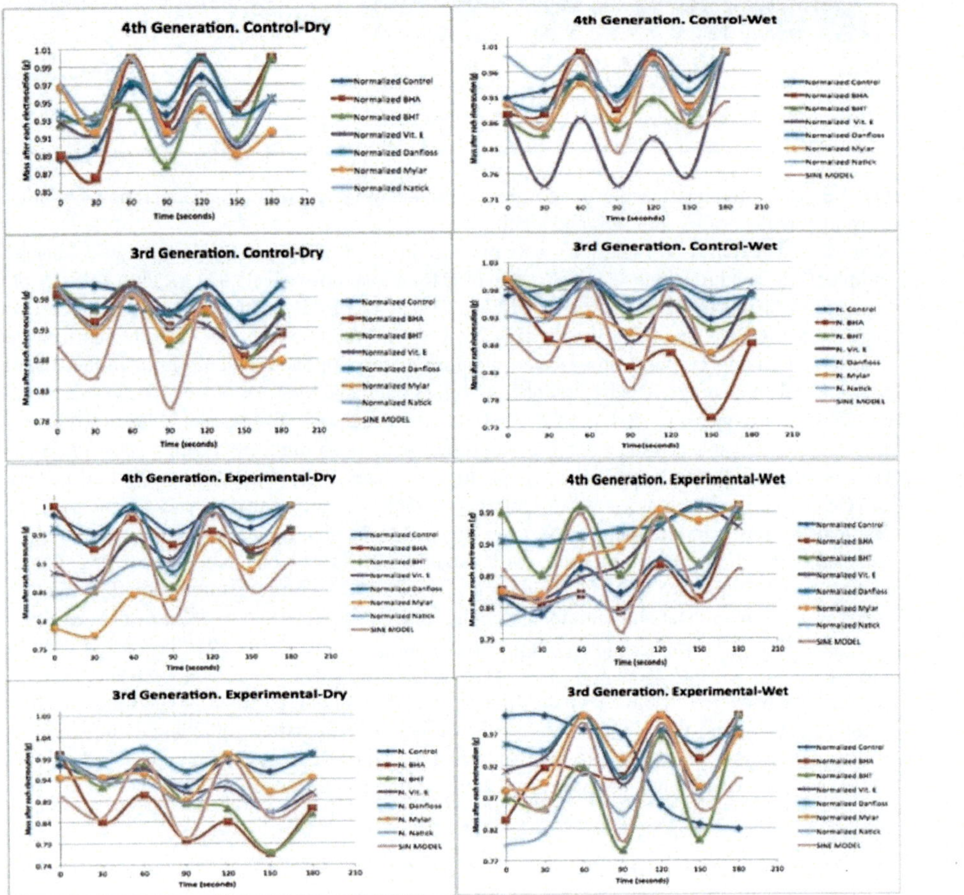

Fig. 1.6 Top left: tabulated results, top right: Cs-137 radiation testing, and bottom left: electroactivity profiles of radiation resistance ground experimentation of Generation 3 and Generation 4 Synthetic Muscle™ with various additives and coatings. (Photo credits: Ras Labs) Note: Sine curve fitting ofelectroactivity data, Mathematical formulation of the model used:y = Asine(kx) +B, where A models the amplitude of the curve, k models the frequency, and B moves the curve along the y-axis

Taylor at PPPL described this as eyeballs, "Eyeballs in or eyeballs out." These soft materials were backed by hard metals. Imagine you are laying down on the floor of a super-fast elevator. If you are facing up, the force during ascent would push your eyeballs into your head. If you were facing down, the force would push your eyeballs out of your head. The space experiment, with the soft EAP samples backed against the

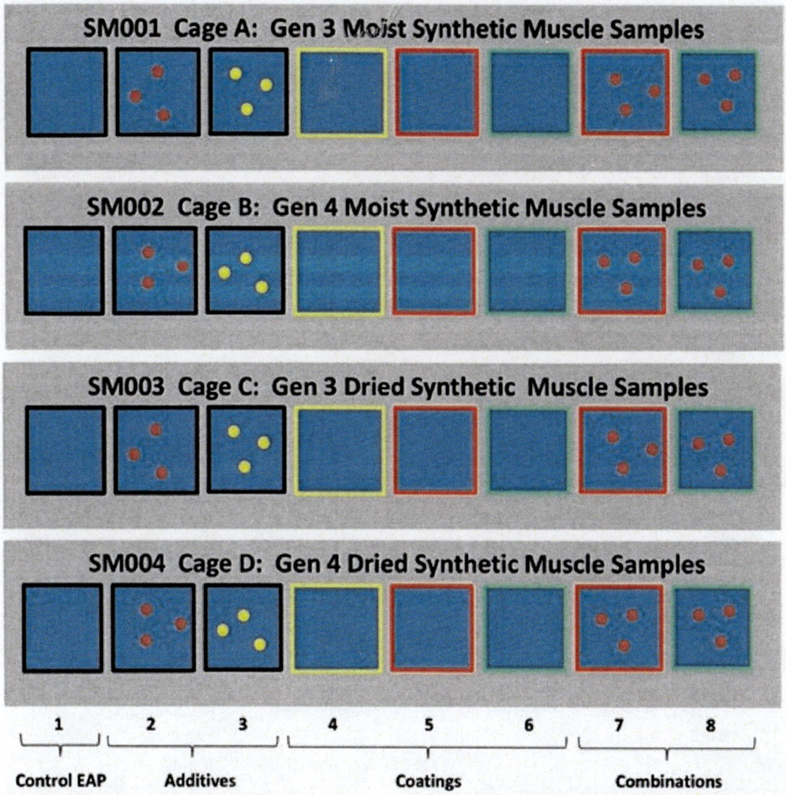

Fig. 1.7 Design of the Ras Labs-CASIS-ISS NL Synthetic Muscle™ Experiment: Resistance to radiation
Note: From the results of the radiation testing on Earth, the following experiment was designed for radiation resistance testing on the ISS National Laboratory. All four cages have the following 8 Synthetic Muscle™ samples: 1. control EAP sample (no additives, no coatings), 2. EAP with BHT additive, 3. EAP with γ-tocopherol (Vitamin E) additive, 4. EAP with Mylar® coating, 5. EAP with US Army Natick Labs meals ready to eat (MRE) 3-ply coating, 6. EAP with US Army Natick Labs MRE 4-ply coating, 7. EAP with combination BHT and Natick MRE 3-ply coating, and 8. EAP with combination BHT and Natick MRE 4-ply coating

titanium coupons in their aluminum cages and with their mesh windows for viewing, were oriented with the mesh screening of the cages facing up, "eyeballs in," to help keep the samples tight and well positioned during the ascent. For all of the bolt points that were in place, these were also glued and then even wired in place [and then taped with adhesive aluminum tape]. This is because with that lateral shaking [directly prior to and during liftoff], what often causes catastrophic failure is the beginning of just one little piece getting loose, like even just one small bolt. That then destroys other things, which weakens parts that then get loose, and the wreckage can continue, exponentially, quickly becoming catastrophic. The cargo transfer bag (CTB) had our experiment plus a bunch of other experiments to make the best use of space available. Good planning was not only essential for our experiment, but to help protect all of the experiments and the astronauts on the ISS NL.

Top: Front and back views of Cage A: G3 EM: S/N SM001. Middle left: All four cages for the Synthetic Muscle™ Experiment bubble wrapped, placed in large Ziplock bag and placed in cargo transfer bag (CTB) ready for transport to the ISS-NL, oriented with mesh viewing windows up, i.e., "eyeballs in." Middle right: Launch of SpaceX Dragon CRS-6 at 20:10:41 UTC from Cape Canaveral on April 14, 2015. Up, up, and away! Bottom: RMS capture of the SpaceX Dragon CRS-6 at 10:55 UTC on April 17, 2015. Photo credits: NASA.

Fig. 1.8 Preparation for the Ras Labs-CASIS-ISS NL Synthetic Muscle™ Experiment UA-2014-119, liftoff of SpaceX Dragon CRS-6, and Remote Manipulator System (RMS) capture at the ISS-NL

Each cage had its own microclimate, where one of the Gen 3 cages was a moist environment and one was a dry environment, with the same for the Gen 4 cages, one moist microclimate and one dry microclimate. The idea was that the materials for this type of technology could be desiccated. Ras Labs' EAPs can be dried and then hydrated when needed, and they work fine. The dry microclimates were for deep space travel concepts. A duplicate to the space experiment, with 32 identical samples, matrix design, and packaging, was the ground control, which remained on Earth for comparison.

In April [2015], I flew down from Massachusetts for the launch and flew down [from West Virginia] one of my biggest supporters throughout my life, my mother Winola Carman, and brought my youngest son who was in middle school at the time. He was his school's reporter for the event. Cape Canaveral, and Florida in general, has a lot of lightning strikes. NASA is very careful about many important things, including the weather. On April 13, the launch was scrubbed less than 3 minutes out because of cloud coverage that entered its 10-mile radius. The launch was rescheduled for the next day, April 14. The observation deck at the NASA center was about a mile from the actual launch site. We were given sunglasses. And it happened [at 4:10 PM], the countdown, 10, 9, 8, 7, 6, 5, 4, 3, 2, 1... and it went off!! Even though we were that distance from the launch, when SpaceX Dragon CRS-6 went up, not only could you see and then hear the rocket taking off, but you could also feel it, and you could feel those pressure waves. It was very bright. Florida is bright, it's very sunny, but even so, the light from the rocket was like looking at the sun.

It is awe-inspiring what us humans can do. The International Space Station is truly an international collaboration, where all us of around the globe are working with military precision, but together, in this quest to explore the cosmos.

One of the reporters asked my mother if this was on her bucket list, and she started crying [in a good way], so then I started crying, because this was so beyond any of our bucket lists. Growing up, we had hard times. We hunted, fished, and trapped to put food on the table. A deer would help get us through the winter. I don't know how she did it; my mother worked nights as a nurse and put me and my brothers through school [Virginia Tech and West Virginia University]. Again, if I had tried to do this, statistically, the odds are so low, but a series of events happened. I am honored and very privileged to have been part of these communities, at my *alma maters* [Virginia Tech and Purdue University], at Princeton University, and with NASA, CASIS, and ISS National Laboratory.

After the ascent, the SpaceX Dragon CRS-6 was in orbit and aligned with the space station over 2.5 days [62 hours, 45 minutes]. The duplicate ground control experiment remained in our lab in Massachusetts on its shelf. Later that week, photos were downlinked from NASA to show that our flown experiment arrived at the ISS NL and was safely in place in its locker. Figure 1.9 shows this first photo event. This was going to be a 90-day exposure up on the ISS to see how well Synthetic Muscle™ withstood the radiation up there. Why is it important for NASA to test things for the space program for radiation? And why on the ISS? Gamma radiation exposure was ground tested on Earth, but in space, there are so many other things going on: high LET [linear energy transfer] particles, plus gamma, beta, microwave, and X-ray radiation, *all at once.* To really duplicate this on Earth is nearly impossible. At the Brookhaven National Laboratory, high LET type particles can be generated. Samples can be exposed to gamma, beta, and alpha radiation and put in a microwave or X-rayed, but to do all of these at once, those broad-spectrum radiation effects encountered in space and hitting the ISS [plus additional effects from periodic solar flares] are almost impossible to duplicate here on Earth and would be prohibitively expensive. The International Space Station National Laboratory is a truly unique lab for space application testing, with its near microgravity, radiation from solar and cosmic effects, extreme temperatures outside the station, and, for human factors, its self-contained environment.

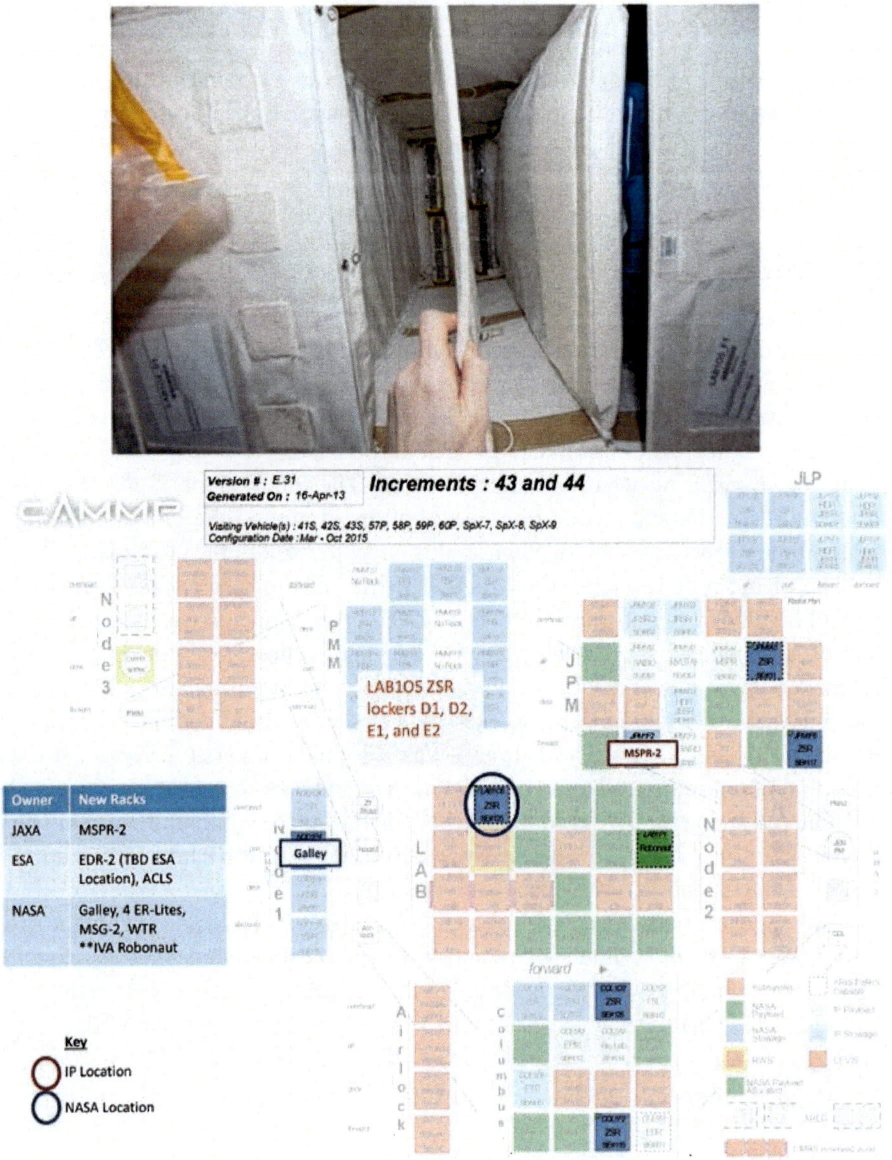

Fig. 1.9 First NASA photo event for Ras Labs-CASIS-ISS NL Synthetic Muscle™ Experiment UA-2014-119 and its location on the ISS NL. (Photo credit: NASA)
Note: Friday April 17, 2015, from www.NASA.gov: While the International Space Station was traveling 257 statute miles over the Pacific Ocean just east of Japan, Expedition 43 Flight Engineer Samantha Cristoforetti of the European Space Agency, with the assistance of Expedition 43 Commander Terry Virts of NASA, successfully captured the SpaceX Dragon spacecraft with the station's robotic arm at 6:55 AM EDT. The Synthetic Muscle Experiment was transferred and placed on the ISS-NL, Monday April 20, 2015, 10:30 AM, 1st photo event. Live NASA feed for the SpaceX missions to the ISS and other events is at www.NASA.gov and on NASA TV

There are various nodes on the International Space Station. Some areas are more protected from radiation, like where the astronauts do most of their living and eating. Other areas are not as protected. We wanted our experiment on one of the more external nodes, in a less protected area, to up the radiation level [that the experiment

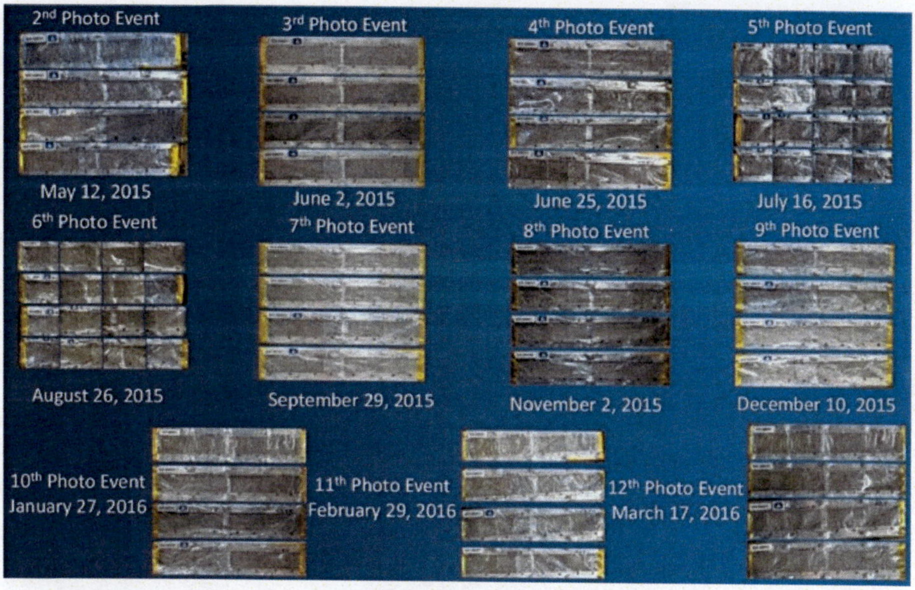

Fig. 1.10 NASA ISS NL photo events of the Ras Labs-CASIS-ISS NL Synthetic Muscle™ Experiment UA-2014-119. (Photo credits: NASA)

would be exposed to]. All of the ISS is made of low z-materials, because things like lead and concrete and other shielding materials are very heavy. Even though the areas for the astronauts are somewhat protected, they still get a considerable amount of [radiation] exposure up there, but our experiment was purposely located in a less protected area of the ISS.

The experiment was supposed to come back on SpaceX Dragon CRS-7, which didn't happen. We weren't sure when the experiment would come back down. NASA continued to do high-resolution photo events every 4 to 5 weeks so that we could keep an eye on the experiment. Figure 1.10 shows the photo events of the larger overall views of the experiment. NASA also provided close-ups of the samples, all downlinked back to Earth. Starting in about November, we began to see something happening around the edges of some samples, but we weren't totally sure what was going on until the experiment returned to Earth [in 2016]. Meanwhile, we were being asked to talk to local schools about the space experiment. I asked the CASIS and NASA folks, "These scientific pictures are great. They're perfect, they're wonderful, but could we just have a few photos of our cages to show the perspective and the feel of space?" This required some coordination. It was Astronaut Scott Kelly [in the twin study with his brother Mark Kelly] who graciously took the beautiful perspective photos for us. It's hard to even pick a favorite; they are all works of art (Fig. 1.11).

The other thing we were mindful of in this space experiment was high LET particles, which are cosmic radioactive subatomic particles that are generated from the cosmos and are heavy hitting [damaging]. We are so protected here on Earth, not just because of our atmosphere but also because of electromagnetic fields that are around and shielding our planet, which Mars does not have [Mars also has a much thinner atmosphere]. Figure 1.12 shows the tracking from high LET particles in astronaut

Fig. 1.11 Perspective photos of the Ras Labs-CASIS-ISS NL Synthetic Muscle™ Experiment UA-2014-119. (Photo credits: Astronaut Scott Kelly, NASA)

helmets from the Apollo 8 and 12 lunar missions [65–68]. ISS is low orbital. On the Moon, you are in deep space. The astronauts, when they closed their eyes, would see sparks [this also occurs on the ISS [69]. Those sparks were from high LET particles. These are really bad for us. This is a big challenge for the Mission to Mars. How are we going to withstand the long-term, and even the short-term, effects of high radiation during deep space travel? When we analyzed the returned experiment using microscopy, high LET tracking was something we kept an eye out for. The concept is that our EAP-based materials could potentially be used in space suits to help assist with

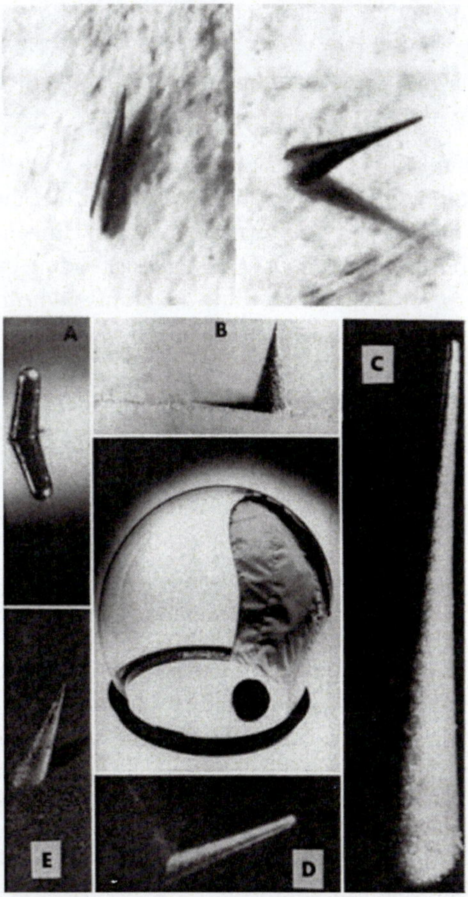

Tracks of heavy cosmic rays inside of an Apollo space helmet made of Lexan®. Left: a track from a particle entering the helmet. Right: an ending track from a particle that passed through the interior of the helmet and came to rest in the opposite side. The tracks are 0.5 mm and 0.7 mm in length [65, 66, 67].

Cosmic ray tracks in Apollo helmets. The Apollo 8 helmet of Astronaut Jim Lovell, one of several used in a personnel dosimetry experiment, surrounded by replicas of etched cosmic ray tracks. Tracks A, D, and E were from the helmets of Apollo 12 Astronauts Pete Conrad and Richard Gordon. The other tracks were from a control helmet exposed to primary cosmic rays. Track C was from a zinc ion. The lengths (in micrometers) of the track replicas are A, 350; B, 300; C, 700; D, 600; and E, 480. These tracks of heavy cosmic rays that penetrated the Apollo helmets provided a measure of the cytologically lethal component of cosmic radiation. Tracks A, C, and D crossed the inside of the helmet [68].

Fig. 1.12 Deep space radiation: Tracks from high LET particles in Apollo 8 and Apollo 12 space helmets [65–68]

motion around the astronaut's joints and to help protect the astronauts by capturing some of the high LET particles. This would lower the amount of radiation that entered astronaut's bodies and thus lessen their cellular damage.

SpaceX Dragon CRS-8 brought our experiment back home to Earth, landing in the Pacific Ocean on May 11, 2016 (Fig. 1.13). The flown Synthetic Muscle™ Experiment arrived by ground transport back in our lab in June [June 1, 2016]. When we opened up the experiment [June 3, 2016], each step of the process was recorded. The radiation exposure from the flown experiment was about 65 mGy, compared to the ground control exposure with less than 1 mGy here on Earth. This was over 75 times the total radiation exposure and over 90 times the daily radiation dose for the mission (Fig. 1.14). This was a long space exposure, which led to a better assessment. Two things became very clear when we opened everything up:

1. The moist microclimate cages had dried out somewhat. Even though they were all sealed for over a year, the ISS atmosphere is on the dry side, so some of the

moisture had leached out over time, on both the flown and ground control moist microclimate cages.
2. The flown moist samples in particular had a noticeable yellowing effect, which can be indicative of accelerated aging in polymers.

For all of us working on the project, this was an eye opener on just how harsh space is, even for an internal platform [on the ISS] at low orbital. Consider that the significant yellowing in these EAPs was observed in the flown wet samples. Humans are organized fluid, and except for our bones, we are about 60% water depending on our level of hydration [and even our bones are 31% water, while the brain and heart are about 73% water] [70]. This experiment visually struck home just how challenging space is on matter and by default on us.

All of the EAP samples from the flown Synthetic Muscle™ Experiment survived. All of the flown samples were in one piece. Ironically, one of the ground control samples on Earth, as it dried out, formed a crack. The EAPs were tested for pH, durometer, and electroactivity (Fig. 1.15). Initially, both the ground control and flown experimental EAPs averaged a little over pH of 5 [average pH of 5.28 for ground control and average pH of 5.53 for the flown samples]. After several weeks of soaking the EAPs in electrolytic solution [swelling to equilibrium], virtually all of the EAPs evened out to a pH of 4, which is typical for these ion-containing smart materials [10, 71]. In the flown experiment, after hydration, there were three EAP samples that were a little harder from the space exposure than the other samples. Nonetheless, after several weeks of hydrating in electrolytic solution, all of the EAPs were substantially softer and were flexible again, which is essential for good electroactivity. From the electroactivity testing:

1. In terms of additives and coatings, the Number position 1 (no additives, no coatings), position 4 (no additives, Mylar coating), and position 5 (no additives, 3-ply MRE coating) showed the most electroactivity.
2. In terms of the EAP formulations and microclimate comparisons, the G3ED, G3EM, and G4EM Synthetic Muscle™ samples showed the most electroactivity, i.e., the flown moist microclimate environments and the Generation 3 EAP formulation were more electroactive [as trends].

Overall, all of the flown samples had similar profiles [material integrity, pH, durometer, and electroactivity] to each other and to the ground control samples. These Synthetic Muscle™-based EAPs were radiation resistant, both to the extreme radiation testing performed here on Earth at PPPL and to the rigors of space from an extended exposure on the ISS NL.

We did look for indications of tracking in the EAPs using light microscopy and scanning electron microscopy (SEM) to see if any [high LET particle] tracks were captured in the EAPs, which was inconclusive. For this long-duration space exposure, some high LET particles certainly passed though the flown EAPs. Because these EAPs are softer and much more flexible than harder materials like polycarbonates [Lexan®], the EAPs may have re-annealed over time, removing any anecdotal evidence of high LET particle tracks. In the flown aluminum cages, track pitting was evident on all sides of the cages, including some on the inside walls of

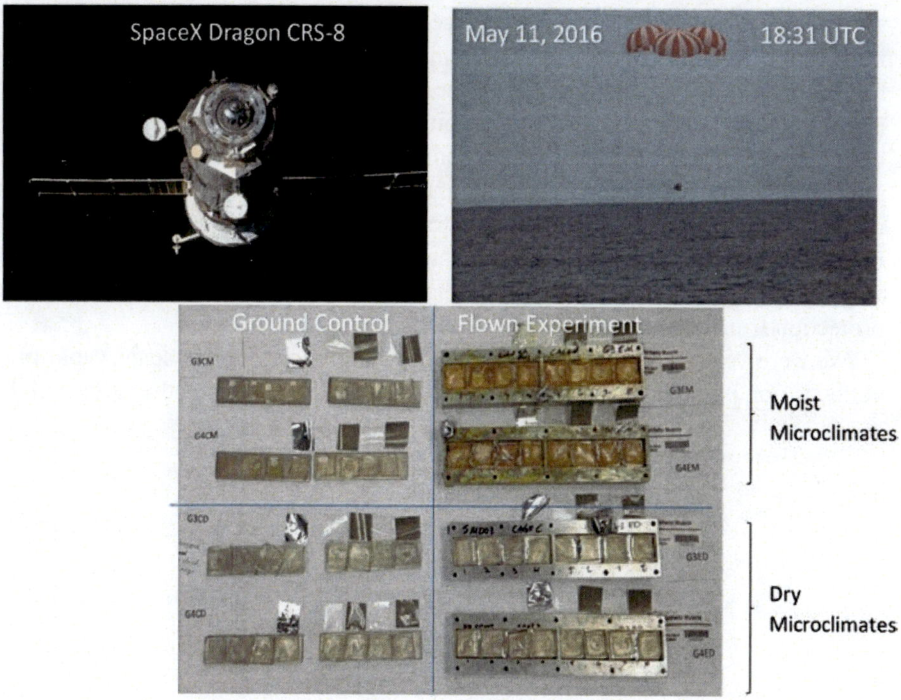

Fig. 1.13 Return and examination of the flown Ras Labs-CASIS-ISS NL Synthetic Muscle™ Experiment. (Photo credits: (*top*) NASA, (*bottom*) Ras Labs)

RAM dosimeter's exposure time periods

RAM Name	RAM S/N	Assembled	Launched SpX-6	Landed SpX-8	Received at SRDL[1]
RAM 1	2752	11/20/2014	4/14/2015	5/11/2016	6/3/2016
RAM 2	2753	11/20/2014	4/14/2015	5/11/2016	6/3/2016
Control 1	2754	11/20/2014	N/A	N/A	5/26/2016
Control 2 (Ras Labs)	2755	11/20/2014	N/A	N/A	6/3/2016

RAM dosimeter's total radiation dose and rates

RAM Name	RAM S/N	Mission Duration (days)	[2]RAM Mission Dose (mGy)	[3]RAM Dose Rate (μGy/day)
RAM 1	2752	393	65.60 ± 1.07	166.9 ± 2.7
RAM 2	2753	393	64.85 ± 1.39	165.0 ± 3.5
Control 1	2754	N/A	0.238 ± 0.005	1.21 ± 0.03
Control 22(Ras Labs)	2755	*481	0.855 ± 0.032	1.78 ± 0.07

[1]Space Radiation Dosimetry Laboratory, NASA Johnson Space Center, Houston TX 77058.
[2]RAM 1 Dose, RAM 2 Dose, and Control 2 Dose values have been already corrected to subtract the ground handling/shipping exposure as measured by Control 1.
[3]RAM Dose Rate values are computed by dividing the RAM Mission Dose by the Mission Duration.
*Control 2 Mission Duration represents the time spent at Ras Labs, Quincy, MA 02169.

Fig. 1.14 Radiation exposure comparisons between the ground control and flown experiment the milled areas that held the EAPs in place (Fig. 1.16). This pitting was not observed in the bottom of the milled areas that held the EAPs in place; however, these areas had housed not only the EAPs but also the titanium coupons backing the EAPs. The flown titanium coupons had much less pitting than the aluminum cages. This may be indicative that the EAPs provided some protection, but keep in mind that titanium is a much harder metal than aluminum. Aluminum is much easier to pit and mechanically damage than titanium, so tracks from high LET particle impacts are much more observable in aluminum than in titanium. Further experimentation would be needed to definitively prove that these EAPs can provide us with some protection from cosmic radiation.

We were very curious about the yellowing in the flown experiment. A lot of the yellow color, particularly the bright yellowing, was found to be on the surface of the

pH of Synthetic Muscle™ Samples — June 3, 2016

Control

				G3CM					Average
Sample	1	2	3	4	5	6	7	8	
pH Level	3.5	6.0	4.0	5.0	3.5	4.0	5.5	5.5	4.63
				G4CM					
Sample	1	2	3	4	5	6	7	8	
pH Level	4.0	6.5	6.0	6.0	4.5	4.5	5.5	6.0	5.38
				G3CD					
Sample	1	2	3	4	5	6	7	8	
pH Level	4.5	6.0	6.5	4.0	4.5	4.0	6.0	6.0	5.19
				G4CD					
Sample	1	2	3	4	5	6	7	8	
pH Level	6.5	6.5	5.5	6.0	6.0	5.5	5.5	6.0	5.94
Average	4.63	6.25	5.50	5.25	4.63	4.50	5.63	5.88	

Experimental

				G3EM					Average
Sample	1	2	3	4	5	6	7	8	
pH Level	5.5	4.5	4.5	5	4.5	5.5	5.5	5.5	5.06
				G4EM					
Sample	1	2	3	4	5	6	7	8	
pH Level	5.5	6.5	6.5	6	6.5	6	6	6	6.13
				G3ED					
Sample	1	2	3	4	5	6	7	8	
pH Level	4.5	5.5	6	4.5	4.5	4.5	4.5	6	5.00
				G4ED					
Sample	1	2	3	4	5	6	7	8	
pH Level	5.5	6	6	6	6	6	6	6	5.94
Average	5.25	5.63	5.75	5.38	5.38	5.50	5.50	5.88	

pH of Synthetic Muscle™ Samples, Rehydrated

	Post-Flight		7 Days Rehydrated		14 Days Rehydrated		32 Days Rehydrated		60 Days Rehydrated	
	Control	Exp	Control	Exp	Control	Exp	Control	Exp	Control	Exp
pH	4.5-6.5	5−6.5	3.5-6	4.5-6	4	4	4	4	4	4

Shore O Durometer of Synthetic Muscle™ Samples after 32 days Rehydration — July 5, 2016

				G3CM					Average
Sample	1	2	3	4	5	6	7	8	
	30	37	36	37	34	36	35	36	35.00
				G4CM					
Sample	1	2	3	4	5	6	7	8	
	36	38	41	41	42	37	36	41	38.71
				G3CD					
Sample	1	2	3	4	5	6	7	8	
	26	31	31	30	26	32	25	33	28.71
				G4CD					
Sample	1	2	3	4	5	6	7	8	
	33	40	40	35	36	36	37	39	36.71
Average	31.25	36.5	37	35.75	34.5	35.25	33.25	37.25	

				G3EM					Average
Sample	1	2	3	4	5	6	7	8	
	41	36	45	48	35	55	57	68	48.125
				G4EM					
Sample	1	2	3	4	5	6	7	8	
	72	40	37	45	39	34	43	30	42.5
				G3ED					
Sample	1	2	3	4	5	6	7	8	
	34	34	36	36	25	33	33	36	33.375
				G4ED					
Sample	1	2	3	4	5	6	7	8	
	43	41	39	38	39	40	34	37	38.875
Average	47.5	37.75	39.25	41.75	34.5	40.5	41.75	42.75	

Fig. 1.15 Tabulated results and electroactivity of ground control and flown experiment for the Ras Labs-CASIS-ISS NL Synthetic Muscle™ Experiment UA-2014-119 with various additives, coatings, and microenvironments. (Photo credits: Ras Labs)

(continued)

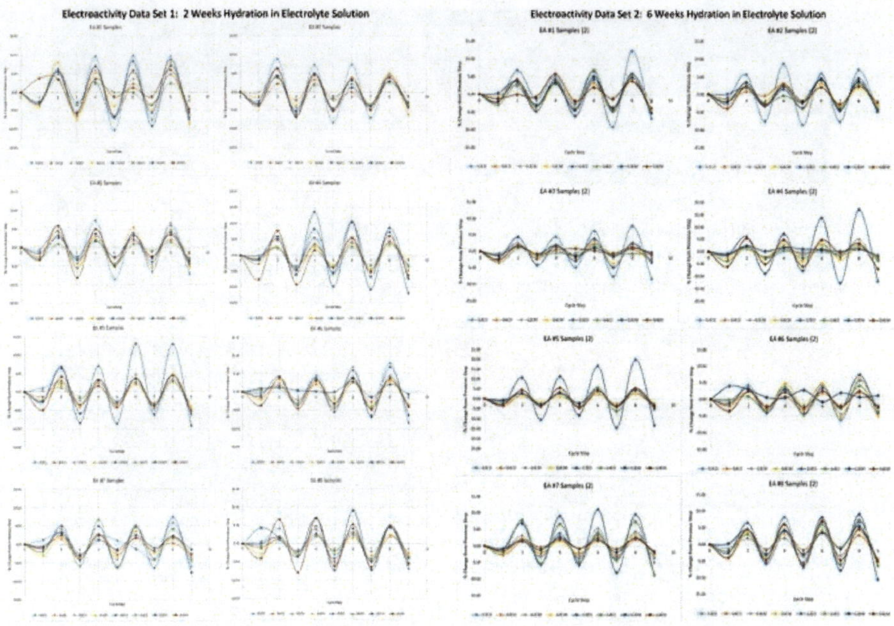

Fig. 1.15 (continued) Note: Peroxide testing was not performed on the Ras Labs- CASIS-ISS NL Synthetic Muscle™ Experiment ground control and flown experimental EAPs. Ionizing radiation creates free radicals, which initiate chain reactions involving cascades of reactions in a moist aqueous environment. A single hydroxy radical can form over 30 peroxide molecules [72, 73]. Free radicals and peroxides are very reactive, thus the potential for short-term and long-term damage to substances. However, they are also very short-lived due to this high chemical reactivity. The time duration for the Synthetic Muscle™ Experiment to return to Earth from the ISS, travel to the NASA Johnson Space Center in Houston, TX, and then travel by ground transport to the Ras Labs Synthetic & Prototyping Laboratory in Massachusetts was 21 days. Any free radical and thus peroxide formation caused from the space exposure would have totally dissipated

EAPs in the flown moist microclimate environments. By zooming in using stereo microcopy, this surface yellowing was observed as areas or patches, i.e., the bright yellow areas were not uniform. There was some slight yellowing throughout the EAPs, but the bright yellowing was a surface phenomenon in the flown EAPs, particularly those EAPs from the flown moist microclimate environments (Fig. 1.17). This could have been imparted from the aluminum screening. Energy-dispersive X-ray spectroscopy (EDS) was used to analyze the elemental composition of the EAPs' surfaces at the microscopic level. The presence of aluminum was found on the surfaces of the flown EAPs. The highest aluminum peaks from the EDS analysis were on the surfaces of the EAPs from the flown moist microclimates, particularly compared to the flown samples in the dry microclimate environments (Fig. 1.18) and to the ground control EAPs. Aluminum was not in the original EAP formulations, but was part of the cage structures that surrounded the EAP matrices. Yellowing in polymers can be indicative of aging; however, the bright yellow color was determined to be primarily a surface effect from the aluminum cage systems that served to protect the EAP matrices. The yellowing did not adversely affect the electroactivity or other characteristics of the EAPs.

Fig. 1.16 High LET particle tracks observed in a flown aluminum cage and possibly in a flown titanium coupon. (Photo credits: Ras Labs)

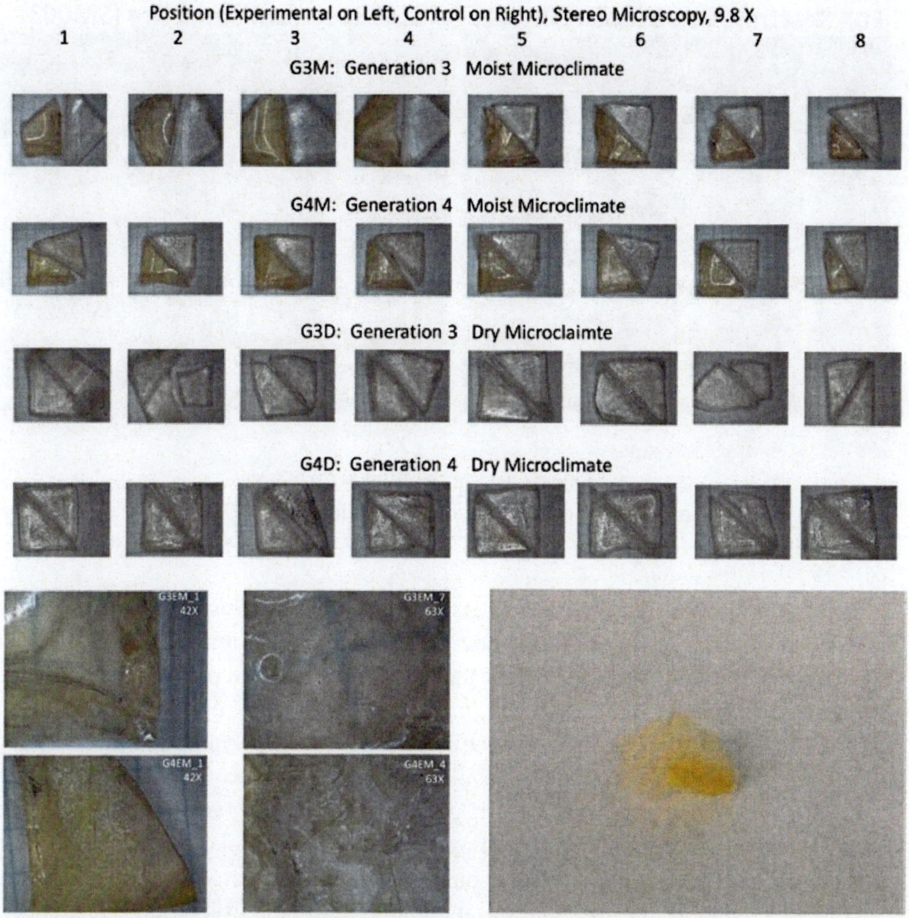

Fig. 1.17 *Top* and *bottom left*: Microscopic analysis of ground control and flown experiment for the Ras Labs-CASIS-ISS NL Synthetic Muscle™ Experiment UA-2014-119. Note the *yellow* coloration in the EAPs from the flown moist microclimate environments. *Bottom right*: Photo of a small piece of G4EM-3 (flown experimental Generation 4 EAP, Position 3, moist microclimate) that still appeared *yellow*, even when hydrated. (Photo credits: Ras Labs)

All of the EAPs in the Ras Labs-CASIS-ISS NL Synthetic Muscle™ Experiment UA-2014-119 were electroactive. The EAPs had good material integrity, good durometer, and typical pH values, particularly after rehydration. There were concerns about the color changes [the observed yellowing]; however, this was determined to be primarily a surface effect from contact with the aluminum, particularly for the flown EAPs caged in the moist microclimate environments. There was no macroscale breakdown, even with this long space exposure on the ISS NL [393 days], which was much longer than originally planned [90 days]. Ras Labs' Synthetic Muscle™-based EAPs were found to be radiation resistant, determined from both the extreme radiation testing performed on Earth at PPPL and the rigors of space from an extended exposure to broad-spectrum radiation on the ISS NL.

The quest is to bridge the gap between form and function. Science influences art, but also art influences science. Below is Luke Skywalker's arm [Star Wars: Episode

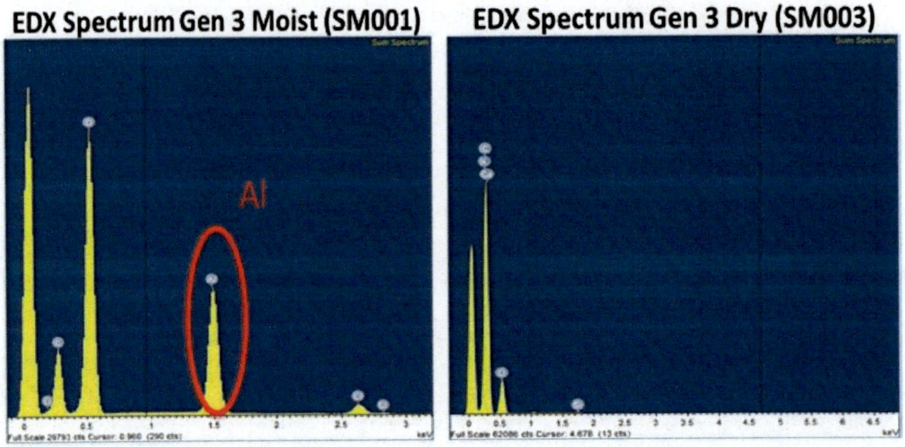

Fig. 1.18 EDS comparison of two flown experimental EAP samples, both Generation 3 and Position 3, from the moist microclimate (*left*) and the dry microclimate (*right*). Note the high level of aluminum on the surface of the flown EAP housed in the moist microclimate environment

V, The Empire Strikes Back, 1980 release] with Synthetic Muscle™ photoshopped into the prosthesis. In the movie, when Luke opens up his arm, it is metal cables inside that are moving, drawn from the technology available in the 1970s. Our vision is biomimetic human-like grasp for prosthetic hands and robotic grippers [also known as end effectors or end of arm tooling (EOAT)] using Synthetic Muscle™ and our next generation of EAPs under development. Ras Labs' compliant FingerTip™ sensors are being integrated into robotic grippers for sensitive tactile touch-like feedback. Instead of long, long ago in a galaxy far, far away, if Luke were alive, here on Earth in the near future, his prosthetic hand just may be powered by smart materials capable of extremely life-like motion, integrated with human-like tactile feedback. All of us in the smart material community are making science fiction a reality.

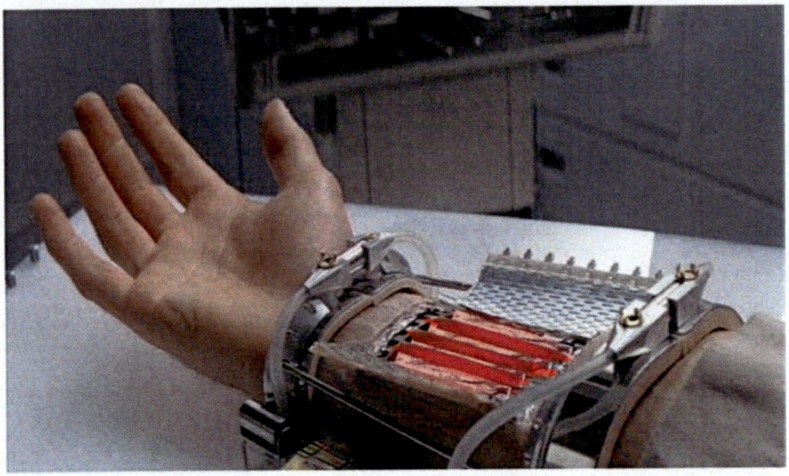

Rendition: Ras Labs from base image of mechno-arm [74]

1.2 Synthetic Muscle™ EAP Shape-Morphing and Pressure Sensing: Considerations for Robotic and Prosthetic Applications

The smart material community will play a role in assisting humanity. Technology pushes humanity forward and saves lives. Synthetic Muscle™ is a class of EAP-based materials and actuators that contract and expand at low voltage (battery levels), sense pressure from gentle touch to high impact, and attenuate force [10]. These EAPs can withstand environments unsafe to humans due to extreme temperatures, radiation, and pressures (Table 1.1), as well as environments with biological agents such as infectious contagions with no cure. Ras Labs-patented Synthetic Muscle™ technology is scalable (Fig. 1.19). For shape-morphing, the dual motion of contraction and expansion is achieved by easily switching the electric polarity of the voltage applied to the EAP back and forth to produce contraction-expansion cycles. The amount of movement can be controlled by adjusting the voltage level, which lends these materials to biofeedback [10]. In addition, these EAPs serve dual use as mechanical pressure sensors, from gentle touch to high impact [17]. Ras Labs' EAPs, synthetized to mimic the unique gentle-yet-firm nature of human tissue, is a potential asset to both robotic and prosthetic applications. Shape-morphing Synthetic Muscle™-based pads were explored in prosthetic liners. Both the shape-morphing aspects and the sensing aspects as soft sensors of these EAPS were applied to robotic grippers.

Robotics We are in the fourth revolution of the industrial era, the robotics revolution, but robots do have their challenges. Part of it is hardware. Most automation and robots are very strong, are very fast, and usually need to be partitioned away from humans for safety reasons. For many repetitive tasks that humans do with delicate or fragile objects, it would be beneficial to use robotics: whether it is for agriculture, for medical surgery, for therapeutic or personal care, or in environments where it isn't safe for humans to be. Another challenge with robots is sensing. Currently, robotic sensing is mainly visual, which is useful up until the point of contact, and then what? To understand how an object is being gripped, tactile feedback is needed. For handling fragile objects, if the grip is too tight, breakage occurs, and if the grip is too loose, the object will slip out of the grasp, also leading to breakage. Rigid robotic grippers using a visual feedback loop can struggle to determine the exact point and quality of contact. Robotic grippers can also get a stuttering effect in the visual feedback loop. By using soft Synthetic Muscle™-based EAP pads as the sensors, immediate feedback was generated at the first point of contact. Because these pads provided a soft, compliant interface, the first point of contact did not apply excessive force, allowing the force applied to the object to be finely controlled. The EAP sensor could also detect a change in pressure location on its surface, making it possible to detect and prevent slippage by then adjusting the grip strength. In other words, directional glide provided feedback for the presence of possible slippage to then be able to control a slightly tighter grip, without stutter, due to both the feedback and the soft gentleness of the fingertip-like EAP pads themselves. The soft nature of the EAP fingertip pad also naturally held the gripped object, even delicate objects like ripe fruit, improving the gripping quality over rigid grippers without an increase in applied force (Fig. 1.20). Analogous to finger-like tactile touch, the EAPs with appropriate coatings and electronics were positioned as pressure sensors in the fingertip or end effector regions of robotic grippers. This development of using Synthetic Muscle™-based EAPs as soft sensors provided for sensors that feel like the pads of human fingertips. Dome-shaped Synthetic FingerTip-like sensors and flat rectangular Synthetic FingerTip™ sensors provided feedback for point of contact, what we know as tactile touch, as well as for glide to detect slippage (Figs. 1.20, 1.21, 1.22, and 1.23).

Ras Labs' newest EAP-based FingerTip™ soft sensors are extremely sensitive, with a broad pressure range from at least 0.005 N up to 50 N, with 1 mm resolution (Figs. 1.20 and 1.21). These FingerTip™ sensors were validated and verified down to 0.005 N, with ongoing testing to capture the limits of the extreme sensitivity and linearity above 50 N. Human touch is around 0.1 N pressure sensitivity [75]. In Fig. 1.20, each pixel corresponds to 1 square millimeter. The 1mm^2 white pixel showing the highest contact pressure, surrounded by shades of red depicting surrounding pressure intensities. The FingerTip™ sensors were also able to pick up heartbeat pulses, including components of the pulse, at the wrist and other locations. From fatigue testing, these sensors were also robust, surviving 1000,000+ cycles of compression force at ~1.3 Hz, with 300,000 cycles of 500 g weight (5 N) followed by 700,000 cycles with 1 kg weight (10 N), with full functionality and no visible breakdown. This testing is

Fig. 1.19 Scale-up batch of Synthetic Muscle™. (Photo credit: Ras Labs)

FingerTip™ Sensors retrofitted to a robotic arm and EOAT gripper, handling a variety of fruit with real time point of contact detection, with no droppage, no slippage, and no damage.

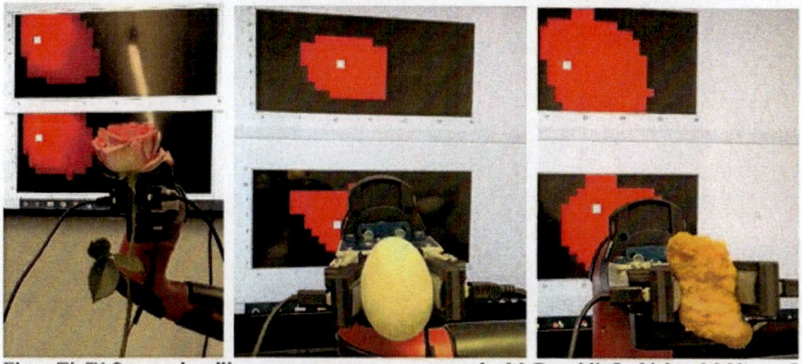

FingerTip™ Sensors handling a rose stem, a raw egg, and a McDonald's® chicken McNugget® with real time point of contact detection and with no droppage, no slippage, and no damage.

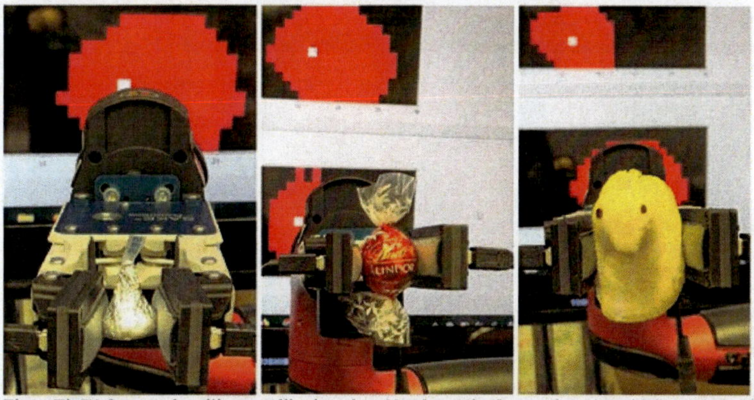

FingerTip™ Sensors handling a milk chocolate Hershey Kiss®, a soft centered Lindt Lindor® milk chocolate truffle, and a Just Born marshmallow Peep® with real time point of contact detection and with no droppage, no slippage, and no damage.

Fig. 1.20 EAP-based FingerTip™ sensors, retrofitted on a Sake Robotics® EOAT gripper with a Rethink Robotics® Sawyer robotic arm, handling a variety of fruits, a rose, a raw egg, a chicken nugget, a variety of candies, a croissant, a paper cup, a thermometer, business cards, and a test tube,

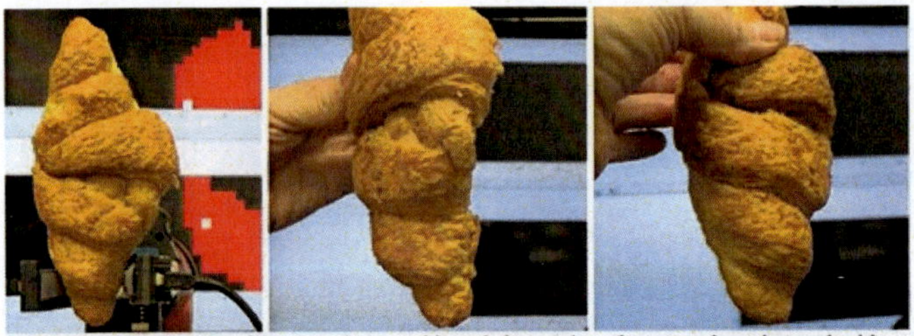

FingerTip™ Sensors handling a croissant with real time point of contact detection and with no droppage, no slippage, and no damage.

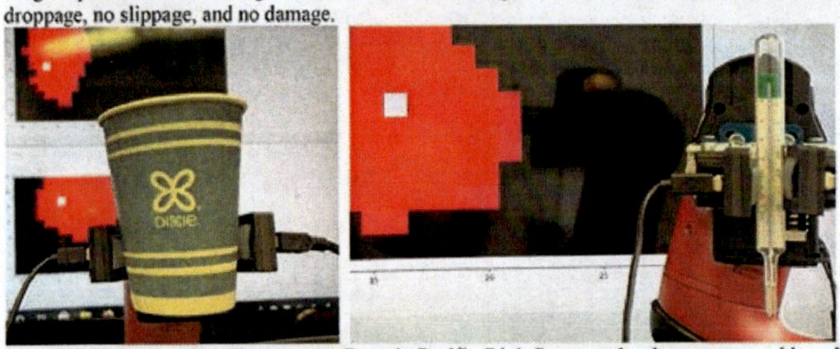

FingerTip™ Sensors handling a paper Georgia-Pacific Dixie® cup and a thermometer with real time point of contact detection and with no droppage, no slippage, and no damage.

FingerTip™ Sensors handling a business card with real time point of contact detection and with no droppage, no slippage, and no damage.

FingerTip™ Sensors, retrofitted to a robotic gripper and arm, showing a ripe cherry tomato being picked off the vine, with no droppage, no slippage, and no damage. Circled in yellow shows the predominantly downward z-direction in near instantaneous real time: at ~0 N force before picking (far left), maximum force prior to tomato and vine cleanly separating (middle), and return to ~0 N force after the tomato was picked from the vine(far right).

FingerTip™ Sensors handling a Pyrex® glass test tube as it was filled with 50 mL of water. Circled in yellow shows the force in the downward z-direction in near instantaneous real time, from ~0 N (far left) to 0.5 N (far right).

Fig. 1.20 (continued) with real-time contact points displayed for both sensors, and with no droppage, no slippage, and no damage. The *white* pixel shows the highest contact pressure, surrounded by shades of *red* depicting surrounding pressure intensities, with 1 mm^2 resolution for each pixel. For the tomato, point of contact and release from the vine were easily detected. For the test tube, water was added and easily detected in real time. (Photo credits: Ras Labs)

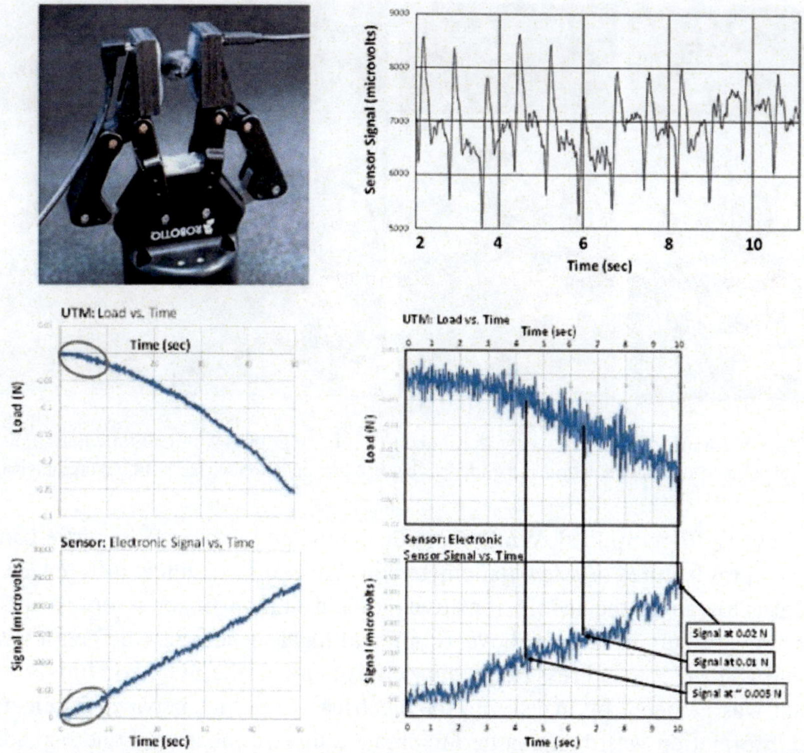

Fig. 1.21 Top left: Ras Labs' extremely sensitive FingerTip sensor retrofitted to a Robotiq® gripper gently handling fruit. Top right: the FingerTip sensor detecting heartbeat at the wrist. Bottom left and right: pressure range and pressure sensitivity of the FingerTip™ soft compliant sensor. Universal Testing Machine (UTM) mechanical analysis using flat platens and a 10-lb load cell in compression for the D-LR50 sensor at 0.25 mm/min to -0.20 mm, where the flat platen of the UTM began to touch the sensor at approximately 2 seconds into the run. On right is expansion of the circled first 10 seconds of the run. (Photo credits: Ras Labs)

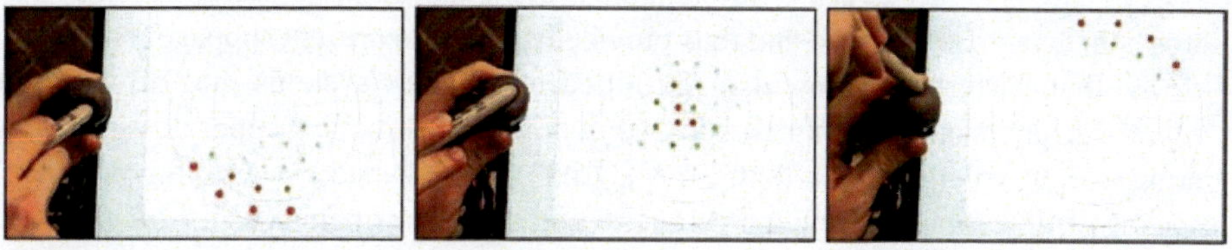

Fig. 1.22 Dome-shaped FingerTip-like sensor with real-time output display, which provides point-of-contact detection and glide feedback to prevent slippage [62]. (Photo credits: Ras Labs)

ongoing, with analysis every 100,000 cycles using the 1 kg load (10 N). Figure 1.21 shows the results of the pressure sensitivity and large pressure range using our most optimized FingerTip™ sensors to date. These soft compliant FingerTip™ sensors are robust, are extremely sensitive, and have a wide pressure range.

The dome-shaped soft sensor has a hemispheric shape. The goal of the Synthetic FingerTip-like sensor was to mimic the form and function of the human finger in terms of tactile touch of mechanical pressure. Human sensitivity is around 0.1 N of

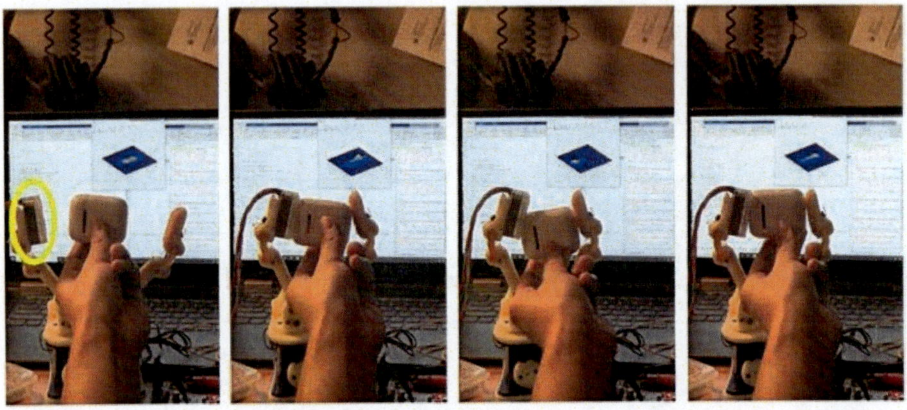

Fig. 1.23 Flat FingerTip™ like sensor retrofitted to a robotic gripper with real-time output display to provide point-of-contact detection and glide feedback to prevent slippage [62]. (Photo credits: Ras Labs)

force over a fingertip [75]. While individual cells on the human fingertip can react to structural features of a few nanometers in size and mechanical differences in the pascal (Pa) range, the tactile sensitivity of the human finger is only capable of detecting structural features above 10 nm and local mechanics (the gentlest touch) around 1 kPa [75, 76]. Pressure sensing information of the Synthetic FingerTip-like sensor was gathered by measuring the electrical properties between its electrodes. This information was then coalited to create a three-dimensional understanding of the pressure profile. Basic pressure position and magnitude tests have been successful, with pressure sensitivity down to 0.2 N using this sensor. Even more sensitive sensors have been developed and were discussed above.

The flat Synthetic FingerTip™ is shown in Fig. 1.23, where the gripper pad using the EAP-based soft sensor is circled in yellow. On the computer, the different touch points of the gripper were again observed in real time. An open-source robotic hand, OpenHand [77], was 3D printed to demonstrate the longer and flatter finger-like pad on the robotic gripper and its ability to sense different areas along the length of that soft sensor. This was beneficial to show glide and thus provide feedback to prevent slippage.

Synthetic Muscle™ was also retrofitted as actuator systems into off-the-shelf robotic grippers and is being considered in novel biomimetic gripper designs, operating at low voltages (less than 50 V). Like the EAP pad systems for prosthetic socket fit, the robotic gripper EAP actuation can also operate using off-the-shelf batteries for locations on Earth and in space where access to a plug-in connection is limited or non-existent. These EAPs contract and expand in a controllable fashion by simply adjusting the voltage [10, 62]. An off-the-shelf sprue gripper was retrofitted with a Synthetic Muscle™-based EAP system. Its actuation was compared to standard pneumatic pressure to handle soft fruit such as a blueberry (Fig. 1.24). This EAP system weighed 9 g, which included the EAP, the wiring, and the coating. First 20 V was applied to actuate to point of contact, and then it was reduced to 5 V to hold the blueberry, with no droppage, no slippage, and no damage to the fruit.

Another Synthetic Muscle™-based EAP system was retrofitted into what is known in the robotics trade as an air hand. This provided for pick-and-place of fragile thin-walled cylinders (Fig. 1.25).

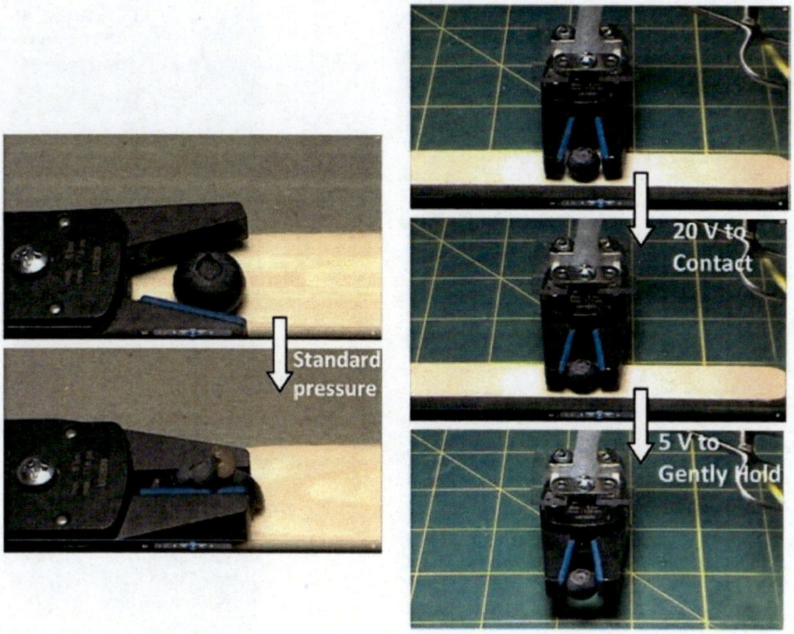

Fig. 1.24 EMI® gripper handling a blueberry using standard pneumatic pressure (*left*) and retrofitted with a Synthetic Muscle™-based EAP actuator (*right*) with 0 V (*top right*), 20 V to quickly make contact (*middle right*), and then 5 V to handle blueberry with no droppage, no slippage, and no damage to the fruit (*bottom right*) [62]. (Photo credits: Ras Labs)

Prosthetics The first application in prosthetics we addressed was the socket fit [17, 78, 79] where the prosthetic device attaches to the residual human limb. Shape-morphing pads incorporating Synthetic Muscle™ is expected to allow amputees to continue their active lives without needing to adjust the fitting of their prosthetic device(s) throughout the day. This technology will potentially resolve major issues facing amputees, most notably tissue breakdown from prosthetic slippage and the inconvenience of adding or removing prosthetic socks to adjust fit. The EAP-based pads actively expand or contract at low voltages while offering impact resistance and pressure sensing, all in one system.

In collaboration with United Prosthetics, amputee testing of these EAP-based pads located in strategic areas of the socket was case tested with customers, both below-knee (BK) and above-knee (AK) amputees, for evaluation and feedback [79]. The AK amputee was tested using 9 V batteries and a three-pad EAP system placed around the circumference of the residual limb in the prosthetic socket, with good results under both strenuous exercise and normal activity conditions. Further iterations to the EAP-based pad system were made, and a BK amputee was tested next using 6 V batteries and a three-pad EAP system strategically placed around the circumference of the residual limb in the prosthetic socket. The BK patient was tested on three separate occasions, under both vigorous exercise (in a heated room, purposely pushing toward dehydration) and normal activities, with very good results. He was able to work out longer and harder than in all of the 12 years since he became

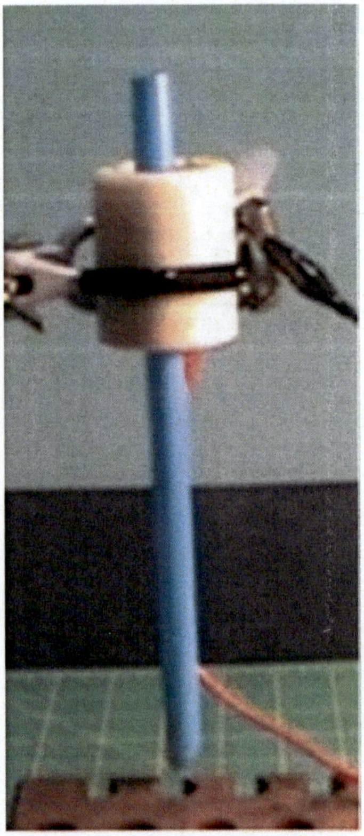

Fig. 1.25 Pick-and-place of a thin-walled cylinder using an off-the-shelf air hand retrofitted with a Synthetic Muscle™-based EAP actuator [62]. (Photo credit: Ras Labs)

an amputee, *comfortably* (Fig. 1.26). An unexpected positive outcome was that he also had very good proprioception while using the EAP pad system. Proprioception, which is a challenge for amputees, is the intuitive perception or awareness of the position and movement of the body. He could walk much more naturally by "feel" without having to rely so much on sight. With this adaptive fit, another feedback response was that he felt that he had his leg back. He wanted to take the device home with him, which would be the next step in a 25- to 30-person clinical trial.

In the prosthetic socket, the ability to withstand mechanical compression force is crucial during walking and running ambulation. Mechanical compression testing was performed on selected EAPs and then the scaled-up EAPs, using a 5 kN or 100 N load cell, depending on the sample, with an Instron® 3345 universal testing machine (UTM), at a compressive rate of 5 mm/min after pre-test load of 3 N, followed by hold at 1.5 mm compression distance for 60 seconds, to EAP samples 1 × 1 square inches by about 5 mm thick. The harder EAPs, such as the EAP with the highest concentration of particulate additives, with a Shore OO durometer of 92, also had the highest compressive load of 3.8 MPa at 1.5 mm compression. The EAPs without additives tended to be the softest EAPs in durometer and compression testing, with Shore OO hardness around 35 and compressive load around 0.03 MPa. These EAPs had some creep, which in a high-pressure area may help with comfort in the socket. Ras Labs' EAPs also have very good, fast rebound when pressure is removed (low hysteresis), which is extremely beneficial for maintaining good fit during ambulation [80].

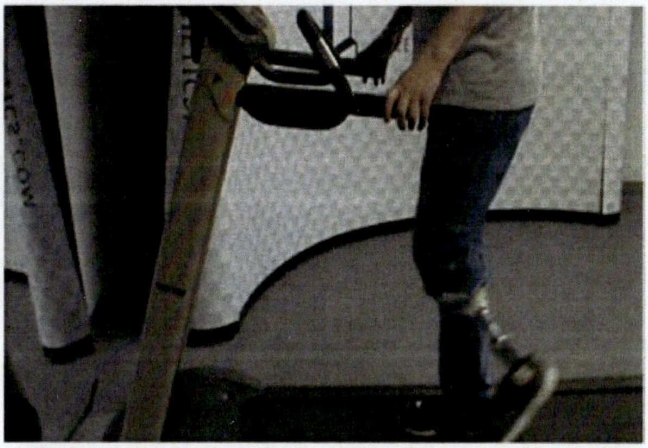

Fig. 1.26 BK customer evaluation and feedback. (Photo credit: Ras Labs)

Humans typically walk between 5000 and 7000 steps per day [81]. For these EAP pad systems to last a reasonable amount of time, at least 6 months, they will have to hold up under the pressures and stresses of ambulation. For the durability testing, long-term mechanical cycle testing was performed from 5 psi to 30 psi at 4 Hz for 3,000,000 cycles, followed by a 50-psi compression. Five selected EAP samples, 1 inch in diameter and about 5 mm thick, were encapsulated in a silicone coating, which added another 2 mm to the overall thickness. All five samples tested, including those with additives, survived the long-term mechanical cycle testing with no observable damage to the EAPs. Three of the samples developed small tears in the silicone coating, and some moisture escaped during the approximately week-long cycle testing of each sample, but these EAPs still held up well to the 3,000,000 cycles of the mechanical pressure testing.

The modulation of the EAP shape-morphing control as needed can be achieved by controlling the voltage level. This technology is expected to provide for an adjustable prosthetic liner or socket that can maintain dynamic perfect fit throughout the day, for an advanced standard of care for amputees. Early evaluation and feedback from AK and BK amputees were very promising. Durable Synthetic Muscle™ has the potential to sense pressure and provide for self-adjusting prosthetic socket fit, with no need for the patient to adjust the fit of their device throughout the day. The maintenance of dynamic perfect fit throughout the day could be handled automatically to prevent tissue damage from even beginning to occur.

Another application in prosthetics is using Synthetic Muscle™ for actuation to produce biomimetic motion by contracting, like human muscles, with electric input and with expansion on the opposite side of the jointed linkage. There are about 105,000 upper limb (UL) loss people in the United States with amputations from diabetes, trauma, or birth defects, and 70% of these amputations are distal to the elbow [82]. Hand injuries account for a third of all injuries at work, a third of chronic injuries, a quarter of lost working time, and a fifth of permanent disability [81]. Often underreported, the most common amputation is a partial hand amputation, representing 61,000 amputations per year [82], which is 33% of all 185,000 yearly amputations [83]. Better reporting is in pediatrics, where nearly 92% of amputations over a 12-year national study among children and adolescents resulted in a partial hand [84]. Prostheses provide a way for amputees to interact with the world with a substitute for the human hand; however, it is difficult to achieve a realistic,

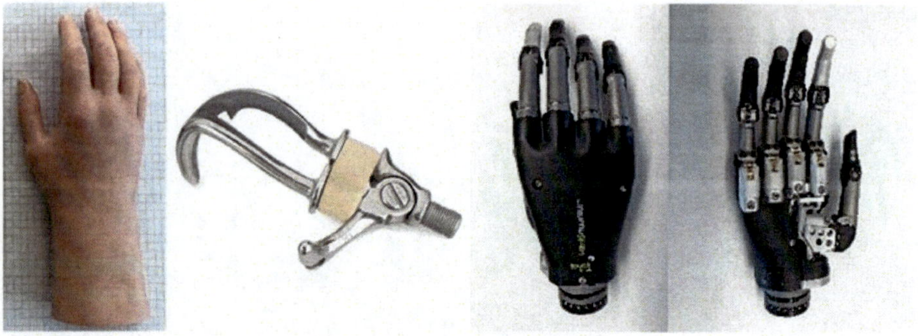

Fig. 1.27 Examples of current prosthetic hand options in the marketplace. (Photo credits: Ras Labs)

functional human hand replacement. Many prosthetic hands either look very realistic with little function or are highly functional without accurately mimicking and appearing like a human hand (Fig. 1.27).

Ras Labs collaborated in a side project with United Prosthetics and the Harvard Medical School MED Science Program to create a 3D printed customized partial prosthetic hand to be intuitive to operate, lightweight, aesthetic, and affordable for a petite quad amputee. The sizing of this hand has applications in pediatrics. Equipping a child with a prosthetic device as soon as possible is crucially important due to the brain-mapping that occurs in early childhood [80]. For children, having any level of prosthetic hand not only helps with the brain-mapping for limbs but also improves all aspects of learning, and this intuitive cognition is expected to be much better with extremely realistic prostheses. In other words, using a limb, prosthetic or otherwise, opens up and unlocks cognitive functions for the brain in early development [80]. Our hands are an integral and defining part of being human. The development of a truly realistic prosthetic hand in form, function, and feel would be a game-changer for people who have lost their hands or who were born without fully developed limbs. This work toward replicating human grasp has implications in robotics as well, particularly for collaborative robots, also known as cobots; for humanoid robots, also known as hubots; and for safe human and work assist robotics.

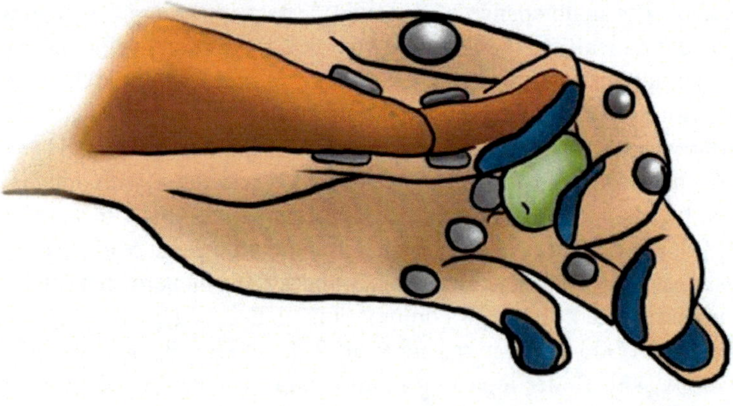

Conceptual rendition with EAP sensing and shape-morphing areas for a hand-like gripper

What Is Needed to Drive This Technology Home Is a Small Robust Fuel Cell Shape-morphing Synthetic Muscle™ hydrogels need to be moist (need water) and need electricity, and above 1.23 V, they release oxygen and hydrogen. Fuel cells require oxygen and hydrogen and then release water and electricity. This tie-in is a perfect loop (Fig. 1.28). Fuel cell inclusion would reduce energy draw and increase overall efficiency. This isn't a perpetual engine due to some energy loss through heat and friction, but nonetheless, this would be an extremely energy-efficient way to produce motion, including biomimetic movement.

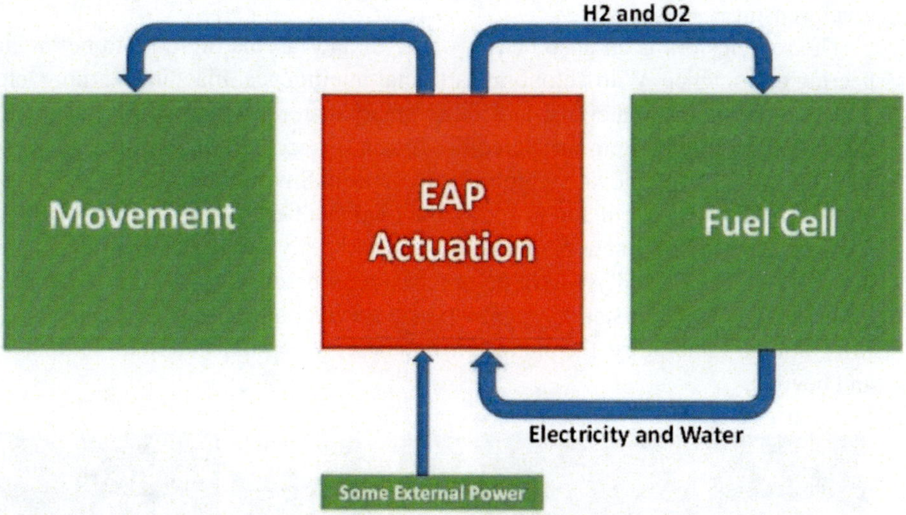

Fig. 1.28 Fuel cell tie-in for extremely energy-efficient motion [10]

Replicating human grasp has implications in robotics, particularly for collaborative robots, also known as cobots; for humanoid robots, also known as hubots; and for safe human and work assist robotics. Synthetic Muscle™ can survive and work in environments where humans cannot safely enter, due to extreme environments or due to contagions that have no cure. In addition to shape-morphing abilities, these EAPs were also used to intrinsically sense pressure due to the correlation between mechanical force applied to the EAP and its electronic signature. Two sensor designs were investigated: a dome-shaped FingerTip-like soft sensor and a flat rectangular prism FingerTip™ soft sensor. Both designs gave good point of contact detection and were able to provide glide information, which is useful for slippage detection when robotic grippers are handling objects, particularly fragile items. Machine learning (ML) and artificial intelligence (AI) can be layered into these sensor systems; however, even with these add-ons for more intuitive feedback, it is highly desirable to have these EAP sensor systems as consistent, reliable, and simple as possible to maintain low latency, i.e., to keep the feedback speed near instantaneous in real time without delays. Concerning glide feedback, to prevent slippage, there could be a fast feedback loop without much ML and AI, while for more sophisticated tasks and motion, ML and AI could be more fully employed. The analogy is when our hand feels pain, we immediately jerk our hand back because of the faster response from the quick spinal cord feedback loop, while the more delayed interpretation of pain and analysis comes from the brain's feedback loop. The immediate feedback loop helps prevent injury, while for more sophisticated tasks like fine craftsmanship, most of the feedback to our hands is cerebral. Compact Synthetic Muscle™-based actuators, operated using off-the-shelf batteries, were retrofitted into traditional grippers to provide controllable gentle grip. The combination of Ras Labs' EAP shape-morphing and sensing features promises the potential for robotic grippers and prosthetic hands with human hand-like dexterous control and tactile touch sensing.

The robotics era is an absolutely wonderful and fascinating time to be in engineering and science. With autonomy, artificial intelligence, machine learning, electroactive polymeric materials, and other smart material technologies all coming together, this is leading to an incredible fusion of mechanical and biological concepts to make truly innovative biomimetic motion. Smart material technology will allow humanity, on Earth and in space, the final frontier, to explore new worlds, to seek out new life and new civilizations, and to boldly go where no one has gone before [85]. To go boldly where no one has gone before includes in research and development, for understanding, growth, and survival on Earth and in space: on the ISS National Laboratory, the planned Moon base, the anticipated Mars settlements, and beyond.

Photo credits and concepts (top L to R): Ras Labs, Ras Labs, E. Starkman PPPL, E. Starkman PPPL, (bottom) NASA.
P. 3 and p. 40 concept credits: Artist C. Foley © 2020.

Acknowledgments We gratefully acknowledge the National Science Foundation, the Center for the Advancement of Science in Space, the Kalenian Award, Breakout Labs, Children's Hospital of Philadelphia/Philadelphia Pediatric Medical Device Consortium, the US DOE, and the US DOD for funding of the synthetic muscle projects. We gratefully acknowledge Livia Rizzo of the Harvard Medical School MedScience Program and interns Curran Dillis, Cole Schreiber, and Jesse d'Almeida for their work with the customized 3D printed prosthetic hand. STEM internships were supported in part through the MLSC Internship Challenge and the PPPL NUF and SULI Programs.

References

1. Q. Pei, M. Rosenthal, R. Pelrine, S. Stanford, R. Kornbluh, Multifunctional electroelastomer roll actuators and their application for biomimetic walking robots. Proc. SPIE **5051**, 11 (2003). https://doi.org/10.1117/12.484392
2. P. Brochu, Q. Pei, Dielectric elastomers for actuators and artificial muscles, in *Electroactivity in Polymeric Materials*, Chapter 1, ed. by L. Rasmussen, (Springer-Verlag, GmbH & Co. KG, © 2012), p. 40
3. M. Shahinpoor, Biomimetic robotic Venus flytrap made with ionic polymer metal composites. Bioinspir. Biomim. **6**(4), 046004 (2011). https://doi.org/10.1088/1748-3182/6/4/046004
4. E.F. Hebling, R.J. Wood, A review of propulsion, power, and control architectures for insect-scale flapping-wing vehicles. ASME App. Mech. Revs. **70**, 010801–010801 (2018)
5. R.J. Wood, 2007, Liftoff of a 60 mg Flapping-Wing MAV, Proc 2007 IEEE/RSJ IROS, 1889–1894. (2007). https://doi.org/10.1109/IROS.2007.4399502
6. J.S. Hyeon, J.W. Park, R.H. Baughman, S.J. Kim, Electrochemical graphene/carbon nanotube yarn artificial muscles. Sens. Acts B: Chem. **286**, 237–242 (2019). https://doi.org/10.1016/j.snb.2019.01.140
7. A. Miriyev, K. Stack, H. Lipson, Soft material for soft actuators. Nat. Commun. **8**, 596 (2017). https://doi.org/10.1038/s41467-017-00685-3
8. K.Y. Ma, P. Chirarattananon, S.B. Fuller, R.J. Wood, Controlled flight of a biologically inspired, insect-scale robot. Science **240**(6132), 603–607 (2013). https://doi.org/10.1126/science.1231806
9. D.A. Wells, *The Science of Common Things* (Palala Press, © 2015), p. 290
10. L. Rasmussen (ed.), *Electroactivity in Polymeric Materials* (Springer-Verlag, GmbH & Co. KG, © 2012)
11. L. Rasmussen, L.D. Meixler, D. Schramm, D. Pearlman, K. Mullally, P. Rasmussen, A. Kirk, Considerations for contractile electroactive polymer based materials and actuators. Proc. SPIE **7976**, 2B1–2B13 (2011)
12. L. Rasmussen, C.J. Erickson, L.D. Meixler, G. Ascione, C.A. Gentile, C. Tilson, E. Abelev, Considerations for contractile electroactive polymeric materials and actuators. Polym. Int. **59**, 290–299 (2010)
13. L. Rasmussen, Electrically driven mechanochemical artificial muscle: For smooth 3-dimensional movement in robotics and prosthetics. Proc. SPIE **6524**, 20 (2007)
14. L. Rasmussen, L.N. Albers, S. Rodriguez, C.A. Gentile, L.D. Meixler, G. Ascione, R. Hitchner, J. Taylor, D. Hoffman, D. Cylinder, L. Moy, P.S. Mark, D.L. Prillaman, R. Nodarse, M.J. Menegus, J.A. Ratto, C.T. Thellen, D. Froio, L. Valenza, C. Poirier, C. Sinkler, D. Corl, S. Hablani, T. Fuerst, S. Gallucci, W. Blocher, S. Liffland, Ras Labs-CASIS-ISS NL experiment for synthetic muscle returned to Earth: Resistance to ionizing radiation. Proc. SPIE **10163**, 1016310 (2017). https://doi.org/10.1117/12.2267716
15. L. Rasmussen, E. Sandberg, L.N. Albers, S. Rodriguez, C.A. Gentile, L.D. Meixler, G. Ascione, R. Hitchner, J. Taylor, D. Hoffman, D. Cylinder, L. Moy, P.S. Mark, D.L. Prillaman, R. Nodarse, M.J. Menegus, J.A. Ratto, C.T. Thellen, D. Froio, C. Furlong, P. Razavi, L. Valenza, S. Hablani, T. Fuerst, S. Gallucci, W. Blocher, S. Liffland, Ras Labs-CASIS-ISS NL experiment for synthetic muscle: Resistance to ionizing radiation. Proc. SPIE **9798**, OP1–O10 (2016). https://doi.org/10.1117/12.2219473
16. L. Rasmussen, C.J. Erickson, L.D. Meixler, The development of electrically driven mechanochemical actuators that act as artificial muscle. Proc. SPIE **7287**, E1–E13 (2009)
17. L. Rasmussen, S. Rodriguez, M. Bowers, G. Franzini, C.A. Gentile, L.D. Meixler, G. Ascione, R. Hitchner, J. Taylor, D. Hoffman, D. Cylinder, L. Moy, P.S. Mark, D.L. Prillaman, R. Nodarse, M.J. Menegus, R. Carpenter, D. Martin, M. Maltese, C. Furlong, P. Razavi, G. Martino, Synthetic muscle electroactive polymer (EAP) based actuation and sensing for prosthetic and robotic applications. Proc. SPIE **10594**, 105942C (2018). https://doi.org/10.1117/12.2297660

18. R.J. Wood, The first takeoff of a biologically inspired at-scale robotic insect. IEEE Trans. Robot. **24**(2), 341–347 (2008). https://doi.org/10.1109/TRO.2008.916997
19. F. Madsen, A.E. Daugaard, S. Hvilsted, A.L. Skov, Review: The current state of silicone-based dielectric elastomer transducers. Macromol. Rap. Comms. **37**(5), 378–413 (2016). https://doi.org/10.1002/marc.201500576
20. R. Pelrine, R. Kornbluh, Q. Pei, J. Joseph, High-speed electrically actuated elastomers with strain greater than 100%. Science **287**(5454), 836–839 (2000). https://doi.org/10.1126/science.287.5454.836
21. F. Carpi, D. De Rossi, R. Kornbluh, R. Pelrine, P. Sommer-Larsen, *Dielectric elastomers as electro-mechanical transducers: Fundamentals, materials, devices, models & applications of an emerging electroactive polymer technology* (Elsevier, © 2008)
22. R. Pelrine, R. Kornbluh, Q. Pei, J. Joseph, High-Speed Electrically Actuated Elastomers with Strain Greater Than 100%, in *Electroactivity in Polymeric Materials*, Appendix B, ed. by L. Rasmussen, (Springer-Verlag, GmbH & Co. KG, © 2012), pp. 151–159
23. C. Bonomo, L. Fortuna, P. Giannone, S. Graziani, S. Strazzeri, Motion Sensors and Actuators Based on Ionic Polymer-Metal Composites, in *Device Applications of Nonlinear Dynamics. Understanding Complex Systems*, ed. by S. Baglio, A. Bulsara, (Springer, Berlin/Heidelberg, © 2006), pp. 83–99. doi.org/10.1007/3-540-33878-0_7
24. M. Shahinpoor, Y. Bar-Cohen, J.O. Simpson, J. Smith, Ionic polymer-metal composites (IPMCs) as biomimetic sensors, actuators and artificial muscles: A review. Smart Mater. Structs. **7**(6), R15 (1998). https://doi.org/10.1088/0964-1726/7/6/001
25. Pneumatic Artificial Muscles (© 2020), https://en.wikipedia.org/wiki/Pneumatic_artificial_muscles
26. G. Wang, N. Wereley, T. Pillsbury, Non-linear quasi-static model of pneumatic artificial muscle actuators. J. Int. Mater. Systms. Structs. **26**(5), 541–553 (2015). https://doi.org/10.1177/1045389X14533430
27. V.L. Nickel, J. Perry, A.L. Garrett, Development of useful function in the severely paralyzed hand. J. Bone Joint Surg. **45A**(5), 933–952 (1963)
28. Scientific Instruments, Biorobotics: Build Your Own Robotic Air Muscle Actuator [DIY McKibben AM] (© 2020), https://www.imagesco.com/articles/airmuscle/AirMuscleDescription01.html
29. mGrip and other grippers (© 2020), https://www.softroboticsinc.com/
30. Y. Osada, Conversion of chemical onto mechanical energy by synthetic polymers (chemomechanical systems), in *Advances in Polymer Science*, ed. by S. Olivé, G. Henrici-Olivé, vol. 82, (Springer, © 1987), pp. 1–46. doi.org/10.1007/BFb0024041
31. Y. Osada, D.E. De Rossi, *Polymer Sensors and Actuators* (Springer, © 2010)
32. Y. Osada, A. Khokhlov, *Polymer Gels and Networks* (Marcel Dekker, © 2002)
33. Y. Osada, *Polymer Sensors and Actuators* (Springer, © 2000)
34. Y. Osada, K. Kajiwara, *Gels Handbook* (Academic Press/Elsevier, © 2000)
35. T. Narita, N. Hirota, J.P. Gong, Y. Osada, Effects of counterions and co-ions on the surfactant binding process in the charged polymer network. J. Phys. Chem. **103**, 6262–6266 (1999). https://doi.org/10.1021/jp990358l
36. M. Uchida, M. Kurosawa, Y. Osada, Swelling process and order-disorder transition of hydrogel containing hydrophobic ionizable groups. Macromolecules **28**, 4583–4586 (1995). doi:0024.9297/95/2228-4583
37. Y. Osada, S.B. Ross-Murphy, Intelligent gels. Sci. Amer. **268**(5), 82–87 (1993). https://doi.org/10.1038/scientificamerican0593-82
38. H. Okuzaki, Y. Osada, Electro-driven chemomechanical behaviors of polymer gel based on reversible complex formation with surfactant molecules, and polymer gels: Intelligent soft materials as new energy transducers, in *Proceeding of the First Conference on Intelligent Materials, ICIM 92*, ed. by T. Takagi, K. Takahashi, M. Aizawa, S. Miyata, (Kanagawa, © 1992)
39. Y. Osada, H. Okuzaki, H. Hori, A polymer gel with electrically driven motility. Nature **355**, 242–243 (1992). https://doi.org/10.1038/355242a0

40. D. De Rossi, K. Kaliwara, Y. Osada, A. Yamauchi, *Polymer Gels* (Plenum Press, New York, © 1991)
41. M. Miyano, Y. Osada, Electroconductive organogel 2. Appearance and nature of current oscillation under electric field. Macromolecules **24**, 4755–4761 (1991). doi:0024.9297/91/2224.4755
42. Y. Osada, Chemical valves and gel actuators. Adv. Mater. **3**(2), 107–108 (1991). doi:0935-9648/91/0202-0107
43. J. Gong, I. Kawakami, Y. Osada, Electroconductive organogel. 4. Electrodriven chemomechanical behaviors of charge-transfer complex gel in organic solvent. Macromolecules **24**, 6582–6587 (1991). doi:0024.0207/91/2224.6582
44. R. Kishi, Y. Osada, Reversible volume change of microparticles in an electric field. J. Chem. Soc. Faraday Trans. 1 **85**(3), 655–662 (1989). https://doi.org/10.1039/F19898500655
45. Y. Osada, K. Umezawa, A. Yamauchi, Oscillation of electrical current in waterswollen polyelectrolyte gels. Makromol. Chem. **189**, 597–605 (1988). doi:0025-116X/88
46. Y. Osada, M. Hasebe, Electrically activated mechanochemical devices using polyelectrolyte gels. Chem. Lett. **14**(9), 1285–1288 (1985)
47. Y. Osada, M. Sato, Conversion of chemical into mechanical energy by contractile polymers performed by polymer complexation. Polymer **21**, 1057–1061 (1980). doi:0032-3861/80/091057-05
48. T. Shiga, Y. Hirose, A. Okada, T. Kurauchi, Bending of a high strength gel in an electric field. Polym. Preprt. **30**(1), 310–314 (1998)
49. T. Shiga, Deformation and viscoelastic behavior of polymer gels in electric fields. Adv. Polym. Sci. **134**, 131–162 (1997)
50. T. Tanaka, I. Nishio, S.T. Sun, Collapse of gels in an electric field. Science **218**, 467–469 (1980). doi:0036-8075/82/1029-0467
51. T. Tanaka, A.Y. Grosberg, Molecular dynamics of multi-chain coulomb polymers and the effect of salt ions. AIP Conf. Proc. **469**, 599–606 (1999). https://doi.org/10.1063/1.58554
52. A.E. English, T. Tanaka, E.R. Edelman, Polymer and solution ion shielding in polyampholytic hydrogels. Polymer **39**(24), 5893–5897 (1998). https://doi.org/10.1016/S0032-3861(98)00106-2
53. E.S. Matsuo, T. Tanaka, Kinetics of discontinuous volume-phase transition of gels. J. Chem. Phys. **89**(3), 1695–1703 (1988). https://doi.org/10.1016/S0021-9606/88/151695-09
54. S. Hirotsu, Y. Hirowaka, T. Tananka, Volume-phase transitions of ionized n-isopropylacrylamide gels. J. Chem. Phys. **87**(2), 1392–1395 (1987). doi:0021-9606/87/141392-04
55. T. Tanaka, Gels, in *Structure and Dynamics of Biopolymers*, NATO ASI Series E, ed. by C. Nicolini, (Martinus Nijhoff Publishers, Boston, © 1987)
56. T. Tanaka, E. Sato, Y. Hirokawa, S. Hirotsu, J. Peetermans, Critical kinetics of volume phase transitions of gels. Phys. Rev. Lett. **55**(22), 2455–2458 (1985). https://doi.org/10.1103/PhysRevLett.55.2455
57. T. Tanaka, Critical dynamics, kinetics and phase transitions of polymer gels. Polym. Preprt. **27**(1), 235 (1985)
58. T. Tanaka, Gels. Sci. Amer. **244**(1), 124–138 (1981). https://doi.org/10.1038/scientificamerican0181-124
59. T. Tanaka, D. Fillmore, S.T. Sun, I. Nishio, G. Swislow, A. Shah, Phase transition in ionic gels. Phys. Rev. Lett. **45**(20), 1636–1639 (1980). https://doi.org/10.1103/PhysRevLett.45.1636
60. V.V. Vasilevskaya, I.I. Potemkin, A.R. Khokhlov, Swelling and collapse of physical gels formed by associating telechelic polyelectrolytes. Langmuir **15**, 7918–7924 (1999). https://doi.org/10.1021/la981057q
61. R.A. Haslam, M. Boocock, P. Lemon, S. Thorpe, Safety Sci. **40**, 625–637 (© 2002)

62. L. Rasmussen, P.N. Vicars, C. Briggs, T. Cheng, E.A. Clancy, S. Carey, B. Secino, Synthetic muscle electroactive polymer shape-morphing and pressure sensing for robotic grippers. Proc. SPIE **11375**, 1137505 (2020). https://doi.org/10.1117/12.2558965
63. NASA Space Radiation Analysis Group, Johnson Space Center (© 2019), https://srag.jsc.nasa.gov/SpaceRadiation/What/What.cfm
64. National Institute of Standards and Technology, Radionuclide Half-Life Measurements (© 2010), http://www.nist.gov/pml/data/halflife-html.cfm
65. R.L. Fleischer, P.B. Price, R.M. Walker, *Nuclear Tracks in Solids: Principles and Applications*, Chapter 2. (University of California Press, © 1975), p. 54
66. P.B. Price, L.R. Fleischer, Identification of energetic heavy nuclei with solid dielectric track detectors: Applications to astrophysical and planetary studies. Annu. Rev. Nucl. Sci. **21**(310) (1971). doi/abs/10.1146/annurev.ns.21.120171.001455
67. G.M. Comstock, R.L. Fleischer, W.L. Giard, H.R. Hart, G.E. Nichols, P.B. Price, Cosmic-ray tracks in plastics: The Apollo helmet dosimetry experiment. Science **172**, 154–156 (1971)
68. R.L. Fleischer, *Tracks to Innovation* (Springer, © 1998), p. 121
69. S. Kelly, *Endurance: My Year in Space, a Lifetime of Discovery* (Knopf Penguin Random House, © 2017)
70. US Geological Survey, US Department of the Interior (© 2020), www.usgs.gov/special--topic/water-science-school/science/water-you-water-and-human-body?qt-science_center_objects=0#qt-science_center_objects
71. L. Rasmussen, L.D. Meixler, C.A. Gentile, Contractile electroactive materials and actuators. Proc. SPIE **8340**(10), 1–14 (2012)
72. M. Chaplin, Peroxide and Oxygen Radicals, in *Water Structure and Science*, Creative Commons. (2020). www1.lsbu.ac.uk/water/o2water.html
73. K. Tomanova, M. Precek, V. Mucka, V. Vysin, L. Jiha, V. Cuba, At the crossroad of photochemistry and radiation chemistry: Formation of hydroxyl radicals in diluted aqueous solutions exposed to ultraviolet radiation. Phys. Chem. Chem. Phys. **19**, 29402–29408 (2017). https://doi.org/10.1039/C7CP05125E
74. Mechno-arm (© 2020), https://starwars.fandom.com/wiki/Mechno-arm
75. B. Patel, Synthetic Skin Sensitive to the Lightest Touch (© 2010), https://spectrum.ieee.org/biomedical/bionics/synthetic-skin-sensitive-to-the-lightest-touch
76. Y. Jiang, U. Mansfield, K. Kratz, A. Lendlein, Programmable microscale stiffness pattern of flat polymeric substrates by temperature-memory technology. MRS Commun. **9**(01), 181–188 (2019)
77. W.G. Bircher, A.M. Dollar, A. Morgan, OpenHand (© 2020), https://www.eng.yale.edu/grablab/openhand/
78. L. Rasmussen, S. Rodriguez, M. Bowers, D. Smith, G. Martino, L. Rizzo, C. Scheiber, J. d'Almeida, C. Dillis, Adjustable liners and sockets for prosthetic devices. Can Prosth. & Ortho J. **1**(2), 1–3 (2018)
79. L. Rasmussen, S. Rodriguez, M. Bowers, D. Smith, G. Martino, L. Moy, P.D. Mark, D. Prillaman, R. Nodarse, R. Carpenter, D. Martin, C. Scheiber, J. d'Almeida, Synthetic muscle electroactive polymer based actuation and pressure sensing for prosthetic and robotic gripper applications. Proc. SPIE **10966**, 1096626 (2019). https://doi.org/10.1117/12.2514429
80. Discussions with prosthetic experts: Mt Sterling, OH, 2016-present; Duderstadt, Germany, 2017; with Dr. Matthew Maltese, Children's Hospital of Philadelphia, Philadelphia, PA, 2015-present; and with Dr. Todd Farrell, LTI, Holliston, MA 2015-present
81. W. Burngardner, Very Well Fit (© 2018), www.verywellfit.com/whats-typical-for-average-daily-steps-3435736
82. Industrial Safety & Hygiene News, Statistics on Hand and Arm Loss, BNP Media (© 2018), http://www.ishn.com/articles/97844-statistics-on-hand-and-arm-loss

83. Access Prosthetics, Living with Limb Loss (© 2017), https://accessprosthetics.com/15-limb-loss-statistics-may-surprise/
84. D.S. Smith, Partial-Hand Amputations. inMotion, **17**(1). (© 2007), https://www.amputee--coalition.org/wp-content/uploads/2015/05/partial_hand.pdf
85. G. Roddenberry, Star Trek: Original Series and Star Trek: The Next Generation, 1966–1969 and 1987–1994

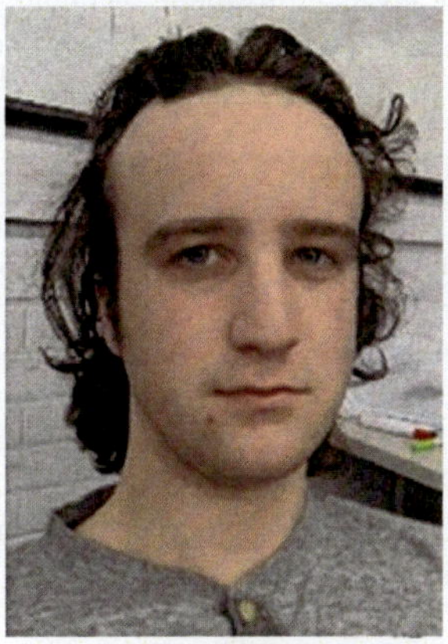

Calum R. Briggs, MS, Robotics Engineer, in addition to robotics engineering, is our hardware-software integrator and a creative force at Ras Labs. He holds a MS in Robotics Engineering from WPI and double major Bachelor of Science degrees in Electrical and Computer Engineering and Robotics Engineering from WPI.

Peter N. Vicars, CEO of Ras Labs, is a serial CEO with a highly successful international and domestic track record. Vicars has led a range of companies with his deep start-up experience and strong operational insights. A visionary and technologist, his leadership has driven growth and turnarounds while providing increased revenues and liquidity to shareholders. Vicars is experienced in all facets of business, from raising capital, M&A, marketing, sales, business development, professional services, software engineering, and finance. He values integrity and growth through problem identification and resolution. His last companies included a telepresence robotics company, an early-stage robotics play in creating right-sized boxes for the E-Commerce world, and a co-robot company in the pick and place world of E-commerce.

Lenore Rasmussen, PhD, is the CTO and Founder of Ras Labs, and she is also an adjunct professor at Worcester Polytechnic Institute (WPI), where she created and teaches the graduate course "Smart Materials and Actuators (and Biomimicry)" through the Robotics Engineering Department. Her corporate experience includes Johnson & Johnson. While in grad school at Purdue, a relative suffered a traumatic farm injury, and she investigated prostheses, but was disappointed with what was available in the marketplace. Her drive to advance prosthetic technology led her to invent Synthetic Muscle™ – electroactive polymers (EAPs) that contract and expand at low voltages, which garnered attention in the robotics community. Dr. Rasmussen is a recognized global leader in the EAP industry with multiple patents and publications, presentations at global conferences, including the SPIE Electroactive Polymers Actuators and Devices (EAPAD) Conferences, Orthotic and Prosthetic Association (AOPA) conferences and world congresses, and robotics meetings and consortia, and multi-media communication, including NASA TV. She holds a PhD in Chemistry, Polymer Chemistry Specialty, from Virginia Tech; MS in Biology, Biophysics Specialty, from Purdue University; and double major Bachelor of Science degrees in Biochemistry and Chemistry from Virginia Tech.

Chapter 2
Q&A with NASA Astronaut Yvonne Cagle, MD Moon and Mars Missions RBE 533 "Smart Materials, Actuation, and Biomimicry," Worcester Polytechnic Institute on October 8, 2019

Yvonne Cagle

Abstract Q&A with Astronaut Yvonne Cagle, MD, about Missions to Moon and Mars, and developments in amplified muscle restoration, plus the challenges for smart materials, robotics, and humanity as we explore the cosmos and consider colonizations in space.

Keywords Accelerated healing · Amplified muscle restoration · Bone density · Spacesuit · Moon · Lunar · Mars · Smart material · Robot · Colony

Yvonne Cagle: Maybe I should start with a little intro. I'll start with who I am and my trajectory, which may touch on the answers to some of your questions and then can fill in with some questions that are outliers and a few of my favorites.

Y. Cagle (✉)
NASA, Mountain View, CA, USA
e-mail: public-inquiries@hq.nasa.gov

© Springer Nature Switzerland AG 2022

L. Rasmussen (ed.), *Smart Materials*,
https://doi.org/10.1007/978-3-030-70514-5_2

I'll start with who I am and a bit about my trajectory, for my profession as an astronaut, my aspiration and my passion are to take it as far off the planet as we can to sustain human life and explore. In order to achieve that type of resiliency for humans as a species as we venture off planet, I spend the majority of my day liaising in Silicon Valley, trying to identify inventions, strategies, and ideas that will help the human system sustain space exploration. For space exploration, that's all systems – vehicles, communication – but for me, the human system is the most exciting. All of that is a fancy way of saying, basically, the "BioMe" is consumed with coming up with ways for the human system to keep up with the robotic systems so we can work collaboratively as we venture off the planet.

How did I get to this place? For me, it started when human, for the very first time, on July 20, 1969, set foot on the Moon, and at that moment my dreams took wing. One day, I, too, wanted to see my footprint on the Moon. It made me feel somewhat like that "Overview Effect",[1] the mysteries behind our place in the universe – on this Spaceship we call "Earth" – and our species occupying it. I wanted to find a way to

[1]The overview effect is a cognitive shift in awareness, reported by some astronauts in spaceflight while viewing the Earth from outer space. It is described as the experience of seeing firsthand the reality of the Earth in space, which is immediately understood to be a tiny, fragile ball of life, shielded and nourished by an extremely thin atmosphere. For example, from Michael Collins, Apollo 11, "The thing that really surprised me was that it [Earth] projected an air of fragility. And why, I don't know. I don't know to this day. I had a feeling it's tiny, it's shiny, it's beautiful, it's home, and it's fragile."

F. White, "The Overview Effect – Space Exploration and Human Evolution," Houghton-Mifflin © 1987.

understand the human body, so we could thrive literally "off-the-planet". I did my undergraduate work at San Francisco State University in biochemistry, then went to the University of Washington at Seattle for my MD [medical doctor], and then served in the Air Force. I spent 22 years, the majority of which I spent flying in a wide variety of high-performance jets. After 15 years as a flight surgeon, I re-engaged my childhood dream and aspired to "get faster" and explore in the Space Shuttle. So, for me, that is what my goal is. And in 2024, every one of your 300+ astronauts are in line to go back to the Moon – for the first time in this century! In 2024, we plan to have humans on the Moon, so we can test out all of our systems that will help get us onto Mars. So, the Moon is not a landing or strip or runway, but actually the Moon landing is a springboard for Mars in 2035. Because the only way we're really going to be able to make it that far in 9 months, and then 6 months on surface, before the 9 months returning home, is we need to validate the expedition systems first. The preliminary plans are in 2020 to do a suborbital insertion, where we test out the heavy vehicles. In 2021, plans are underway to send the first uncrewed vehicle, around the Moon – a lunar orbit. In 2024, we land that first crew back on the Moon. These are 30-day mission, where we will validate: propulsion, robotics, rovers, propellants, communications, and of course, human systems, because we will do a lot of science and engineering for these validations.

K. Chang, The New York Times, ISSN 0362-4331 © 2019, www.nytimes.com/2019/07/16/science/michael-collins-apollo-11.html

That brings us to the adaptation in the altering gravity environments of space. Many things happen to the human body when we go to space. For example, if exercise goes down or when our physiologic systems become stressed, we're more vulnerable to changes in the environment and our microbiome. Radiation is always a concern, [because of higher likelihood of] cancer, so we're looking into creating environmental materials and wearables to protect us. [In microgravity], fluid continues to accumulate in our head and in some cases, we become more farsighted. So there's headaches, pressure in our eyes, and in our neural networks. We are not sure to what degrees and who might be more susceptible. We have some exciting noninvasive, wearable equipment that can allow us to do research in the near future and better understand human adaptations ranging from evolutions in our DNA and genes, to the brain's compliance, to accelerating muscle and bone restoration.

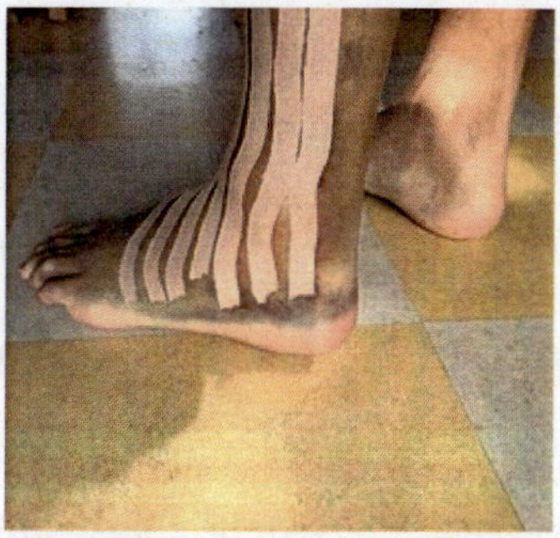

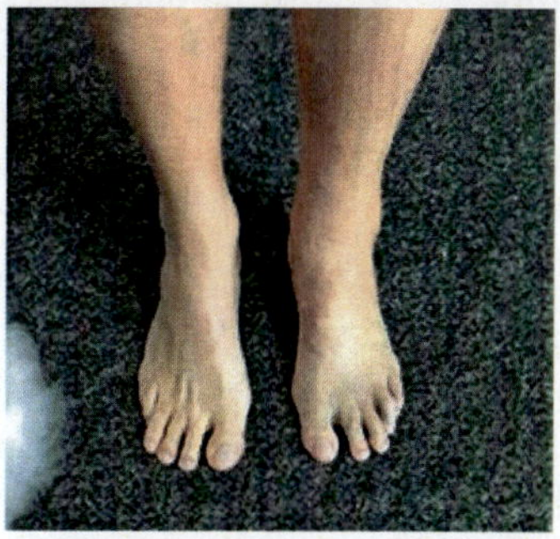

One example is an inventive approach that I designed capable of placing loads back on the body and has been shown to amplify and accelerate restoration of a deconditioned body or soft-tissue injury, shown to reduce weeks of therapy into days, and days of discomfort into minutes. What you're looking at is the before picture of an ankle sprain by an athlete who's training for a triathlon in three weeks. He rolled his ankle 2 days before the photo [top image], and his pain exceeds 10 out of 10 on the pain scale. Typically, thus injury takes 10 to 12 weeks to rehabilitate, and even then, he would be susceptible to reinjury and most likely won't be as competitive over the years. It might even affect his career. Let's put up the after picture [bottom image]. After applying the wearable guide, he now has no pain, he's able to bear weight, and even climb stairs. Anyone want to hazard a guess on how long it may have taken him to get from the before picture to the after picture?

Audience: It was his left leg that was injured?

YC: Yes, it was his left leg, and you can see stripped marks from the injury tape on his leg he had been applying in the initial days after his injury that we have since removed after this session. Any guesses?

Audience: 3 days? 10 days? 10 weeks? 10 hours?

YC: Great guesses! The time from the before photo to the after photo was only – 10 minutes! Let me create a little more insight. For you, we're going to up the game. So now you're this triathlon athlete, but you're not training for a race, you are on Mars, and you don't have 6 weeks – let alone 10 weeks – to rehabilitate on space station. You have been sleeping. You have become so deconditioned that you may have as little as only 6 days before you might no longer be able to actually rehabilitate your limb, and you may be left functionally incapacitated.

So that [photo on right] was at the end. This is just a reflection how fast the body can heal or be activated to heal itself. You can see the restoration after the first session, [photo on right], and he very rapidly regained strength and stability in the injured ankle within a week after the intervention of the invention. He was running again in 2 days and ran his race in 3 weeks. This could apply to a family member or someone who is injured, or advanced age, or a veteran, or "alternately-able", or first-responder, or the working wounded, or someone who just wants a chance to re-engage with their vocations or recreations, or be able to hold their baby or attend their child's wedding. The important point is that this is an example of how we prep on Mothership Earth before we step off the planet. And that's why we want to push for space exploration. Because of what we can learn and apply to help our environment, our climate, and our lives right here on Earth!

Audience: For the Mars mission landing party, when they [the astronauts] arrive, my understanding is that they will be as weak as kittens?

YC: Yes, very much so. Unless they do some sort of rapid reconditioning on board. So, you've got the trip down there – 9 months. With nuclear based propulsion, we could get that to 6 months. Right now, the ticket to ride is about 9 months. That's one way to get there and land. Then 6 months on surface to do science and then 9 months to return. Although to be honest with you, 9 months in microgravity, to the body, is equivalent to running six 25 K marathons back-to-back. How ready would you be to get up after one 25 K marathon and immediately after crossing back over the finish line, line up at what is now your new starting line, let alone six more races! You're going to be deconditioned for 6 to 9 months. That gives us two options: (1) is to go and take that chance and come back, (2) is my option – go and stay to be early expeditioners – all in!!

Audience: How does your therapy work with microgravity?

YC: The next step is to test the load resistant distribution of my invention/approach in microgravity. The KC-130, zero-Gravity aircraft is a parabolic aircraft and every time it dives down, like a roller coaster, for 26 seconds, you're weightless. You're experiencing microgravity. During that dive would be a great time to test my invention out. But 26 seconds isn't going to be the ideal environment. The only way to really test it in microgravity is to take it to take it first to the ISS and/or to the Moon. It's not the same as Mars, but at least it gives us some preliminary space validation.

Someone asked about the pressures encountered and the pressure suits on Mars. The Martian atmosphere is about 0.085 to up to 0.110 psi – very thin [water vaporizes at most temperatures at this pressure]. Our bodies as you know needs to be around 1 atmosphere to keep from bubbling and boiling away. The spacesuit is at 4.3 psi [Earth's atmosphere at sea level is 14.7 psi (1 atmosphere), the ISS maintains the same pressure]. That's not quite as high as what we need to have a good, pressurized environment. That's why we have to quick breathe oxygen, in order to blow off a lot of the nitrogen, so that we don't bubble up and get the bends. Even though we don't have it as high as what would be ideal, if the pressure suit is too high, then outside the station [ISS], then they wouldn't have the flexibility and agility in the fingers and joints – it would be really cold and tight outside. As you can imagine, these suits will be pretty stiff, and slow down the work. The extra effort and work may help the muscles, and over time, your bones. On Mars the pressure suits will be 5.8 psi to help us deal with the atmosphere, so the suit will be pretty stiff, which will slow us down quite a bit. When you're working against that type of pressure, the resistance on your limbs and extra effort and that work out may have a benefit to make your stronger, which hopefully over time will help your bones.

Audience: Exoskeletons may help with pressure suits? And smart materials? Helping with assistance?

YC: Absolutely! I was so excited to hear about the robotics students in the mix [audience]. The future of space flight is going to require us working in cooperation with robotic capabilities and for exactly that reason: to be able to do different tasks that we may not be able to do physically or stamina-wise or because we are being exposed to an [extreme] environment. The other thing I find most exciting about using robots, is when you're working in reduced gravity or microgravity, you're taking three times as long, and people forget that. When we timeline, we have to make sure we have enough time because it's like you're moving in slow-motion. Robots not only actuate in real time, but robots don't have to take the additional time to pull off tasks in order to suit up or suit down!

Audience: For your device, can it be modified to maintain bone density by load?

YC: With compression devices, at very high loads, there is a limit where the neurovascular system can't tolerate pressures over a certain threshold with vascular supply issues to the nerves. You have to be careful with compression devices, if they are too tight or if they are worn too long. If you can get these [loads] high enough and for long enough [wear], you very well may be able to affect the bone. On the other hand, we may not need [high loads] – 'more is better' – may not necessarily be the answer. It may be strategically how, where, and how long the loads to the bones are applied. It really depends why and where in the human body these signals arise from as to how we can titrate and moderate the stimulation and modulation of those areas being "effected".

Audience: For the lunar and Mars missions, what is the training of the crew? How big will the crews be? What are the skill sets needed?

YC: Couple of things to keep in mind: Is that when we go to the Moon and Mars and beyond, these are no longer considered lower orbit missions. These will go for several weeks, several months, upwards to a year. [Most previous missions] were "tethered" space missions, so if something goes awry, we're still in real-time communication with mission control and not too far from getting back home. But when we do Moon and Mars, those are going to be untethered. Those are going to be deep space missions. We will have delays in our communications with latencies ranging from 3 to 20 minutes each way. We will need to take with us most everything we need or hopefully find strategic ways to resource in situ what we can either innovate, 3D print or otherwise additive manufacture, or find AI [Artificial Intelligence] that can sustain operations.

That applies to the crew mix as well for a diverse set of crew skills. Fortunately, we've already worked towards that on our lower orbit missions. And we certainly are going to need someone with piloting skills to get us there and back.

Communication skills, scientists, engineering skills, robotic skills, computational science skills, etc. – but these are the sort of technical STEM skills that we are all very familiar with. We're going to need, in addition, artistic skills as well. It's in our DNA, it's in our wiring, to be able to have things that can create entertainment, diversion, creativity, and stimulate our curiosity. Hence, this is why I prefer to use the more collaborative term: TEAMS, for Technology-Engineering-Art-Math-Science. Currently we don't fly a physician, on every LEO mission. You can now begin to appreciate just how critical the mix of skill sets are that will need to be allocated. The more important question is not only the well-known technical skills that we're familiar with or currently have a position for on crews, but there are some other disciplines that we need to think about as well: Professions such as dentistry, such as nursing – someone who can prepare a drip or care for a wound over an extended period of time and to be able to know the tissue well enough to detect early warning conditions that might be going awry. You might need some very strong vocational skills: agriculture, blacksmithing, sewing, textiles, fabrics. Celestial or Compass Navigation! Many of the skills that you may not get in an academic university, but they may get on a farm or ranch or in the boiler room of a submarine having signed on as an early recruit coming out of high school or endemic to one's culture. It's going to be interesting to mix those skill sets that we're going to need for really long duration.

Crew size: Ideally you want enough to create society. Society is defined basically by skill mix and redundancy. So not only do you have a good variety of skills and capabilities, but that if someone is incapacitated, others can take over and adapt. You would need at least 3 astronauts in the mission for redundancy. You saw that with the Apollo mission. But you could look at it with skills approved to upwards of 12 or 14 or 15 on your way to 20 would be needed to build a site. In actual practice it would be dependent upon how many people can be provided with enough to eat and with enough room and supplies in an escape vehicle. You never want more people on a crew than you can feed and rescue. Right now, that is what will determine and drive the crew size. You can get creative with that. Many people have said to try to have an odd number of crewmates. If you have an even number, you can split the vote. Odd numbers are better for community decision making. Fortunately, in addition to orbital, ISS, and lunar missions we have a number of terrestrial analog simulators operating out of remote stations that can help us assess and validate our resources and approaches such as polar, biodome, and even deep and submariner habitats. It is crucial to also recognize deep space missions with untethered spacecrafts and communities. Moreover, it is crucial to consider not just the technical, but also the behavioral norms, and how important [these norms are] for us to learn about how to live on a planet. This expeditionary community is a mini-society—and any mini-society needs to be community, resourced with a diversity of skill sets and life experiences, in order to not just survive – but to thrive!

Audience: In one of the recent space walks, you could see a blue bar, where the [smaller] astronaut, she could fit under the bar. Is there in any part of this training where you have gender inclusivity? In terms of precautions to take or advantages of one gender over the other?

YC: Excellent question. There are a lot of questions in this day and age about gender parity, even in space. This is less about gender and more about proportions, because the human system has some basic capabilities and after a certain threshold, it doesn't matter what your gender is. It's human resiliency and human survivability. And at what point do we cross into cyborgs and robotics.

If you step back a little bit, it's less a question of gender and more about strategy and strategic matching up a of size, proportions, loads, and platforms. Over the years, it's been "Let's get a big strong person out there to do a spacewalk," and there are definite advantages for that, like stamina and duration, but there are a lot of cubby spaces that are all around the International Space Station and will be in the habitats that we build, that require a diverse set of sizes, skills, capabilities, and dexterity considerations. You name it, whatever it takes. The more we can be inclusive, the more we're going to be safe, robust, and resilient, because we open up options for problem solving. That's not just something that's physical – it's strategic, too. Diversity brings together a lot of life experiences, problem solving, different lenses, and points of view. If everyone thought the same and had only the same experiences, then they might find that the solution is not looking for one more round hole, when this time you need a square peg. Diversity enables a broader conversation and tapestry of problem-navigating at the table. You'll be surprised by the solutions and how resilient you are.

Audience: What is NASA looking for? Is industry or academia favored by NASA?

YC: Industry and academia are really beginning to blur. As part of the government and space, NASA has more relationships with public-private partnerships. It's less about where your pedigree comes from and who your corporate sponsor is and is much more about your capabilities and your ability to innovate. It may or may not be a traditional academic pathway and may not be the same sort of industry infrastructure that we're used to seeing. I would encourage you to unbox yourself. Think more about skill sets, think about do-it-yourself projects. Think about intermingling and integrating industrial thinking in a way to come up with new and creative ways to problem solve and to grow capabilities within yourself and the environment you're going to be working in or constructing or building out. I am so excited to see that we're not settling for only teaching to the test, but rather teaching people how to think!

Audience: By 2035 [or sooner], human level AI is anticipated. How comfortable are we with robots equipped with AI on Mars? Like a humanoid [hubot]?

YC: It's happening! We're moving in that direction. Our level of comfort is not just a luxury, our level of preparation is a crucial part of the conversation and how we prepare for that inevitability. Most important to keep in mind as we prepare for those types of constructs and integrations is that we are only as safe as our reliability and reproducibility. Tech is wonderful. We can discover and do lots of things, but we're only as safe as that reliability and reproducibility. Hence, that's how we should posture ourselves as we begin to integrate humans. We are only as strong and robust as what we program. With machine learning and AI to become what we need is continual machine learning when we program our robots, which is great, but it's also vitally important to understand that in the same way that the human mind and technology can give us great capabilities, technology is advancing in a very exponential way, and we are vulnerable to also perpetuating errors. So it's not only about how far we go. With human error, we still have the luxury of scale to oftentimes being able to fix it. When an error propagates exponentially, there can be catastrophic consequences. We need to be more error intolerant and be very cautious as we program from the very beginning with the principles of AI and machine learning, to make sure that we minimize our exposure to error and bias as we move forward. I was a Girl Scout, "On my honor, I will try: To do my duty… To help people at all times, and to live by the Girl Scout Law." A Girl Scout is prepared! I think in our preparation as a species we need to humble ourselves and recognize that we are only as good as our weakest link or biggest assumption.

Audience: What are the sensors in the spacesuits and in the clothes astronauts wear for biofeedback and to check health status and such?

YC: There are external sensors and right now we are looking at undergarments and to get sensors into the gear, clothing, and the suit, especially the EVA [extravehicular activity] suit itself. We are working in that direction for basic biodesign information, such as variabilities in respiration, temperature, and biometrics. We would love to get loads and muscle perfusion, tissue oxygenation, oxygen consumption rates, galvanic skin resistance [GRS], and more automatic sensors to look at the ANS [autonomic nervous system] to give us early warnings of parasympathetic and sympathetic tradeoffs. [The sympathetic nervous system controls the body's responses to a perceived threat and is responsible for the "fight or flight" response, while the parasympathetic nervous system controls homeostasis and restoration of the body to the state rest or calm and is responsible for the body's "rest and digest" function.] There are increasingly more wearable sensors coming online that use materials conducive to early indications about what's going on with the human system. The human system, by being its own "weathervane" or thermostat, could help us as we walk around and do things. There might be environmental sensors, in and out of the vehicle or the habitat. The sensors could tell us about and sense planetary surface exposures or sense radiation levels starting to rise or build up. If we're about to encounter some kind of GCR [Galactic Cosmic Radiation] or solar flares or change in atmospheric quality, the sensors could warn us to "check and protect" ourselves. This could enable an exciting extension of our habitat.

LR: Thank you for talking with us tonight. Thank you!! [Applause!]

YC: And thank you! I realize that I'm planting some seeds for some amazing future scenarios that will take us not only to the Moon and Mars, but beyond and returns. Wishing all the very best. Remember there's space for all!

Yvonne Cagle, M.D., Ph.D., NASA Astronaut

Col USAF, Ret. Senior Flight Surgeon

Aerospace Professor

Photo and visual concept credits: p. 1 photos, NASA; p. 2 photos, Astronaut Y. Cagle, MD; p. 4 photo, NASA, p. 4 visual concept, Shutterstock, p. 7 visual concept, Shutterstock, p. 8 visual concept, Shutterstock

Chapter 3
Shape Control of Large Lightweight Mirrors with Dielectric Elastomer Actuation

R. D. Kornbluh, D. S. Flamm, H. Prahlad, K. M. Nashold, S. Chhokar, R. Pelrine, D. L. Huestis, J. Simons, T. Cooper, and D. G. Watters

Abstract Space-based astronomy and remote sensing systems would benefit from extremely large-aperture mirrors that can permit greater-resolution images. To be cost-effective and practical, such optical systems must be lightweight and capable of deployment from highly compacted stowed configurations. Such gossamer mirror structures are likely to be very flexible and therefore present challenges in achieving and maintaining the required optically precise shape. Active control based on dielectric elastomers was evaluated in order to address these challenges. Dielectric elastomers offer potential advantages over other candidate actuation technologies including high elastic strain, low power dissipation, tolerance of the space environment, and ease of commercial fabrication into large sheets. The basic functional element of dielectric elastomer actuation is a thin polymer film coated on both sides by a compliant electrode material. When voltage is applied between electrodes, a compressive force squeezes the film, causing it to expand in area. We have explored both material survivability issues and candidate designs of adaptive structures that incorporate dielectric elastomer actuation. Experimental testing has shown the operation of silicone-based actuator layers over a temperature range of –100 °C to 260 °C, suitable for most earth orbits. Analytical (finite element) and experimental methods suggested that dielectric elastomers can produce the necessary shape change when laminated to the back of a flexible mirror or incorporated into an inflatable mirror. Interferometric measurements verified the ability to effect controllable shape changes less than the wavelength of light. In an alternative design, discrete polymer actuators were shown to be able to control the position of a rigid mirror segment with a sensitivity of 1800 nm/V, suggesting that sub-wavelength position control is feasible. While initial results are promising, numerous technical challenges remain to be addressed, including the development of shape control algorithms, the fabrication of optically smooth reflective coatings,

R. D. Kornbluh (✉) · D. S. Flamm · H. Prahlad · K. M. Nashold · S. Chhokar · R. Pelrine
D. L. Huestis · J. Simons · T. Cooper · D. G. Watters
SRI International, Menlo Park, CA, USA
e-mail: roy.kornbluh@sri.com

© Springer Nature Switzerland AG 2022
L. Rasmussen (ed.), *Smart Materials*,
https://doi.org/10.1007/978-3-030-70514-5_3

consideration of dynamic effects such as vibration, methods of addressing large numbers of active areas, and stowability and deployment schemes.

Keywords Electroactive polymers · Dielectric elastomers · Adaptive optics · Telescope · Mirrors

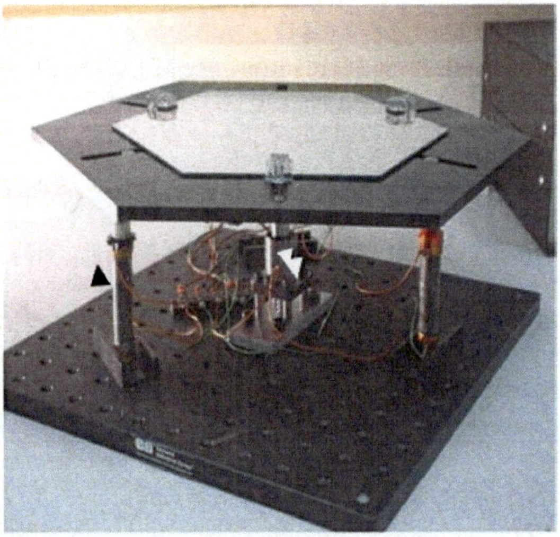

3.1 Introduction

Large-aperture optics are required to achieve the high imaging resolution desired of future space-based sensing systems. For example, the Hubble space telescope is 2.4 m in diameter and has an areal density of 180 kg/m^2 [1]. This telescope is of the largest diameter that can be lofted into space in a single piece. For many national security and scientific reasons, it is desirable to look at earth from a geosynchronous orbit, so that the telescope can remain stationary over an area of the earth. To achieve the same resolution in geosynchronous orbit as a 2.4-meter telescope can achieve in a low-earth orbit would require a telescope with a diameter of on the order of 100 times that of the Hubble. If these large-aperture optic systems are to be cost-effective and practical, the optical systems must be lightweight and capable of deployment from highly compacted stowed configurations. Optical systems based on membrane mirrors or other lightweight structures can address these needs. However, such gossamer structures present many challenges, particularly the need to achieve and maintain precise shape (at the level of a fraction of the wavelength of light used). It is doubtful that such shape control can be achieved without active control. Since active control will eventually be needed, the choice of actuator technology becomes crucial.

Traditional actuation technologies for shape control such as electromagnetics, piezoelectrics, and magnetostrictives are relatively bulky and heavy. Figure 3.1 shows a graphic comparison of these candidate actuation technologies and

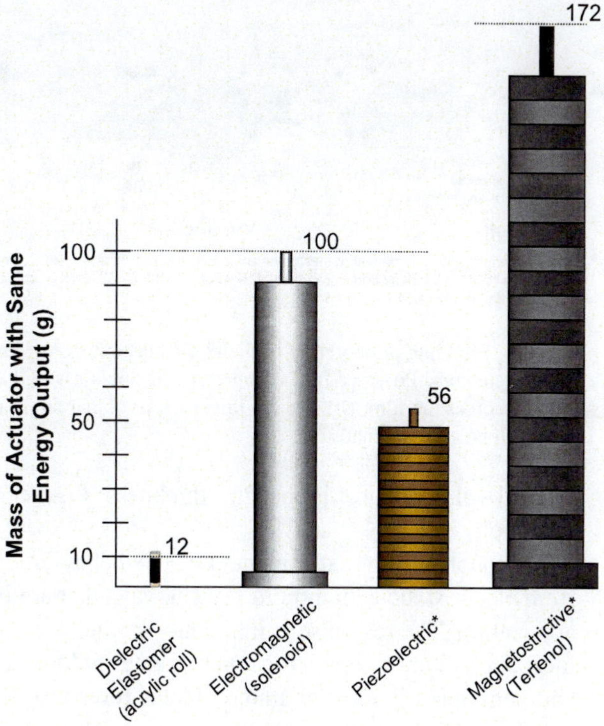

Fig. 3.1 Comparison of actual performance of dielectric elastomer actuators with that of commercially available actuators on an energy density basis. (Sources: Giurgiutiu et al. 1996 [2]; Pei et al. 2002 [3])

compares them with dielectric elastomers. Other actuation technologies that are based on thermal or electrochemical response have also been proposed for shape control (e.g., shape memory alloys). While they have exceptional energy density, these approaches are too slow to respond and difficult to accurately control.

In addition to performance differences, electroactive polymers (EAPs) in general and dielectric elastomers in particular have advantages over other actuators in ease of fabrication, versatility of form, and ease of integration with flexible materials. These and other potential advantages of dielectric elastomers can best be understood by considering their basic principle of operation.

Dielectric elastomers are a type of EAP in which a shape change is induced in the polymer by the imposition of an electric field. The basic functional element is a thin polymer film that is coated on both sides with a compliant electrode material. When voltage is applied to the top and bottom electrodes, an electrostatic compressive force squeezes the film. The polymer is essentially incompressible, so the compressive force simultaneously reduces film thickness and increases film surface area. The induced forces are related to the square of the applied voltage. Thus, by controlling the voltage, the amount of deformation can be precisely controlled. Using this simple electrostatic model, we can derive the effective pressure produced by the electrodes on the film as a function of the applied voltage [4].

This pressure, p, is

$$p = \varepsilon_r \varepsilon_o E^2 = \varepsilon_r \varepsilon_o \left(V / t\right)^2 \tag{1}$$

where ε_r and ε_o are the permittivity of free space and the relative permittivity (dielectric contact) of the polymer, respectively, E is the applied electric field, V is the

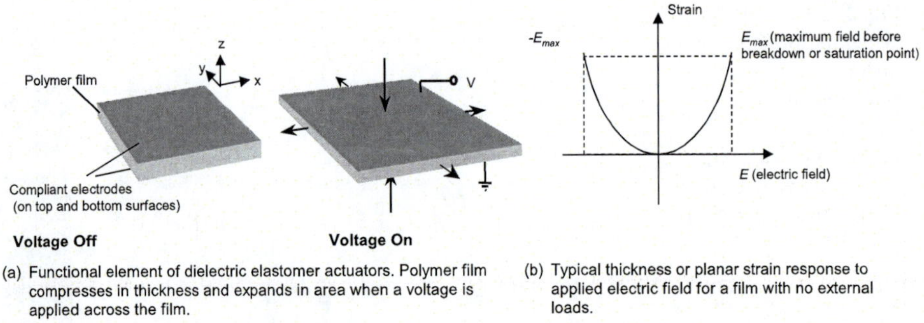

Fig. 3.2 Principle of operation of dielectric elastomers. (**a**) Functional element of dielectric elastomer actuators. Polymer film compresses in thickness and expands in area when a voltage is applied across the film. (**b**) Typical thickness or planar strain response applied electric field for a film with no external loads.

applied voltage, and t is the film thickness. Figure 3.2 illustrates this principle of operation.

Although electrostatic in nature, the behavior of dielectric elastomers differs from electrostatic actuation in several ways that are important for the shape control application. The response is inherently stable, due to the elastic nature of the polymer. The polymer allows for much greater fields before electric breakdown occurs and consequently enables much greater forces to be applied than if air or a near-vacuum was between the two electrodes. The dielectric constant of the polymer multiplies the force output for a given applied voltage.

A great many polymer materials can function as dielectric elastomer actuators [5]. The best- performing materials (those with greatest actuation strains) are based on commercially available formulations of silicone rubber (polydimethylsiloxane) and acrylic elastomers [6]. In fact, both acrylic and silicones have produced maximum strains of more than 100%. The corresponding specific work densities are greater than those of any other field-activated material, including piezoelectric ceramics. Because of this exceptional performance, we focused on these two classes of materials. As will be discussed below, silicones, in particular, can be expected to operate in the extreme environment of space. In addition to the dielectric elastomer film itself, compliant electrodes are also critical to the effective operation of dielectric elastomer actuators. Electrodes based on carbon particles in polymer binders have been the most effective and are expected to be able to survive space environments. In some cases where one of the electrodes does not require large expansion, metallic electrodes such as gold can be used.

This basic functional element can be incorporated into a variety of actuator configurations. Indeed, many configurations have been explored for a large variety of applications [7]. Figure 3.3 shows several configurations that are of particular interest for the actuation of space structures. Only a single addressable electrode area is shown for each actuator configuration. When integrated with the mirror structure, a single actuator could include several individually addressable actuation areas.

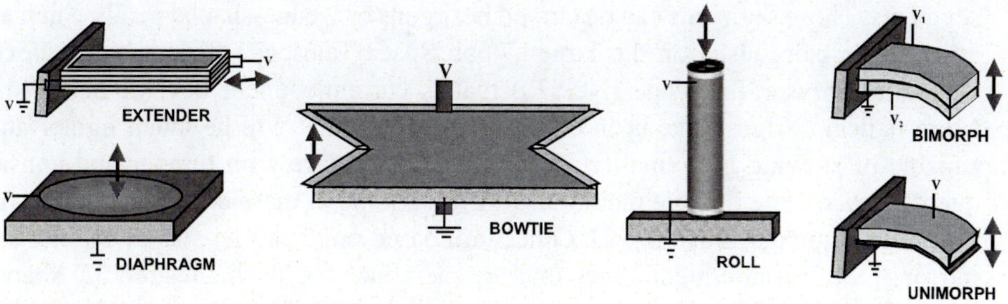

Fig. 3.3 Actuator configurations applicable to space structure control

Such an actuator can be laminated to membrane mirrors, can be incorporated into the boundary of a mirror, or can be a discrete device attached to a mirror or membrane. The films can be manufactured in large sheets at relatively low cost (the best-performing dielectric elastomers have all been based on commercially available polymer formulations). Dielectric elastomers can also be fabricated in a wide variety of shapes and sizes, including large arrays on a single sheet.

A significant difference between dielectric elastomers and other EAP technologies is the wide variety of candidate polymers, which will allow the selection of polymers that offer good performance and can withstand the rigorous environment of space. Polymers such as silicones have shown good performance as dielectric elastomers as well as good performance in space environments. This variety of available polymers further distinguishes the proposed approach from approaches using other electric field-activated polymers such as piezoelectric PVDF materials [8] or electrostrictive polymers [9, 10].

With these advantages in mind, we undertook an effort to explore how dielectric elastomers can help overcome the challenges of large-aperture space optics. This paper describes the results of this study and addresses several aspects of our research:

- Material operation and survivability in the space environment.
- The selection of an actuator configuration and its integration with a stowable, deployable mirror design.
- The gross and fine shape control capabilities of the actuated mirror (both analytical and experimental).
- The surface quality of mirrors with integrated actuation.

3.2 Integration of Dielectric Elastomers with Lightweight Space Mirrors

Several configurations of lightweight space mirrors have been proposed for telescopes that will be used for the observation of space or earth. The most straightforward approach is to make the mirror from an array of smaller, essentially rigid segments. These segments can be curved hexagons or wedge-shaped petals. Such an approach is being used on the James Webb Space Telescope (formerly the Next-Generation Space Telescope (NGST)) that is currently under development [11]. More radical designs have been created for mirrors that will be much lighter and capable of stowing into smaller areas. These designs rely on thinner and consequently much more flexible membrane mirrors. Some of these designs are based on inflatable lenticular mirrors [12]. Others are based on "near-net-shape" membrane mirrors [13, 14], i.e., membranes that are manufactured in the desired 3D shape. Figure 3.4 shows these mirror configurations.

Figure 3.5 shows how these basic mirror designs can incorporate the different dielectric elastomer actuators shown in Fig. 3.3.

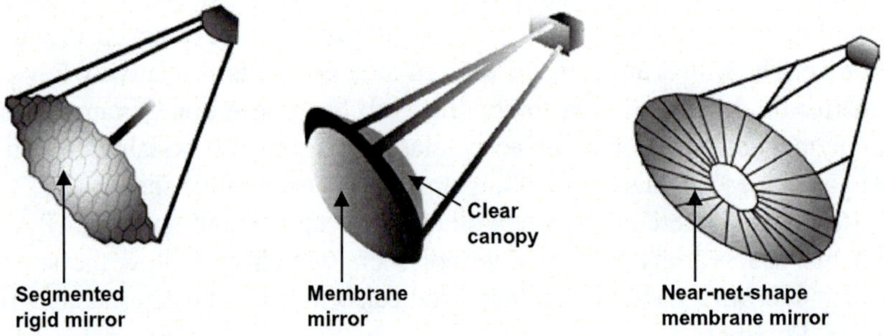

Fig. 3.4 Configurations of existing and proposed large space mirrors. L, proposed NGST segmented rigid mirror [11]; C, inflatable lenticular mirror (microwave antennae) [15]; and R, near-net-shape membrane mirror [13]

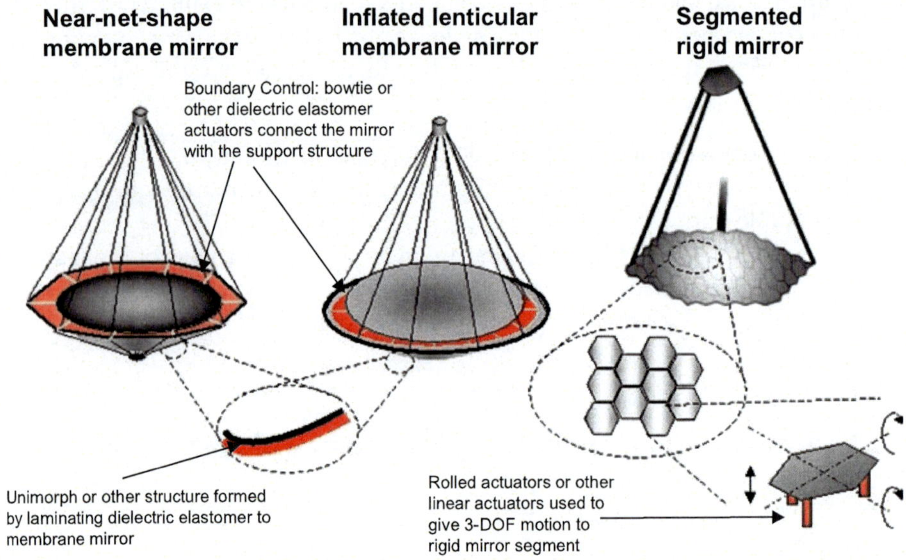

Fig. 3.5 Candidate approaches to shape control of membranes via dielectric elastomer actuation (dielectric elastomer shown in red)

Table 3.1 Design screening criteria

Criteria	Sample parameter(s)
Minimum launch weight/maximum diameter	Areal density
Shape controllability	Degrees of freedom (Zernike coefficients, etc.)
Dynamic control	Bandwidth, settling time
Nominal shape accuracy	Maximum deviation, percentage of wavelength
Stowability/deployment	Ratio of stowed dimension to deployed dimension
Surface continuity and quality	RMS roughness

The designs with boundary control exploit the large-strain capability of dielectric elastomers. Other designs emphasize the elastomers' ability to form large-area arrays, including seamless arrays on a monolithic substrate. In all cases, the elastic nature of the dielectric materials allows for low overall mass, stowability and wrinkle-free deployment, the ability to control shape in the presence of large thermal fluctuations, and the ability to dampen vibrations. Our work focused on the ability of dielectric elastomers to refine and maintain the static mirror shape; deployment and dynamic control issues were not considered. Table 3.1 summarizes important criteria that might be useful in selecting one approach over another. Note that different designs excel in different parameters.

The primary optics, which are most in need of improved adaptation systems, do indeed present by far the greatest challenges due to their large size and low mass requirements. However, we note that the design selection methodology could be applied to secondary (or tertiary, etc.) optics or systems that include both primary and secondary adaptation.

3.3 Material Selection and Survivability

Space is an extremely demanding environment for materials. Major environmental effects include extreme temperatures and thermal cycling, atomic oxygen, UV radiation, micrometeoroids, and debris. While all of these factors are important and must be addressed in the future, we focused on the ability to operate over extremes of temperature as the key technological issue involved in selecting a suitable material.

Solar arrays used in interplanetary missions must withstand temperature fluctuations from −184 °C to 121 °C [16]. We are primarily interested in operation in earth orbit, where the earth itself provides some heating. In the extreme case of geosynchronous orbit, the sensor is roughly 36,000 km from earth, and the time in earth's shadow when cold temperatures are an issue is roughly 1 h. The minimum temperature experienced in geosynchronous orbit is roughly −100 °C. In low earth orbit, earth dominates essentially half the sky seen from the satellite, and earth is radiating as a blackbody at 300 K. In this situation, the equilibrium temperature of an earth-facing mirror is about 210 K or – 60 °C, well within the expected operating range of silicone EAPs.

A variety of commercially available silicone rubbers have shown excellent performance as dielectric elastomers, as noted previously. Certain silicones (methyl phenyl silicones) are known to remain flexible to very low temperatures; generally speaking, if the silicone is flexible, it can be expected to actuate. To experimentally verify this actuation, we stretched a film of the silicone on a circular frame and patterned a small circular active area in the center with powdered graphite as an electrode. This expanding circle actuator was then placed in an environmental chamber. The chamber used a liquid nitrogen coolant to cool the chamber down to as low as −120 °C. A video camera aimed through a window in the chamber was used to measure the amount of strain produced in the circular active area. One silicone in particular, NuSil CV 2287 (NuSil Corporation, Carpinteria, CA), was found to have an exceptional range of performance. As shown in Fig. 3.6, this material produced nearly full actuation at temperatures of nearly −100 °C. Separate tests confirmed that the same material could actuate at more than 200 °C.

Thus, it appears that silicone can withstand the temperature demands of earth orbit. Acrylic was also tested and its operation is limited to roughly −10 °C to 80 °C. Acrylic, therefore, could not be used over 24 consecutive hours without some sort of heat input. A possible solution would be to use a combination of heat storage and insulation to keep the temperature of the polymers in their operating range at all times. For actuators that are laminated to the mirror, around 100 W/m^2 would be required to keep the mirrors at a temperature of −65 °C in geosynchronous orbit. This areal power density would be greatly decreased for discrete or compact actuators. In fact, with suitable insulation and a small amount of joule heating (either intentional or as a side effect of actuation), it should be easy to maintain a temperature of −10 °C to 80 °C, enabling the use of acrylic materials for compact discrete actuators.

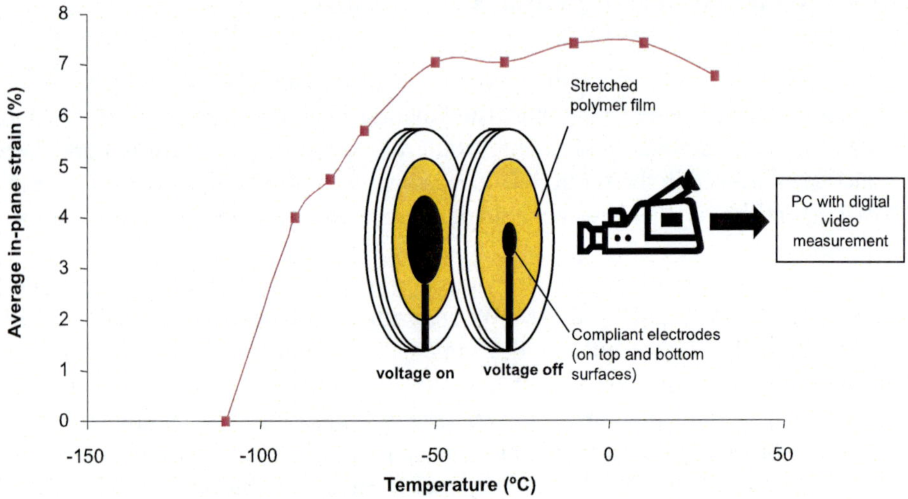

Fig. 3.6 NuSil 2287 silicone radial strain produced with 4 kV actuation as a function of temperature (measurement setup is shown in inset)

3.4 Experimental Evaluation of Designs

Proof-of-principle mock-ups of several of the designs presented in Fig. 3.5 were evaluated experimentally. These designs represent a wide variety of both actuator configurations and mirror designs. A variety of measurement techniques provided shape data ranging from gross shape changes of several millimeters to nanometer-range motion. The measurements allowed us to begin to compare different approaches based on the criteria set forth in Table 3.1.

3.4.1 Laminated Membrane Mirror

Membrane mirrors offer the greatest advantages in terms of areal density and stowability. However, all approaches based on membrane mirrors must be considered high risk because it is extremely difficult to accurately control their shape and ensure good surface quality.

A simple laminated membrane mirror is shown in Fig. 3.7. This basic design is a type of unimorph, since a single actuation phase is laminated to the membrane. We concentrated on the unimorph design because it allows a free surface of the membrane to act as the mirror. For purposes of evaluation, we considered the membrane to be made from strong space-qualified materials such as Kapton (polyimide), while

Fig. 3.7 Laminated membrane structure: test device (top) and basic structure (bottom)

the dielectric elastomer would be a silicone such as that which provided the good temperature range. Good-quality mirrored surfaces can be produced by coating such membranes with metallic finishes.

Before testing the proof-of-principle devices, we performed a computational study shape control by means of the unimorph-type structure. Because of the simplicity of the device's structure and its radial symmetry, we believed that we could use an analytical solution. All versions of the analytical models we have located in the literature are based on assumptions of pure bending (the Kirchhoff hypothesis) of a thin plate with small out-of-plane deflections. In the case of the laminated plates with which we are concerned, the values of the bending moduli of the laminae must be close in order to achieve pure bending. This is not the case with the Kapton-silicone configuration that we have been analyzing; consequently, the assumption of linear variation in strain through the thickness of the laminae is not satisfied, most notably due to a steep discontinuity in slope at the interface. Therefore, we focused on finite element methods.

The same strain discontinuities that invalidated the analytical model also presented challenges for finite element method (FEM) codes. We calculated the response of the disk to a uniform load across the elastomer, using a NIKE-2D FEM code [17]. Unlike the classic parabolic shape predicted by an analytical model with pure bending, the FEM code produced a "Frisbee" shape, with most of the curvature near the edges. Figure 3.8 shows this shape. Qualitative observation of the operation of the laminated mirror test device described above (Fig. 3.7) supports this shape prediction. We further supported the validity of the FEM prediction by using a second type of FEM code. With software that numerically solves partial differential equation models (FlexPDE, PDE Solutions Inc., Antioch, CA), and using automatic and adaptive grid generation and finite element techniques, we again produced the Frisbee.

The FEM model was used to determine whether or not a laminated membrane mirror could, with a patterned array of electrodes, correct for arbitrary errors in shape on the order of the wavelength of light. Figure 3.8 shows the shape produced by the response of a rectangular mirror segment when a narrow strip is actuated. The smoothness of the cross-sectional shape (out-of-surface deflection) is significant and allows us to make some estimates about the feasibility of using a patterned array of active areas for shape control.

A simple heuristic analysis suggests that, even with roughly Frisbee-shaped areas, general 3D mirror shapes are attainable. In this analysis, we assume that the actuated area consists of a flat central portion with a constant-slope wall or lip. For simplicity, we assume that the actuated areas are square. To the zeroth order, actuating a given area causes the lip to slope, depressing the central square some distance below the plane of the mirror. To the first order, the central square of adjacent areas also slopes, because of the moment produced by continuity conditions of the dielectric elastomer at the area perimeters. Thus, actuating a single area produces a piecewise linear approximation of a smooth dimple, and actuating adjacent actuators can make the dimple deeper or more curved, as well as asymmetric. This scheme produces a piecewise, flat approximation to a surface of an arbitrary smooth shape. In

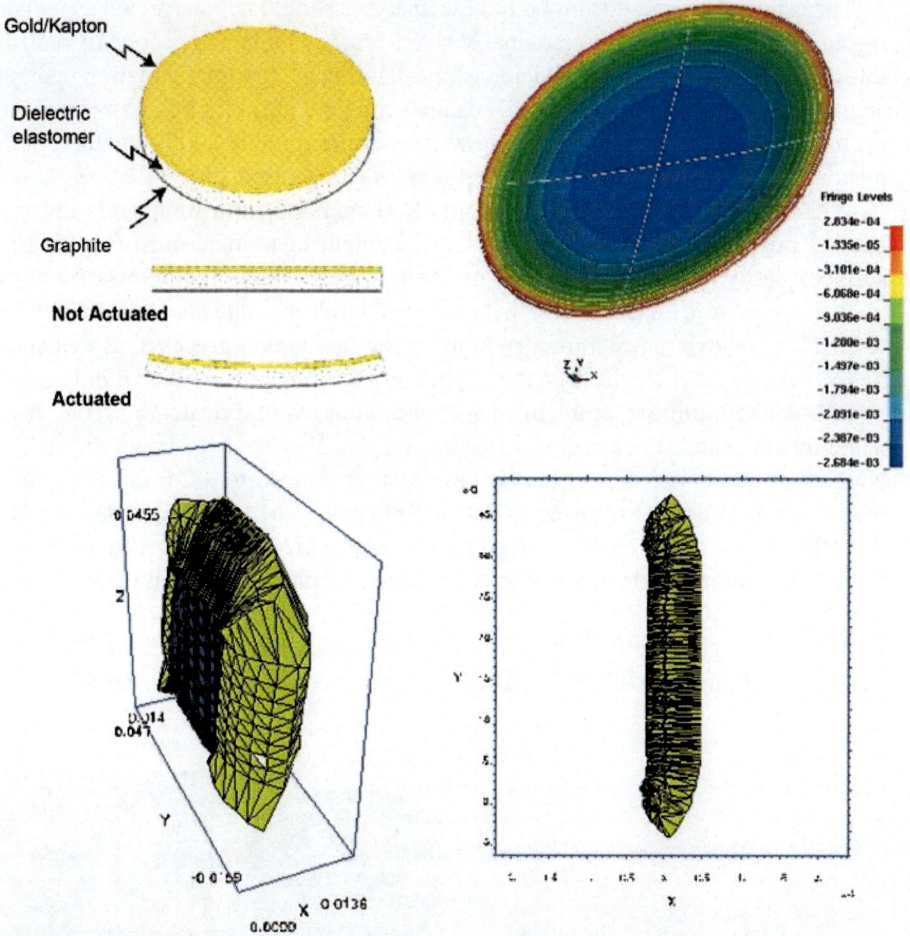

Fig. 3.8 FEM-predicted shape of an actuated laminate membrane: basic model (top L), resultant shape for uniform actuation of a circular disk (top R), and actuation of a narrow strip in a rectangular patch (bottom L and R)

this scheme, an upper bound on the shape approximation error as a function of actuator area size is the maximum deviation of the desired surface from secants of a fixed length equal to the size of the actuator area. We can now apply this analysis to a 5-m parabolic mirror with a laminated dielectric elastomer layer. The dielectric elastomer layer is divided into square actuator regions. A 1-mm square flat actuator produces a shape error of approximately 20 nm, which represents $\lambda/20$ for the minimum wavelength of visible light (about 400 nm). A 10-cm square flat actuator produces a 0.2-mm shape error. If we start with a spherical section instead of an initial flat area, the error is greatly reduced. A spherical-section actuator shape produces a 2×10^{-5} nm error with a 1-mm actuator patch, and a 10-cm spherical-section patch produces a 3.2-nm shape error.

The estimates derived from both these analyses should be interpreted as indicating an order of magnitude for attainable shape errors. Our tentative conclusion from this discussion is that if we can demonstrate significant curvature variation upon the actuation of a 10-cm actuator patch, adequate mirror control for a 5-m or larger mirror may well be feasible. Such control would require roughly 2000 actuators for the mirror or possibly fewer if larger actuator areas can be used.

While the control problem with a laminated membrane mirror appears solvable, another problem remains: to produce and maintain membrane mirrors with good surface quality and continuity. Working with Kapton films roughly 50–100 μm in thickness, we were unable to eliminate obvious wrinkling (due both to actuation and to wrinkles inadvertently introduced during the lamination process), as evidenced by the wrinkles that can be seen in the photograph of the device shown in Fig. 3.7.

In order to minimize problems due to wrinkling, we fabricated a second laminated mirror with a thicker and stiffer membrane. This device, shown in Fig. 3.9, consists of an acrylic dielectric elastomer that is bonded to a 2.6-mil-thick polymeric mirror (Vikuiti Enhanced Specular Reflector, 3 M Corp., St. Paul, MN). A thin silicone elastomer layer provided the bonding. This device was sufficiently stiff to be free-standing without a frame. A single actuation area covered the entire mirror surface.

A 3D photogrammetry system was used to measure the deflection of the laminated membrane (ARAMIS 3D image correlation system, GOM mbH,

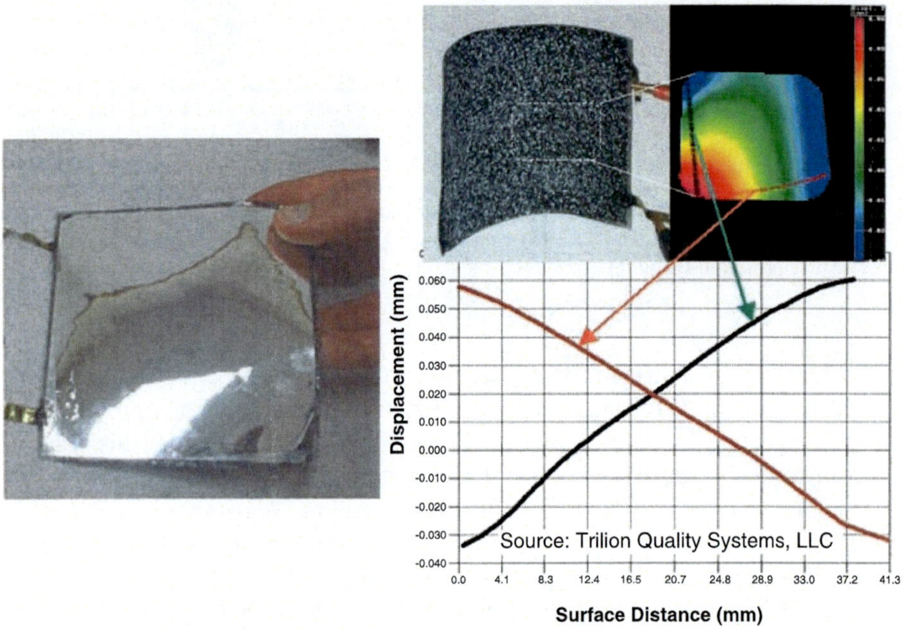

Fig. 3.9 Laminated free-standing polymer mirror: photo of mirrored side (L), backside of device showing dot pattern with out-of-plane deflection map generated by 3D photogrammetry (top R), and out-of-plane deflection as a function of distance in two orthogonal directions (bottom R)

Braunschweig, Germany [18]). The measurements were performed by Trilion Quality Systems, LLC (West Conshohocken, PA).

This system uses 3D image correlation techniques to determine the motion of any point on a surface. To enable the system to function, a random dot pattern was applied to the back (non-mirrored) surface of the device in the form of lightly sprayed black and white paint (see Fig. 3.9). Variations in the dot patterns in stereo images are used to calculate deflection. The out-of-plane deflection of the device generated by this system is shown in Fig. 3.9. The expected smooth curvature over the central portion of the surface of the device is evident.

3.4.2 *Inflatable Membrane Mirror*

An inflatable mirror could overcome problems with wrinkling of membrane mirrors and still offer the benefits of an extremely lightweight structure with the ability to control its shape over a wide range of motion. Inflated lenticular structures such as that flown in the inflatable antenna experiment (Fig. 3.4) have been flown successfully in space. Unlike these structures, we used a much more elastic membrane, the dielectric elastomer itself, in place of the stiffer Kapton or similar film. We hoped that stretching the elastomer would could avoid wrinkling as well as offer greater range of motion. To allow for such a stretchable membrane mirror, a mirrored surface was produced by sputtering gold onto the silicone elastomer surface. This device is shown in Fig. 3.10. Positive or negative pressure differential was used to produce a nominal initial surface curvature. Actuation of the device varied this curvature.

This device showed an ability to undergo large shape changes. A single actuation area covered the entire mirror surface, so its ability to compensate for asymmetric aberrations was not explicitly tested. However, as with the laminated membrane mirror, we expect that such shape control is feasible. With precise control of the

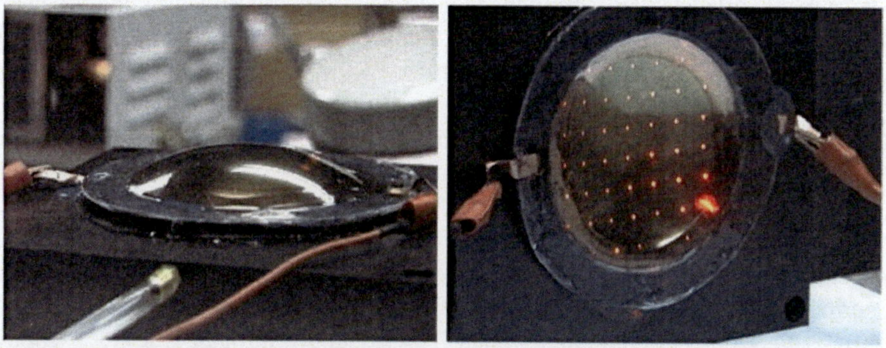

Fig. 3.10 Inflatable active mirror testbed, silicone film with gold surface; The mirror's diameter is 4 in.: mirror under inflation (left) and projected light grid (right)

voltage and patterning of the surface into a sufficient number of areas, it appears that precise shape control could be achieved.

Unfortunately, the mirror quality in our prototype device was too rough to be acceptable, although we succeeded in eliminating wrinkles and achieving good continuity. We explored several combinations and formulations of electrodes on top and bottom of the silicone, as well as variations in the thickness of the gold. We feel that further development and possibly the use of a near-net-shape membrane as the mirrored surface could solve this problem. Additionally, it may be possible to create a truly elastic mirror with good surface quality by stacking a large number of nanometer-thick birefringent elastomeric layers. Such mirrors have been demonstrated with nonelastomeric polymers [19], and there is no reason to think that they could not be made with elastomeric polymer layers.

3.4.3 Positioning of Rigid Mirror

While not nearly as low in areal density as membrane mirrors and therefore limited in size, large mirrors formed from rigid segments may be feasible with today's technology. The 6-m James Webb Space Telescope is an example. Such mirrors still require lightweight materials and support structures and therefore need active positioning of the segments to achieve the necessary shape accuracy. For example, a preliminary designs of James Webb Space Telescope requires actuators with the following specifications [20]:

- Resolution: <20 nm
- Stroke: 6 mm
- Outer diameter (OD): <5 cm
- Mass: <40 g

As shown in Fig. 3.1, dielectric elastomer actuators can offer significant mass savings over other actuators of materials, such as electromagnetics, piezoelectrics, and magnetostrictives, that are capable of producing similar stroke and forces. We have made rolled actuators from the dielectric elastomers. Rolled actuators can produce significant forces and strokes in a compact package: in particular, rolled actuators made from an acrylic dielectric elastomer have produced strokes in excess of 6 mm with ODs of less than 1.5 cm [3]. Their mass is about 15 g.

In order to demonstrate how such linear actuators could be used with a segmented mirror, we designed and constructed a device consisting of a rigid, first-surface mirror mounted on a base plate with a central pivot (Fig. 3.11). Three roll actuators mounted in a tripod configuration were connected to the base plate at equal distances from the central pivot. The central pivot post was loaded with a spring in compression to prevent backlash errors. Each of the roll actuators was connected to a custom-built high-voltage power supply through a joystick control,

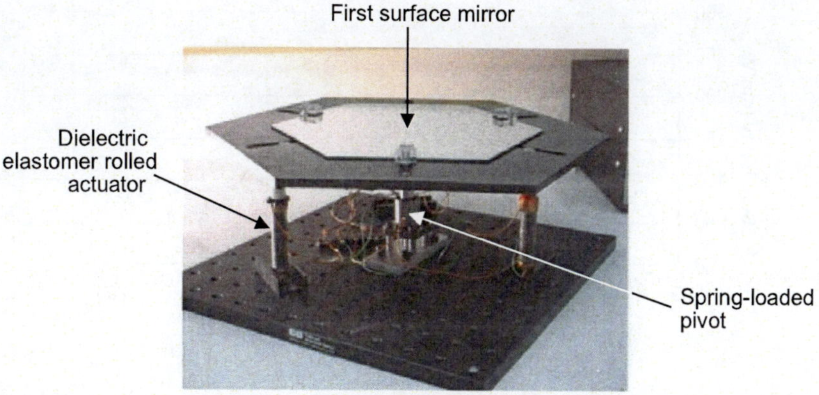

Fig. 3.11 Components of the rigid mirror positioning demonstration device

which was used to command the position of the mirror. Under joystick control, this device could tilt the mirror in any direction.

The resolution of dielectric elastomer rolled actuators, which had not been measured, is a key criterion for the mirror positioning application. An MTI-2000 Fotonic noncontact displacement sensor (MTI Instruments Inc., Albany, NY) with a measurement resolution of 1 microinch (25 nm) and a range of 1.5 mm was used to measure displacements. A Northeast Scientific DC power supply with a range of 0–1600 V and a resolution of 1 V was used to drive the actuators for this test. Displacement measurements (Fig. 3.12) indicate a trade-off in the achievable range and actuator sensitivity for a given power supply.

As seen in Fig. 3.12, trade-offs can be made between the maximum displacement and the drive voltage sensitivity in a given design. In addition, the actuator does not involve any moving parts besides the dielectric elastomer itself. The deflection of the dielectric elastomer is related to the applied voltage, and so the resolution of the system is determined primarily by the resolution of the drive electronics. Therefore, when power supplies are used that have both a high drive voltage and high sensitivity, the acrylic rolls can be used in applications requiring both high range and high resolution. Commercially available power supplies can drive up to 6 kV with a sensitivity of 25 mV. Extrapolating the data from Fig. 3.12 (assuming that sensitivity goes with the square of the applied voltage), we can expect a sensitivity of 1800 nm/V at 3 KV (it should be easy to build a rolled actuator that can produce a stroke of 6 mm at 3 kV). A 25 mV resolution translates to a displacement resolution of 45 nm – close to the goal. Thus, it seems that we can simultaneously meet both the maximum-stroke and the fine-resolution needs with dielectric elastomer actuators.

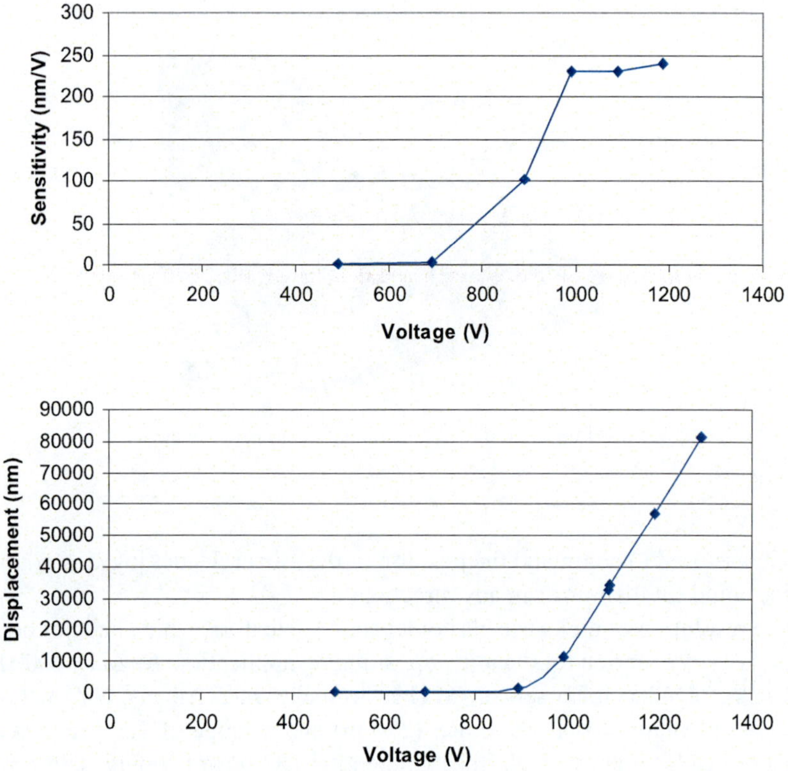

Fig. 3.12 Measurements of displacement at low voltages with an MTI-2000 Fotonic sensor: voltage sensitivity (L) and displacement (R)

3.4.4 Laminated Thin Mirror

The laminated membrane mirrors and the segmented rigid mirror are two extremes of how actuation can be used for shape control. A hybrid approach might be able to offer some of the benefits of low areal density of membrane mirrors with the good surface quality and simpler controllability of rigid segmented mirrors. To this end, we laminated dielectric elastomer actuators to the back of thin mirrors to form a unimorph structure. These mirrors were made out of thin glass or silicon wafers that were approximately 0.5 mm thick. Gold was sputtered onto the front surface to form a first-surface mirror. These mirrors could hardly be called "membranes" because they cannot be obviously deflected in the same way as a thin sheet of Kapton; but they also could not be called "rigid" because they can easily be deformed by more than several wavelengths of light.

We evaluated the performance of these thin mirrors with a Zygo interferometer (Zygo Corporation, Middlefield, CT). As may be seen in Fig. 3.13, these rigid mirrors, whose diameters are 4 in, are flat to within about 20 wavelengths of light and do not appear to have significant higher-order defects (based on the concentricity of

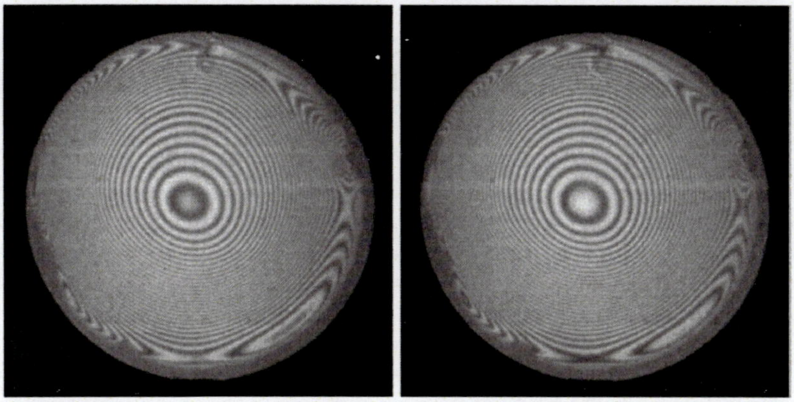

Fig. 3.13 Interferogram of mirror surface of a 4-inch-diameter thin silicon wafer mirror with a laminated acrylic dielectric elastomer actuator: voltage off (L) and voltage on (R)

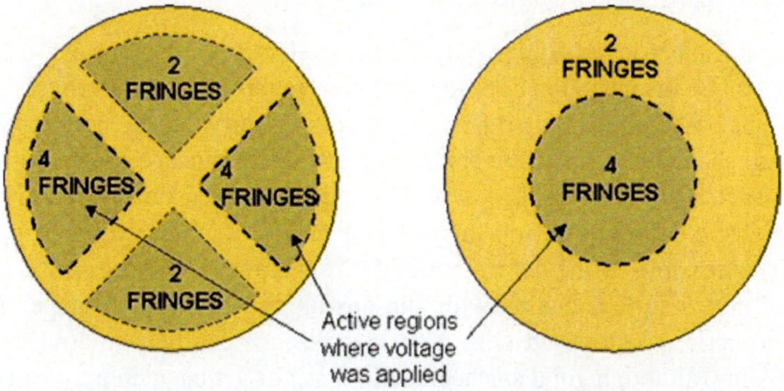

Fig. 3.14 Configuration of active areas and resultant change in the number of fringe patterns resulting from actuation

the interference pattern). This surface shape and quality, while not good enough for an actual imaging system, are surprisingly good in view of the fact that the wafers were not made to optical specifications. Laminated mirrors were made with both silicone and acrylic dielectric elastomer materials. A sputtered gold surface on one side of the wafer formed one electrode, while the other electrode was powdered graphite applied through a stencil on top of the silicone or acrylic elastomer layer. Laminated mirrors were made with both circular active areas (approximately 2 inches in diameter) and 4-quadrant patterns (see Fig. 3.14). The quadrants could be individually actuated.

These mirror-actuator configurations allowed us to demonstrate three important things. First, the application of a voltage did produce a smooth shape change in the mirror surface as evidenced by changes in the number and locations of the fringe lines (similar to the changes evidenced by comparing the two photographs of Fig. 3.13). Second, the shape change was greater within the active region than outside it. For example, the application of a 5 kV produced an out-of-plane displacement of approximately 4 wavelengths directly above an active area and just 2 wavelengths above an inactive area. Third, the shape change could be precisely controlled by controlling the applied voltage. When the applied voltage was varied smoothly by means of a potentiometer, the resultant change in the displacement could be controlled to within a fraction of a wavelength of light. Even at the large voltage of 5 KV, the estimated sensitivity is roughly 0.001 wavelengths per volt.

These results suggest that it is indeed possible to control the shape of a mirror to within a fraction of a wavelength of light.

3.5 Conclusions and Discussion of Future Research

The supposition that dielectric elastomer actuators can enable large high-resolution lightweight optical systems has been supported by several key experiments and analyses. In particular, we have shown that:

- Certain dielectric elastomer polymers and electrodes can survive the range of temperatures expected in both low and geosynchronous earth orbits.
- Designs based on laminated flexible mirrors can produce the necessary effect on overall shape, with sufficient spatial resolution and surface continuity to achieve shape accuracy over large areas, with active areas of reasonable quantity and size.
- Fine shape deformation (on the order of much less than wavelength of visible light) is feasible by the deformation of thin flexible mirror segments.
- Good mirror surface quality with thin flexible mirror segments appears feasible.
- Elastomeric mirrors used with lenticular designs can help minimize wrinkling problems (although good surface quality has not yet been demonstrated).
- Fine positional accuracy (on the order of one-tenth the wavelength of light) can be achieved with rigid mirror segments.

While these results are encouraging, many critical technical issues must still be addressed before the benefits of the technology can be realized.

- The development of control algorithms, including the sensory feedback system, that will generate the needed voltages to the array of actuators to produce and maintain the desired shape (most challenging with membranes and thin flexible mirrors) over the entire mirror surface. Considerable theoretical and analytical work has been done in this area by other researchers with systems based on other actuators. We believe that many of these results are applicable to systems that use dielectric elastomer actuation.
- An analysis of trade-offs among actuation configuration, mirror stiffness, mass, surface quality, shape error, system cost, number of active areas, control complexity, etc., in order to develop the most effective approach for the desired mirror size. A 100-m-diameter mirror would likely behave very differently from a 6-m-diameter mirror.

- The use of distributed actuation for the correction of shape errors resulting from mechanical connections to the mirrors at the boundaries (most critical with membrane approaches).
- The compensation of dynamic effects such as vibration from onboard instruments and the inertial effects pointing the mirror. Dielectric elastomer response times are expected to be fast enough to account for these dynamic effects, but this issue has not been specifically addressed.
- The effects of thermal fluctuation on the shape of the mirror and verification that dielectric elastomers can compensate for these fluctuations. We expect that the large strains available from dielectric elastomers will enable us to compensate for these fluctuations.
- Electrically connecting and addressing the individual actuator segments by methods that minimize complexity and mass and do not affect the surface shape or quality. Approaches based on noncontact methods such as electron beams and light beams may work well with high-impedance dielectric elastomers (in fact, the ability to use light beams to switch dielectric elastomers has already been demonstrated [21]). Other approaches involving time multiplexing may also be possible. Dielectric elastomers operate at low currents, so thick wiring is not needed.
- The development of efficient driving electronics.
- An analysis of the effects of UV, atomic oxygen, micrometeors, static charging, and other factors on the operation of dielectric elastomers (most challenging with laminated membrane designs).
- Stowage and deployment schemes.
- The development of membrane mirrors that do not suffer from wrinkling or poor surface quality. Other research efforts have focused on these factors (e.g., Marker et al. [13]). Mirrors made from elastomeric materials would be compatible with dielectric elastomer actuation and could be inflated or stretched to eliminate wrinkles.
- The longevity of actuators made from specific space-capable dielectric elastomers such as silicones.

The hybrid approach based on the shape control of mirrors consisting of thin (and therefore flexible) mirror segments appears to offer a good compromise between the need to realize mirrors of the necessary surface quality and the shape controllability and minimization of mass and stowed size.

We believe that future efforts could address these remaining challenges and thus enable dielectric elastomers to be used successfully for very large, lightweight space optics.

The dielectric elastomer technology may also be well suited to the shape control of radio and microwave antennae as well as much smaller optical and other high-precision devices for aerospace and terrestrial applications.

Acknowledgments The authors wish to acknowledge the support of the National Reconnaissance Office for this work (Contract NRO000-02-C-0386). The Defense Advanced Research Project

Agency (DARPA) provided support for the building of the rolled actuators and rigid mirror testbed (Office of Naval Research Contracts N00014-02-C-0252 and N00014-00-C-0497). Interferometric measurements of the laminated flexible mirror were made with the assistance of Messrs. Dennis Duneman and Dan Marker of the Air Force Research Laboratory, Kirtland AFB, Albuquerque, NM. Tim Schmidt and John Tyson of Trilion Quality Systems LLC performed the 3D photogrammetry measurements on the shape of the laminate membrane mirror. Dr. D. Laurence Thomsen of the NASA Langley Research Center provided information about materials survivability in space. The authors also wish to acknowledge the help of numerous SRI employees in carrying out the work.

References

1. L. Dye, *High-Tech Reflector: Scientists Developing Superlight Material for Giant Space Mirror* (2002), http://media.abcnews.com/sections/scitech/DyeHard/dyehard011024.html
2. V. Giurgiutiu, C.A. Rogers, Z. Chaudry, Energy-based comparison of solid-state induced-strain actuators. J. Intell. Mater. Syst. Struct. **7**(January) (1996), pp. 4–14
3. Q. Pei, S. Stanford, M. Rosenthal, R. Pelrine, R. Kornbluh, K. Meijer, R. Full, 3-D multi-functional Electroelastomer actuators and their application for biomimetic walking robots, in *Smart Structures and Materials 2002: Industrial and Commercial Applications of Smart Structures Technologies*, ed. by A. McGowan, Proc. SPIE, vol. 4698, (2002), pp. 246–253
4. R. Kornbluh, R.E. Pelrine, Q. Pei, V. Shastri, Application of dielectric EAP actuators, in *Electroactive Polymer (EAP) Actuators as Artificial Muscles – Reality, Potential and Challenges*, ed. by Y. Bar-Cohen, (SPIE Press, Bellingham, Washington, 2001), pp. 457–495
5. R. Kornbluh, R.E. Pelrine, J. Joseph, R. Heydt, Q. Pei, S. Chiba, High-field electrostriction of elastomeric polymer dielectrics for actuation, in *Smart Structures and Materials 1999: Electroactive Polymer Actuators and Devices*, ed. by Y. Bar-Cohen, Proc. SPIE, vol. 3669, (1999), pp. 149–161
6. R. Pelrine, R. Kornbluh, Q. Pei, J. Joseph, High-speed electrically actuated elastomers with over 100% strain. Science **287**(5454), 836–839 (2000)
7. R.D. Kornbluh, R.E. Pelrine, Q. Pei, R. Heydt, S.E. Stanford, S. Oh, J. Eckerle, Electroelastomers: Applications of dielectric elastomer transducers for actuation, generation and smart structures, in *Smart Structures and Materials 2002: Industrial and Commercial Applications of Smart Structures Technologies*, ed. by A. McGowan, Proc. SPIE, vol. 4698, (2002), pp. 254–270
8. J. Wagner, G. Agnes, Optical metrology of adaptive membrane mirrors, in *Smart Structures and Materials 2001: Smart Structures and Integrated Systems*, ed. by L. P. Davis, Proc. SPIE, vol. 4327, (2001), pp. 13–23
9. Q. Zhang, V. Bharti, X. Zhao, Giant electrostriction and Relaxor ferroelectric behavior in Electron-irradiated poly(vinylidene fluoride-trifluoroethylene) copolymer. Science **280**, 2101–2104 (1998)
10. P. Sommer-Larsen, R. Kornbluh, Polymer actuators, in *Proc. Actuator 2002*, (Bremen, 2002, June), pp. 371–377
11. J. Crocker, C. Atkinson, D. Ebbets, L. Gilman, P. Lightsey, D. Neam, S. Texter, TRW/ball: Next generation space telescope (NGST), in *UV, Optical, and IR Space Telescopes and Instruments*, ed. by J.B. Breckinridge and P. Jakobsen, SPIE, vol. 4013, Paper 4013-89, (2000)
12. NASA Goddard, *Preliminary Mission Report, Spartan 207/Inflatable Antenna Experiment, Flown on STS-77*, Spartan Project Code 740.1, NASA Goddard Space Flight Center, Greenbelt, Maryland,14 February 1997
13. D.K. Marker, J.M. Wilkes, R.A. Carreras, J.R. Rotge, C.H. Jenkins, J.T. Ash, Chapter 4: Fundamentals of membrane optics, in *Gossamer Spacecraft: Membrane/Inflatable Structure Technology for Space Applications*, AIAA Progress in Astronautics and Aeronautics Series, Reston, VA, USA, ed. by C. H. Jenkins, (2001)

14. C.H. Jenkins, Chapter 20: Shape control of precision gossamer apertures, in *Electroactive Polymer Actuators*, ed. by Y. Bar-Cohen, SPIE, (2001)
15. R.E. Freeland, G.D. Bilyeu, and G.R. Veal, Validation of a unique concept for a low-cost, lightweight space-deployable antenna structure. IAF-93-1.1.204, 1993
16. E.M. Silverman, *Space Environmental Effects on Spacecraft: LEO Materials Selection Guide*, Contract NAS1–19291 (NASA CR-4661, Parts 1 and 2) (TRW Space & Electronics Group, Redondo Beach, 1993–1995)
17. J.O. Hallquist, LS-NIKE2D: A vectorized, implicit, finite deformation, finite element code for Analyzing the static and dynamic response of 2-D solids, in *LSTC Report 1006*, (Livermore Software Technology Corp, Livermore, 1992)
18. J. Tyson, T. Schmidt, and K. Galanulis, Advanced photogrammetry for robust deformation and strain measurement, in *Proceedings SEM 2002 Annual Conference,* Milwaukee, WI, June 2002.
19. S.R. Carlo, J.S. Shirk, S.R. Flom, R.G.S. Pong, A. Ranade, H. Tai, E. Baer, A. Hiltner, Materials for nanolayered polymeric optical limiters, in *Proceedings of the 224th ACS National Meeting*, Boston, MA, 18–22 August 2002.
20. JWST Actuator Requirements (2003). http://ngst.nasa.gov/Hardware/text/actuator_reqs.html
21. S. Lacour, R. Pelrine, S. Wagner and H. Prahlad, Photoconductive high-voltage switches of thin-film amorphous silicon for EAP actuators, To Be Published in *Smart Structures and Materials 2003: Electroactive Polymer Actuators and Devices*, ed. Y. Bar-Cohen, *Proc. SPIE,* vol. 5051, 2003.

Published with permission from Smart Structures and Materials 2003: Electroactive Polymer Actuators and Devices, ed. Y. Bar-Cohen, Proc. SPIE, Vol. 5051, 143-158, 2003, doi: 10.1117/12.484405.

R.D. Kornbluh is a Principal Research Engineer in SRI International's Robotics Laboratory. Kornbluh has more than 30 years of experience at SRI and has contributed to and led a wide variety of projects to develop new electromechanical transducers for walking, crawling, and flying robots as well as for other systems. He is one of the principal inventors of the dielectric elastomer type of electroactive polymer artificial muscle transducer. Kornbluh is also the principal inventor of the electrolaminate type of mechanical metamaterial that is being applied to a variety of applications in aerospace, robotics, and biomechanics. He is the author or coauthor of more than 60 publications in the areas of polymer actuators and robotics and holds more than 30 patents in these areas. He served on the IPC of the SPIE Electroactive Polymer Actuators and Devices Conference and was an associate editor for the *Journal of NeuroEngineering and Rehabilitation*. Kornbluh has an M.S. in Mechanical Engineering from MIT and a B.S. in Mechanical Engineering from Cornell University. In the early 1990s, he served as a Peace Corps volunteer in Ecuador where he helped improve rural water and sanitation systems. What was the motivation to develop electroactive polymeric actuators and transducers? It started with the desire to create robotic actuators that were more like natural muscle. That work led to more than two decades of innovation in electroactive polymers based on electrostatic forces. Experience with such "soft" electrostatics led to new types of mechanical metamaterials with the ability to modulate stiffness or damping by orders of magnitude, as well as actuators for deformable mirrors and lenses, wall-climbing robots, and improved grippers for materials handling. Such soft active materials are also helping to create wearable robotics for human motion augmentation and even solid-state cooling for improved thermal comfort. Take a look at www.sri.com/core-competency/robotics/ and www.sri.com/bios/roy-kornbluh/.

R. Pelrine is a chief scientist at SRI International. He oversees the development of electroactive polymer actuators, generators, and materials, electrically controlled adhesion, micromachines, magnetics, robotics, magnetic-based sensors, and thermal and fluid modeling. In addition, he has experience in geophysics, especially seismic oil and gas exploration. Since joining SRI in 1989, Pelrine has worked on projects such as pulse rate sensor signal processing, design of a magnetically clamped pipeline robot, design of diamagnetic sensors, micromotors, and cleanroom devices. He is also responsible for the discovery, design, and development of electroactive polymer actuators and generators, polymer-actuated robotics, wall-climbing robots, and electroadhesion. He is a principal inventor of dielectric elastomers, a type of electroactive polymer often referred to as artificial muscle. In 2006, Pelrine received SRI's Fellows Award, which recognizes exceptional staff members for their outstanding accomplishments, and is SRI's highest recognition for technical, scientific, or professional contributions. Pelrine has numerous research papers and holds 70 patents issued in the fields of electroactive polymer actuators and generators, microdevices, magnetics, robotics, sensors, displays, and signal processing. He has numerous additional patents pending. Pelrine has a B.S. in Physics from MIT, a M.S. in Physics from the University of Washington, and a Ph.D. in Mechanical Engineering from the University of Texas. Please see https://www.sri.com/bios/ronald-e-pelrine/.

Chapter 4
The Hydro Muscle and CRFC Valve: An Efficient and Compact Fluidic Robotic System

Marko Popovic, Shannon Moffat, Julia D'Agostino, Ellie Clarrissimeaux, Eric Carkin, and Matthew Bowers

Abstract Hydro Muscle is fluid-actuated artificial muscle with top-rated strain and energy-efficient properties that can closely mimic biological muscle dynamics. In contrast, popular, 60-year-older McKibben muscle is not very efficient and cannot support a biologically realistic muscle strain. Currently, there are no low-cost commercially available fluid control valves that are suitable for a wide range of robotic applications. To address this market shortcoming, the Compact Robotic Flow Control (CRFC) Valve has been recently introduced. The IP-protected CRFC Valve can work in junction with the IP-protected Hydro Muscle. When integrated, Hydro Muscles and the CRFC Valve may be utilized as modular building blocks for robots that can be rapidly assembled and utilized as either perform-alone or wearable robotic systems. The synergy of the CRFC Valve with the cost-effective, energy-efficient, and excellent strain properties of the Hydro Muscle opens a door into a new age of fascinating, useful, and accessible/affordable fluid-operated wearable robotic solutions both in space and on Earth.

Keywords Artificial muscle · Hydro Muscle · Valve · Flow control · CRFC Valve · Robotics · Wearable robotics · Space applications · Space robots

4.1 Introduction

There are a number of biologically inspired muscles and corresponding systems, including those that utilize conventional position-controlled actuators in series with elastic elements, like Series Elastic Actuators [1–3], cable-driven systems also with added elastic elements [4–9], and inherently soft and compliant fluid-actuated

M. Popovic (✉) · S. Moffat · J. D'Agostino · E. Clarrissimeaux · E. Carkin · M. Bowers
Robotics Engineering, Worcester Polytechnic Institute, Worcester, MA, USA
e-mail: mpopovic@wpi.edu

© Springer Nature Switzerland AG 2022

L. Rasmussen (ed.), *Smart Materials*,
https://doi.org/10.1007/978-3-030-70514-5_4

muscles like McKibben [10–12] and Hydro Muscles [13–15]. For a detailed review of biologically inspired muscles, see [16].

Many robotic researchers in academic settings avoid hydraulically and pneumatically operated systems in particular because of the nuisance of leaks, need for custom parts, and complexities associated with fluid circulation systems where the entire system may be affected by local changes or disturbances. However, fluid-operated systems also provide many advantages especially for wearable robotics where size and mass really matter. For example, for hydraulic systems, instead of having one strong and heavy dedicated electric motor *per* actuated degree of freedom, only one strong and heavy electric motor (e.g., pump) is needed for *all* actuated degrees of freedom.

Unfortunately, flow control valves are also required, and those that are available on the market are fairly limited. Valves that are cost-effective, lightweight, compact, and electronically controlled, have short response times, and support a range of pressures are not commercially available.

To address this serious drawback, the patent pending Compact Robotic Flow Control (CRFC) Valve has been recently introduced [17–19]. Based on observations and testing results so far, this novel valve holds potential to improve pneumatic and hydraulic applications.

The simple single-legged robot introduced here represents a research platform for study of a closed hydraulic system with Hydro Muscle actuation controlled with the CRFC Valve.

4.2 Hydro Muscle

In contrast to the pneumatic artificial muscle (PAM), or McKibben artificial muscle, left in Fig. 4.1, the Hydro Muscle, right in Fig. 4.1, can achieve strains identical to and even beyond a biological muscle [14, 15, 20, 21]. Hydro Muscles elongate axially, stiffen radially, and do not bulge when pressurized. When depressurized, they contract axially and soften radially. Hydro Muscles exhibit better efficiency than McKibben muscles [14, 22]. Substantial radial expansion and deflation of the McKibben elastomer introduce substantial energy loss in transfer from the source to the actuated load. Furthermore, Hydro Muscles can easily reach strains of 3.0, while McKibben muscles typically have a maximum strain of about 0.4. Biological muscles have an average maximum strain of 1.0 [23]. According to ref. [20], the Hydro Muscle is ranked as the top synthetic muscle in terms of combined efficiency and strain, and it can fully mimic the actuations of real biological muscle.

Overall, the Hydro Muscle has many favorable characteristics. It is lightweight, small, efficient, soft, fast, similar to biological muscle, and cost-effective. It can be constructed with off-the-shelf materials in less than 10 min, and the materials for each muscle cost less than $10. The muscles can provide substantial force even when pressurized with ordinary tap water at standard household pressures of 0.59 MPa (85 psi). During lab tests, the muscles were pressurized either

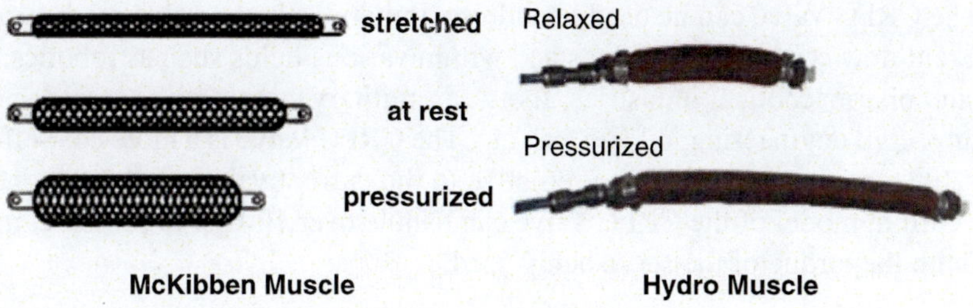

Fig. 4.1 Function of McKibben muscle and Hydro Muscle

pneumatically with compressed air or hydraulically with water pressurized at around 0.69 MPa (100 psi) and typically not in excess of 0.97 MPa (140 psi). Multiple muscles can be actuated from one source using flow and/or pressure control [14].

4.3 The Liquid Actuated Hydro Muscle at Low Temperatures

Some space and planetary applications will require performance at low temperatures. The liquid-actuated Hydro Muscle has not yet been extensively tested at low temperatures. However, as is well known, water freezes at 0 °C, and latex loses its elastic properties at approximately −55 °C [24]. The freezing point may be lowered to −65 °C if a larger percent of ethylene glycol is used as an antifreeze [25]. Still further, the elastomer's elastic properties can be sustained even at temperatures as low as ~ −75 °C, if different rubber compounds are used [26]. Hence, for temperatures greater than −65 °C, the original, or perhaps only slightly modified, Hydro Muscle can be used.

But for deep cryogenic conditions, for example, for temperatures as low as ~ −269.15 °C, different fluids (e.g., liquid helium) could be used, and instead of a rubbery inner tube, one may consider using novel materials like graphene-based foam [27]. Moreover, the inner tube can be exchanged with a simple spring if the outer sheathing can securely contain the pressurized fluid.

4.4 Compact Robotic Flow Control Valve

When coupled with the patent pending Compact Robotic Flow Control (CRFC) Valve, the Hydro Muscle actuation can be finely controlled in real time. The CRFC Valve is a simply operated flow control mechanism, with the ability to manipulate liquid and gas, and uses a servo motor attached to a choking mechanism that controls an entry and exit port for fluids (Fig. 4.2).

The CRFC Valve can be used to address limitations in the valve market in many different flow control applications and within various fields such as robotics, medical and pharmaceutical industries, footwear, military, automotive, aerospace, agriculture, civil engineering, oil and gas, etc. The CRFC Valve is a more cost-effective, versatile, compact, and lighter alternative to the current valves within the industry. The current model of the CRFC Valve can handle over 100 psi of pressure, increasing with the torque of the servo being used.

4.4.1 Design

The 3D printed casing holds the servo motor, which operates as the main actuator of the device, and two tubes, which allow for bi-directional fluid flow (Figs. 4.2 and 4.3). One tube is the entry port, where fluid can be pumped into an attached device, and the other is the release tube, where the fluid can be pumped out. For one-directional flow operations, only one of these tubes would be necessary. On the anterior side of the valve, the two tubes merge with a Y-connector to attach to the

Fig. 4.2 Compact Robotic Flow Control Valve

device. On the posterior side of the valve, the tubes are separate, allowing one to be connected to a pressurizing device, while the other is able to release the fluid when the tube is opened.

String looped around each tube is pulled by the rotating servo to open and close the ports. When tension is added to the string, it pulls the tubing upward against the proximal portion of the casing, which has a slightly curved profile. Pulling the tube around this element seals the tube and prevents fluid from escaping. The semi-circle element also creates a more gradual decline in the angle of choking, taking some pressure off of the tube component. Each tube is made of 5-mm-wide, 1-mm-thick surgical tubing encased in kite fabric. The kite fabric prevents the ballooning and possible bursting of the tube, as well as adds additional protection against friction from the string.

To prevent excessive tension in the string, a 3D printed curved attachment was made to fit over a standard servo horn for the strings to loop around (Fig. 4.4). The curved attachment includes a track for small metal beads to roll along to minimize friction in the system. When the servo angle changes, the string tension will change to alter the size of the inlet opening of the inner latex tubing.

The curved profile was approximately calculated to prevent excessive string tension when the tubing is closed, but to allow sufficient slackening of the string to allow the tubing to open fully. Equations 4.1–4.5 and Fig. 4.5 display the parameters used and calculations performed to customize the valve. These equations yield an

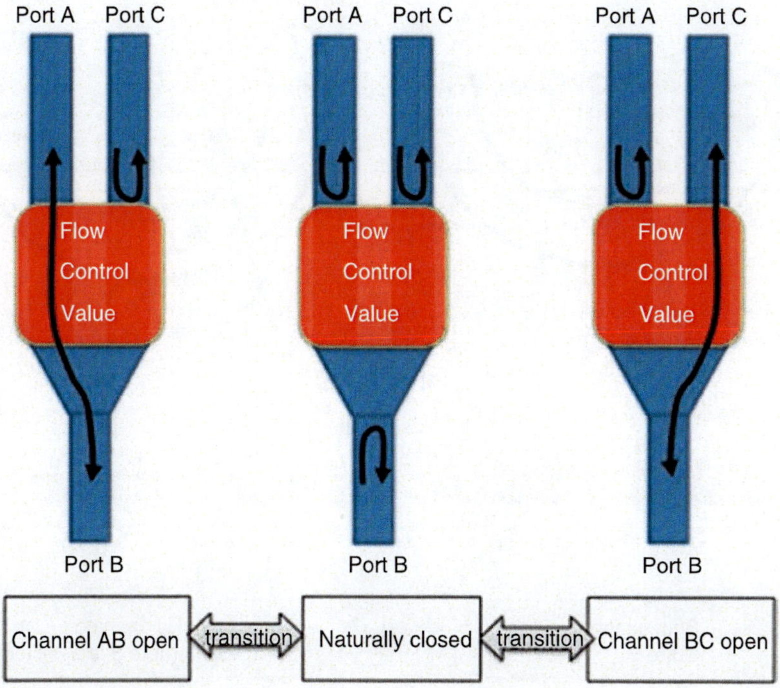

Fig. 4.3 CRFC Valve states

expression for the profile of the curved component, with an arc radius of 20 mm and a total angle span of about 50 degrees.

$$l_0 = l_c - d \tag{4.1}$$

$$s = \sqrt{(h - l_0)^2 + w^2} \tag{4.2}$$

$$\theta_c = 2\sin^{-1}(s/2r) \tag{4.3}$$

$$\theta_0 = \sin^{-1}\left[(h - l_0)/s\right] \tag{4.4}$$

$$\Delta\theta = \theta_c - \theta_0 \tag{4.5}$$

In the resting state, the servo is at a neutral, symmetrical position, with Tube A and Tube B both closed, depicted as State 2 in Fig. 4.4. Rotating the servo counter-clockwise will cause Tube A to open (State 1), while rotating the servo clockwise will cause Tube B to open (State 3). Different angles that the servo moves can control the size of the tube inlet/outlet, introducing intermittent stages.

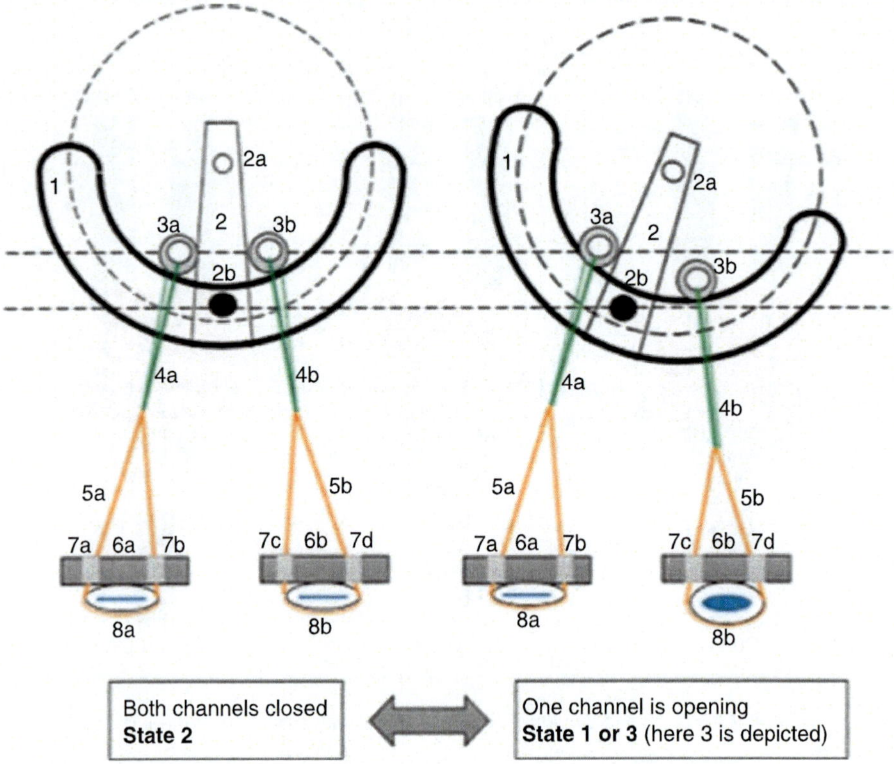

Fig. 4.4 Inlet opening due to servo rotating curved component

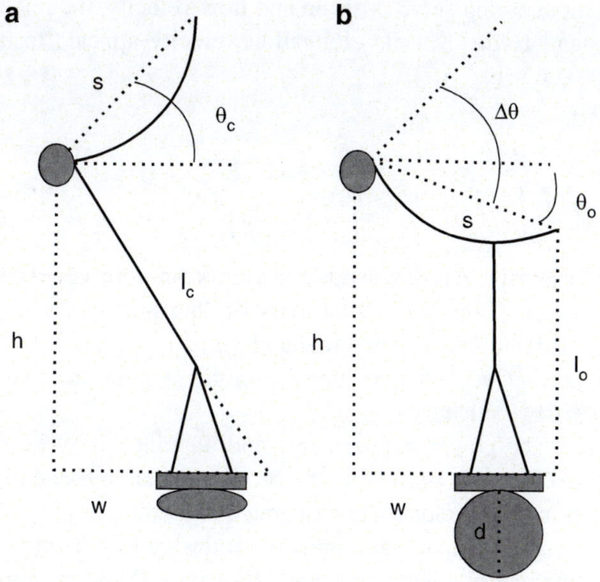

Fig. 4.5 Simplified valve geometry for curved profile, with valve closed (**a**) and open (**b**)

4.4.2 Proportional Flow Control

The flow characteristics of the CRFC Valve were determined through experiments with the valve independently and with the valve interfaced with the Hydro Muscle.

First, a CFM/CMM Thermo-Anemometer was used to measure the speed of air flow while the servo motor moved to different positions. To keep each test consistent, the valve was kept at the same distance from the sensor in a controlled environment that would not affect the flow reading. The valve was tested with 0.69 MPa (100 psi) of air pressure, the air pressure typically used with Hydro Muscles.

At a neutral position, with the servo pointing vertically, no air flow was detected, as was expected given the closed state of both the inlet and outlet tubes. Rotating the servo in a positive direction created a linear trend of flow release, with an R^2-value of 0.969 (Fig. 4.6). The data obtained in this test can be used to control the flow rate based on the angle of the lever component on the servo. From the results, each degree of movement corresponds to an approximate increase of 0.186 m/s of flow velocity.

The CRFC Valve was originally synthesized to work in conjunction with the Hydro Muscle; therefore, an experiment was conducted to determine the relationship between the servo angle and the elongation rate of the Hydro Muscle. This experiment was performed by turning the valve's servo motor to various angles and timing the full elongation of a Hydro Muscle with a length of 10.4 cm in its contracted state.

The rate of elongation was collected with the curved valve attachment being rotated to various degrees, from 0 to 35 degrees in increments of about 5 degrees. There was no air flow at the neutral angle. From this test and the data above concerning servo position and flow velocity, the relationship between flow velocity and Hydro Muscle elongation velocity appears to be linear, with an R^2-value of 0.967 (Fig. 4.7).

4.4.3 Specifications

The valve has dimensions of 6 cm × 5 cm × 2 cm, though it only occupies 2/3 of that volume, and has a total mass of 28.5 grams. For a comparison, a standard on-off solenoid valve with a manually tunable orifice [28] that is readily available on the market occupies six times a larger volume and weighs ten times more than the CRFC Valve.

The power requirements and angular speed are dependent on the servo being used; in this prototype, the servo used had a speed of 0.08 s per 60 degrees of rotation (13.09 rad/s) when operated at 6 volts [29].

The compact and lightweight design of the valve allows for a variety of unique applications. There is potential to expand Hydro Muscles and soft robotics within a variety of fields, and the CRFC Valve enables these technologies to be more compact and efficient. The CRFC Valve is incredibly easy and cost-effective to manufacture. The cost of producing the valve with a standard high-torque servo is around $14 USD.

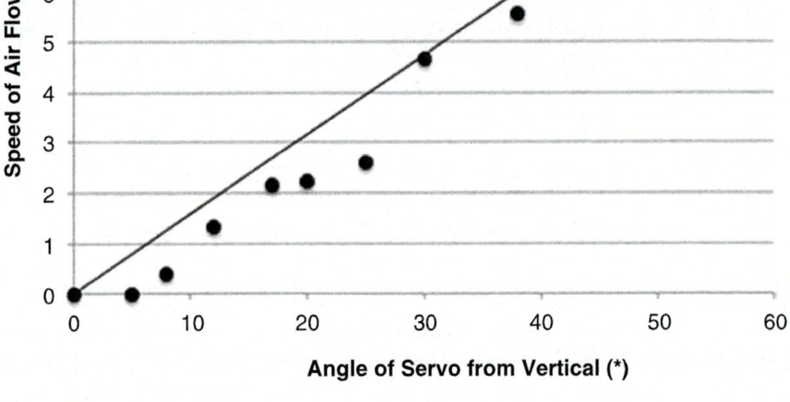

Fig. 4.6 Servo angle vs. air flow velocity

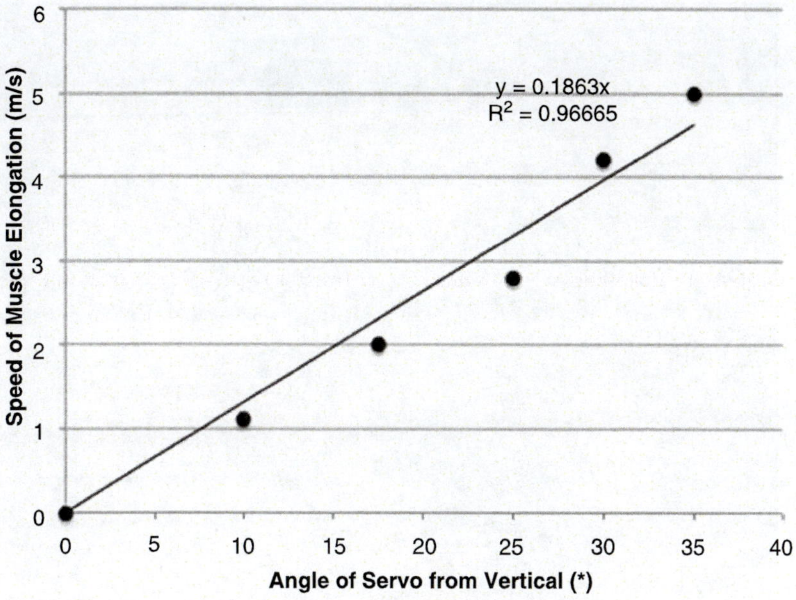

Fig. 4.7 Servo angle vs. average elongation rate of Hydro Muscle

4.5 Proof-of-Concept Leg Model in the Context of a Hydraulic System

To demonstrate the capabilities of the Hydro Muscle and Compact Robotic Flow Control Valve in the context of a hydraulic system, a leg model was built and tested. This model was a simple 3D printed structure connected to a single revolute joint that would act as the leg. A Hydro Muscle was attached to the leg and body so as to provide a substantial moment arm to the joint. The Hydro Muscle was connected to the CRFC Valve, which subsequently connected with a small, 1 lb. water pump and a reservoir of water (Fig. 4.8).

4.5.1 Hydraulic System

The hydraulic system was powered by a pump with pressure ranging between 70 and 140 psi. A short length of tubing connected a small reservoir of water to the inlet of the pump. The outlet of the pump led to the inlet tube of the CRFC Valve, which would control the water flowing into the Hydro Muscle. The outlet tube of the CRFC Valve led back to the reservoir to be circulated again by the pump.

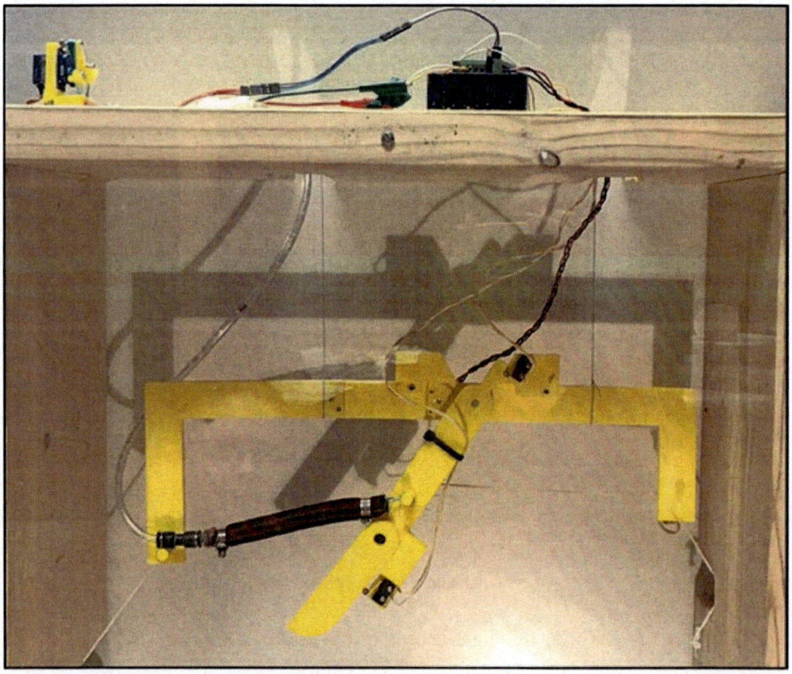

Fig. 4.8 Proof-of-concept leg model setup

4.5.2 Control System

The CRFC Valve was controlled by a simple Arduino program. The objective of the model was to continuously cycle the motion of the leg as if it were walking, meaning the Hydro Muscle would alternate between expanding and contracting at a rhythmic pace. The valve's servo motor was instructed to rotate to allow the inlet tube to open fully so that the Hydro Muscle expands. Since the Hydro Muscle on its own cannot be used to exert pushing forces, a passive elastic element was used to pull the leg forward. A delay (1 s) was set in place to wait for the leg to go past the vertical position. The valve's servo motor then rotated the other way to open the outlet, and the Hydro Muscle contracted to pull the leg back. A limit switch was placed at the joint to determine when the leg should reenter the swing phase and cycle through again.

4.5.3 Work Cycle

The strain of the Hydro Muscle and the hip angle over several cycles is displayed in Figs. 4.9 and 4.10.

4.6 Discussion

As illustrated with Figs. 4.6 and 4.7, the CRFC Valve exhibits interestingly linear behavior in terms of fluid flow velocity and elongation rate of the Hydro Muscle at constant inlet pressure. On the other hand, the Hydro Muscle strain and dynamical leg platform hip movements are rather nonlinear phenomena (Figs. 4.9 and 4.10), in the context of a hydraulic system. That is mainly due to substantially varying inlet pressure caused by a readily available low-cost diaphragm pump [30] and lack of an

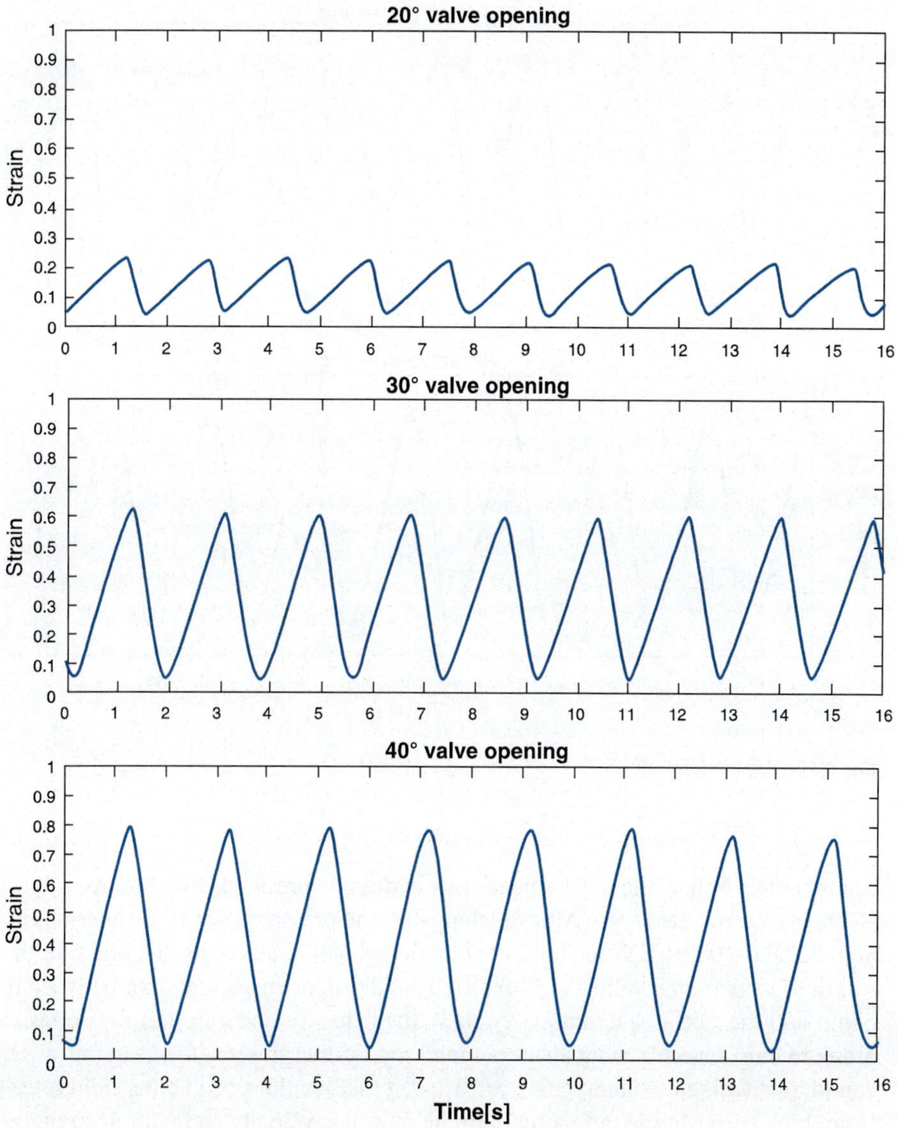

Fig. 4.9 Hydro Muscle strain vs. time

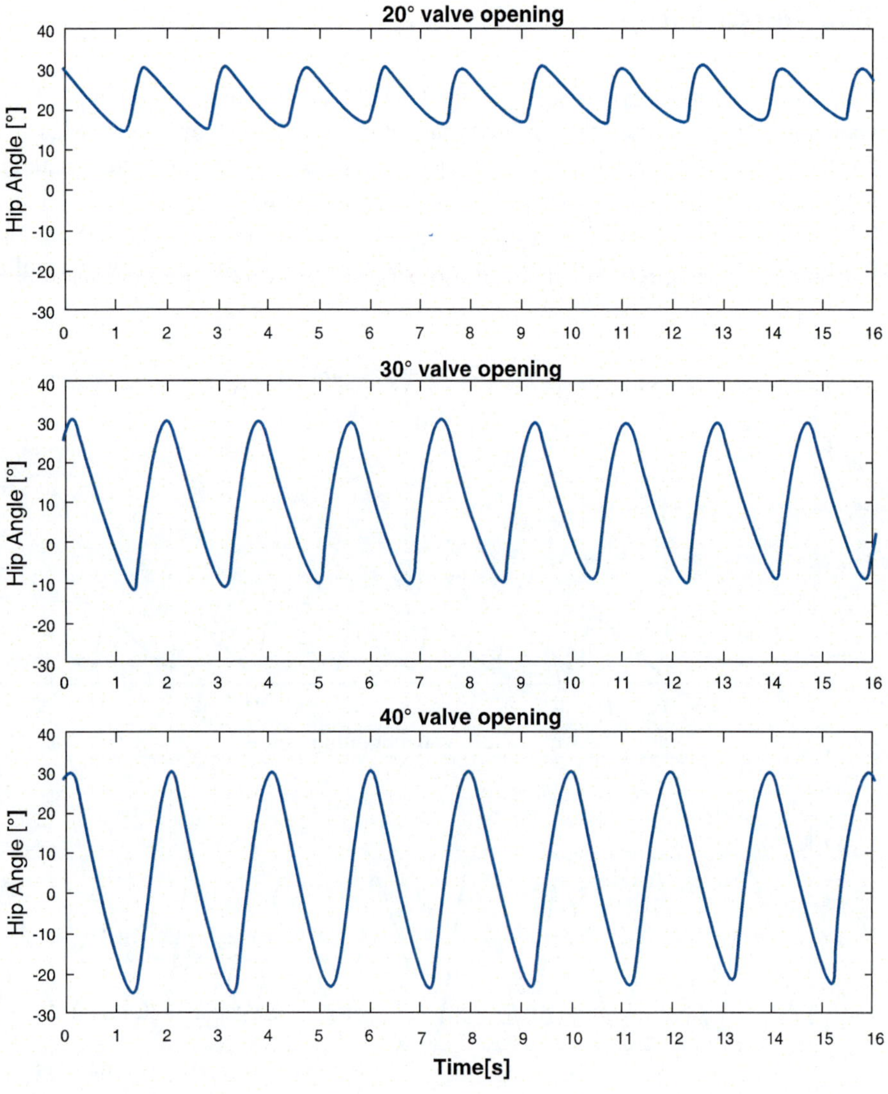

Fig. 4.10 Hip angle vs. time

accumulator. When the inlet opens, the hydraulic pressure drops to its minima (~70 psi), and as the Hydro Muscle elongates, the pressure rises to its intermedium high value (~100 psi). When the inlet closes and the outlet opens, the pressure then spikes to its maximal value (~140 psi). The addition of an accumulator between the pump and the CRFC Valve may regulate this behavior and help mimic a constant pressure fluid source as with an air compressor. Further improvements of the system would be a variable volume quick release bag that would speed up the fluid release from the Hydro Muscle, affecting both the muscle's velocity and pulling strength.

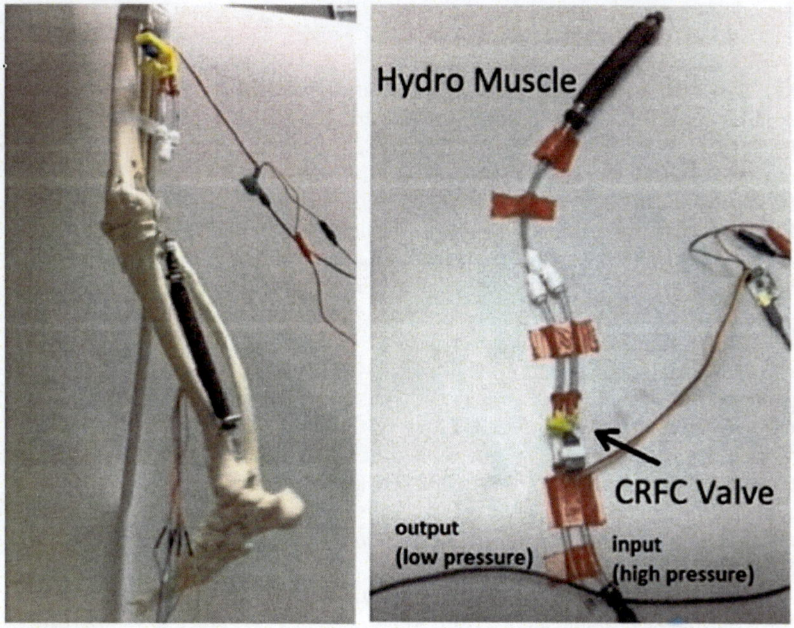

4.7 Conclusion

The CRFC Valve is overall lighter, more compact, controllable, and less expensive than any other valve currently on the market. The CRFC Valve can have numerous applications in a wide variety of fields that all require fine control of fluid flow and that clearly may benefit from a lightweight, compact, and cost-effective valve, e.g., medical and pharmaceutical industry, military, automotive, aerospace, agriculture, civil engineering, oil and gas, etc. The synergy of the CRFC Valve with the cost-effective, energy-efficient, and excellent strain properties Hydro Muscle opens a door into a new age of very interesting, useful, and accessible/affordable fluid operated robotic solutions.

Acknowledgments The authors thank Felix Sanchez for his assistance with CAD for 3D printed leg platform and leg platform tests. The authors thank Thane Hunt for constructive discussion of flow control systems and his suggestions regarding 3D printing and servo control software.

References

1. J.A. Blaya, H. Herr, Adaptive control of a variable-impedance ankle-foot orthosis to assist drop-foot gait. IEEE Trans. Neural Syst. Rehabil. Eng. **12**(1), 24–31 (2014)
2. H. Herr, J.A. Blaya, and G.A. Pratt, Massachusetts Institute of Technology, 2012. Active ankle foot orthosis. U.S. Patent 8, 287,477
3. H. van der Kooij, E. van Asseldonk, and G. van Oort Projects I lopes I Department of Biomechanical Engineering. [online] Universiteit Twente. (2017). Available at: https://www.utwente.nl/en/et/be/research/projects/lopes/. Accessed 15 Feb. 2019
4. S.B. Kesner, L. Jentoft, F.L. Hammond, R.D. Howe, and M.B. Popovic, Design considerations for an active soft orthotic system for shoulder rehabilitation, in *33rd Annual International IEEE EMBS Conference* (Boston, August 30–September 02 2011),
5. I. Galiana, F.L. Hammond, R.D. Howe, and M.B. Popovic. Wearable soft robotic device for post-stroke shoulder rehabilitation: identifying misalignments. *2012 IEEE/RSJ International Conference on Intelligent Robots and Systems* (Portugal, October 7–12 2012)
6. Y. Mao, S.K. Agrawal, Design of a cable-driven arm exoskeleton (CAREX) for neural rehabilitation. IEEE Trans. Robot. **28**(4), 922–931 (2012)
7. A.T. Asbeck, R. Dyer, A. Larusson, C.J. Walsh, Biologically inspired soft exosuit, in *2013 IEEE International Conference on Rehabilitation Robotics (ICORR)*, Seattle, Washington, USA (2013), pp. 1–8
8. A.T. Asbeck, S.M.M. De Rossi, I. Galiana, Y. Ding, C.J. Walsh, Stronger, smarter, softer: Next-generation wearable robots. IEEE Robot. Autom. Mag. **21**(4), 22–33 (2014)
9. E. Saint-Elme, M.A. Larrier, Jr., C. Kracinovich, D. Renshaw, K. Troy, and M.B. Popovic, Design of a biologically accurate prosthetic hand, in *IEEE RAS International Symposium on Wearable & Rehabilitation Robotics* (Houston, 5–8 November 2017)
10. J. Ueda, D. Ming, V. Krishnamoorthy, M. Shinohara, T. Ogasawara, Individual muscle control using an exoskeleton robot for muscle function testing. IEEE Trans. Neural Syst. Rehabil. Eng. **18**(4), 339–350 (2010)
11. Y. Park, C. Bor-rong Chen, N.O. Pérez-Arancibia, D. Young, L. Stirling, R.J. Wood, E.C. Goldfield, R. Nagpal, Design and control of a bio-inspired soft wearable robotic device for ankle–foot rehabilitation. Bioinspir. Biomim. **9**(1), 016007 (2014)
12. S. Kurumaya, K. Suzumori, H. Nabae, S. Wakimoto, Musculoskeletal lower-limb robot driven by multifilament muscles. Robomech J **3**(1), 18 (2016)
13. G. McCarthy, D. Effraimidis, B. Jennings, N. Corso, C. Onal and M.B. Popovic, Hydraulically actuated muscle (HAM) exo-musculature, in *Robot Makers: The future of digital rapid design and fabrication of robots* (RoMa) Workshop, the 2014 Robotics: Science and systems conference (Berkeley July 12, 2014)
14. S. Sridar, C.J. Majeika, P. Schaffer, M. Bowers, S. Ueda, A.J. Barth, J.L. Sorrells, J.T. Wu, T.R. Hunt, M. Popovic, Hydro Muscle – a novel soft fluidic actuator. IEEE Int. Conf. Robot Autom (ICRA), 4104–4021 (2016). https://doi.org/10.1109/ICRA.2016.7487591
15. M. Bowers, C. Harmalkar, A. Agrawal, A. Kashyap, J. Tai, and M.B. Popovic, Design and test of biologically inspired multi-fiber hydro muscle actuated ankle, in *Proceedings of 2017 IEEE International Workshop on Advanced Robotics and its Social Impacts* (University of Texas at Austin, Austin, 8–10 March 2017)
16. M. Popovic, *Biomechatronics*, 1st edn. (Elsevier Academic Press, London, UK, 2019)
17. M. Popovic, S. Moffat, J. D'Agostino, & E. Clarrissimeaux (2019). U.S. Provisional Patent Application No. 62/802,933
18. D'Agostino Julia, Clarrissimeaux Ellen, Moffat Shannon, Florez-Castillo Juan D., Sanchez Felix, Bowers Matthew, and Popovic Marko "Development of Bioinspired Exosuit Actuated with Hydro Muscles and Novel Compact Robotic Flow Control Valve", Biodevices 2020 - The 13th International Joint Conference on Biomedical Engineering Systems and Technologies, BIOSTEC, Valletta, Malta, February 24–26, (2020)

19. S. Moffat, *Bio-Inspired Legs and Novel Flow Control Valve towards a New Approach for Accessible Wearable Robotics* (Worcester Polytechnic Institute, Worcester, MA, USA, 2019)
20. A. Miriyev, K. Stack, H. Lipson, Soft material for soft actuators. Nat. Commun. **8**(1), 596 (2017)
21. C.P. Chou, B. Hannaford, *Measurement and Modeling of McKibben Pneumatic Artificial Muscle* (University of Washington, Department of Electrical Engineering, Seattle, IEEE/ ASME Atlanta, GA, USA, 1992)
22. G. K. Klute, J. M. Czerniecki, and B. Hannaford, McKibben artificial muscles: pneumatic actuators with biomechanical intelligence, in *Proceedings of IEEE/ASME International Conference on Advanced Intelligent Mechatronics*, IEEE/ASME Atlanta, GA, USA, (1999)
23. D.B. Bertoti, P.A. Houglum, *Clinical Kinesiology*, 6th revised edn. (Philadelphia, Davis Company, F. A. 2011)
24. S. Johnson Latex Specifications and Properties (02 March 2019). Retrieved from https://sciencing.com/latex-specifications-properties-7678387.html
25. Antifreeze. (11 April 2019). Retrieved from https://en.wikipedia.org/wiki/Antifreeze
26. Low Temperature Compounds. (2015). Retrieved from https://catalog.precisionassoc.com/asset/PAI%20Low%20Temperature%20Compounds.pdf
27. M. Temming, A New Graphene Foam Stays Squishy at the Coldest Temperatures (01 May 2019). Retrieved from https://www.sciencenews.org/article/new-graphene-foam-squishy-coldest-temperatures
28. Pneumatic Electric Solenoid Valve. U.S. Solid. ¼" 5 way 2 position DC 24 V, (2019)
29. MG90D HIgh Torque Metal Gear, Adafruit. Micro servo, (2019)
30. 130 PSI 6L/Min Water High Pressure Switch Diaphragm Self Priming Pump DC12V 70W, GOOD PUMPS, CE ROHS MOTOR, (2019)

Marko Popovic is a faculty member of the Physics Department, Biomedical Engineering Department, and Robotics Engineering Department at the Worcester Polytechnic Institute (WPI). He is the founder and director of WPI Popovic Labs. Previously, Popovic held research scientist position at the Massachusetts Institute of Technology (MIT). Popovic received his PhD in Physics from Boston University, followed by post-doctoral positions at both Harvard University Physics and MIT Biomechatronics group. He single-authored 'Biomechanics and Robotics', published in 2013 by Pan Standard Publishing. He authored and edited 'Biomechatronics', published in 2019 by Elsevier Academic Press. Popovic is an avid and successful inventor. His Hydro Muscle invention received the largest compensation for intellectual property from an industry partner in WPI history. The Boston Patent Law Association (BPLA) program "Invented Here!" selected Popovic's Hydro Bone invention among top 12 patents from more than 12,000 patents granted in New England in 2018 as part of the BPLA's initiative to honor the region's newest and most innovative technologies. Results of his work were showcased by Reuters, The Boston Globe, The New York Times, Channel News Asia, Business Insider, CNBC, Yahoo! News, Toronto Sun, Daily Times Pakistan, UK Eurosport, CNN, TIME Magazine, National Geographic, and many other domestic and overseas media. WPI Popovic Labs: http://users.wpi.edu/~mpopovic/

Shannon Moffat holds her Degree of Master of Science in Robotics Engineering from Worcester Polytechnic Institute (2019). She is currently employed as Systems Test Engineer at iRobot in Bedford, MA.

Julia D'Agostino is senior undergraduate student at Worcester Polytechnic Institute double majoring in mechanical engineering and robotics engineering.

Ellie Clarrissimeaux is senior undergraduate student at Worcester Polytechnic Institute majoring in mechanical engineering.

Eric Carkin holds his Degree of Bachelor of Science in Robotics Engineering from Worcester Polytechnic Institute (2019).

Matthew Bowers is doctoral candidate in Robotics Engineering at Worcester Polytechnic Institute.

Chapter 5
Contractile Pneumatic Artificial Muscle Generates Extension by Actuating an Integrated Pushrod

Norman M. Wereley, Erica G. Hocking, Benjamin K. S. Woods, and Curt S. Kothera

Abstract Pneumatic artificial muscles (PAMs) are simple mechanical actuators that consist of an elastomeric bladder within a braided mesh sleeve. Upon pressurization of the bladder, the actuator may either contract or extend axially, with the direction of motion dependent on the orientation of the braided sleeve fibers. PAMs that have been constructed for contractile motion are well known for their excellent actuator characteristics, including high specific power, specific work, and power density. Extensile PAMs, on the other hand, are unable to demonstrate comparable performance, in particular demonstrating much reduced blocked force, and are prone to buckling under axial compressive loading. For applications in which extensile motion and compressive force generation are desired, this chapter presents a novel actuator, known as the push-PAM, that harnesses the operational characteristics of a typical contraction/tension PAM, but changes the direction of motion and force with a simple conversion package that demands only a small increase in friction, weight, and cost. Quasi-static testing on this new actuator variant revealed the push-PAM's full evolution of force with displacement for operating pressures ranging from 138 to 621 kPa (20–90 psi in 10 psi increments), in addition to the blocked force and free contraction capabilities at each pressure, confirming that the push-PAM was able to achieve forces and strokes comparable to a contractile PAM tested under the same conditions. Furthermore, it was shown that the performance of both types of actuators could be captured with the same modeling efforts. A refined force balance analysis was applied to these actuators that utilized additional modeling terms to account for three key nonlinear phenomena that were observed in both types of actuators: nonlinear PAM stiffness, hysteresis of the force vs. displacement

N. M. Wereley (✉) · E. G. Hocking
Department of Aerospace Engineering, University of Maryland, College Park, MD, USA
e-mail: wereley@umd.edu

B. K. S. Woods
Department of Aerospace Engineering, University of Bristol, Bristol, UK

C. S. Kothera
InnoVital Systems Inc., Beltsville, MD, USA

© Springer Nature Switzerland AG 2022

L. Rasmussen (ed.), *Smart Materials*,
https://doi.org/10.1007/978-3-030-70514-5_5

response for a given pressure, and a pressure dead-band. This nonlinear model was shown to reconstruct the actuator response with minimal error for both the contractile PAM and the push-PAM.

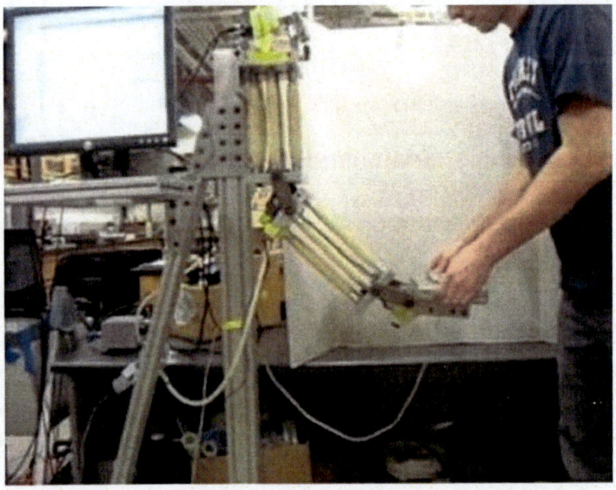

Keywords Pneumatic artificial muscle · Blocked force · Free contraction · Pressure dead-band · Hysteresis · Nonlinear elastic analysis

Nomenclature

A_B	Cross-sectional of the bladder
B	Length of sleeve fiber
D	Outer diameter of the bladder
E	Elastic modulus
F	Force generated by PAM
F_{AVG}	Pressure-averaged force
F_B	Pressure-averaged blocked force
F_F	Friction force
F_G	Gaylord force
F_{LFB}	Linear elastic force balance force correction
F_{NLFB}	Nonlinear elastic force balance force correction
k_F	Friction factor
L_0	Initial (active) length of PAM
L	Instantaneous length of PAM
L'	Corrected length of PAM
M	Summation upper limit

N Number of turns of sleeve fiber
P Actuation pressure
P' Corrected actuation pressure
P_c Pressure offset for PAM actuation (threshold pressure)
R_0 Initial outer radius of the bladder
R Instantaneous outer radius of the bladder
T Tension in sleeve fiber
t Instantaneous thickness of the bladder
V_B Volume of the bladder
v Velocity
α Braid angle (radial)
ε Strain in the bladder
ΔL Change in length of PAM
Δl Tip shape length correction
ΔR Change in radius of the bladder
σ Stress in the bladder

5.1 Introduction

Pneumatic artificial muscles (PAMs) are simple mechanical actuators that harness pressurized air to generate significant forces and deflections while offering several advantages over other actuation technologies. They are known for their high power-to-mass, power-to-volume, and work-to-mass ratios, making them more attractive than electromechanical actuators, solenoids, piezoelectric and magnetostrictive materials, and several other types of actuators in a variety of applications [1–6]. Furthermore, they boast a very simple structure and high durability and fatigue life and are compliant, which makes them safe for human interaction and tolerant of mechanical misalignment and impulsive loading [7]. As a result of these characteristics, they have attracted interest in a wide array of fields, including prosthetics/orthotics [8–10], robotics [3, 11, 12], and recently aerospace engineering [4, 7, 13, 14]. A more comprehensive review of some of the applications of PAMs has been undertaken by Zhang and Philen [15].

These simple, yet effective, devices are comprised of a flexible elastomeric bladder surrounded by a braided sleeve, with two end fittings that serve to seal the ends of the actuators and create a channel for the injection of the working fluid. The operating principle of PAMs is as follows. The inner elastic bladder is pressurized with fluid (air), causing an inflation and expansion of the bladder, which causes a subsequent expansion in the surrounding braided sleeve. Because the braided sleeve fibers are much stiffer than the bladder material, the braided sleeve is unable to stretch as the bladder does and instead contracts or extends axially by an amount ΔL to allow for the expansion in the bladder. In addition to the relative motion between the two end fittings, the stiff sleeve fibers generate either a tensile or compressive force along the main axis of the actuator, corresponding to either an axial contraction or extension, respectively. For a contraction/tension actuator, the bladder expansion is primarily radial. Conversely, for an extension/compression PAM, the bladder expansion is primarily axial [16].

The direction of force and motion are dependent on the initial angle between the filaments of the braided sleeve, which is inherent to the construction of a given actuator and cannot be changed once the actuator is made. PAMs with the filaments aligned primarily along the longitudinal axis of the actuator will be contractile and generate a tensile force. If the filaments of the sleeve are initially aligned primarily along the radial axis of the actuator, then the resulting motion is extensile, and the force generated will be compressive. There is a neutral braid angle between these two extremes that is the crossover point between these regimes. This angle has been found to be 35.26° as measured from the radial axis to the braid filaments or more commonly reported as 54.74° as measured from the longitudinal axis [16–18]. The vast majority of PAMs studied by previous researchers have the braid configured to generate tensile forces and contractile motions [1–14, 19–26]. Although extensile PAMs are capable of producing large axial deformations, they generate low compression forces and are prone to buckling when subjected to axial compressive loading [7, 16, 17, 27, 28] and are thus seldom used as actuators.

Despite the apparent advantages of contractile PAMs, current and future uses of these actuators would benefit from a configuration that allows for comparable forces and stability in an extensile scheme. For example, an industrial stamping process where an extensile actuator is needed to push two plates together, thereby creating a compressive force between them. As such, the goal of this chapter is to develop and characterize a novel PAM design, known as the push-PAM, which takes advantage of the higher forces of the contractile PAM configuration and converts contraction to extension without sacrificing actuator performance or simplicity.

To this end, the remainder of the chapter is organized into four sections. First, the design and fabrication of the push-PAM will be discussed, highlighting its simplicity and similarity to the contractile PAM. The push-PAM will then be characterized through quasi-static testing to reveal the evolution of force with displacement associated with this device at a range of discrete internal pressures, and the results will be compared to those of a contractile PAM to demonstrate the push-PAM's analogous performance. Next, existing quasi-static models for contractile PAMs will be validated against the experimental data from the push-PAM, showing that the output from both types of actuators may be captured through the same modeling scheme. Furthermore, these results will showcase the precision of a refined force balance model that includes corrections to accurately capture the nonlinear quasi-static actuation behavior of the PAMs, incorporating a nonlinear stress vs. strain relationship, friction effects that contribute to hysteresis, and an actuation dead-band at low operating pressures. The chapter will conclude with a summary of the findings from this study and a brief discussion of the significance of the work.

5.2 Design and Fabrication

As alluded to in the Introduction, the design goal of the proposed push-PAM was to harness the operational characteristics of a typical contraction/tension PAM, but to change the direction of motion and force with a simple conversion package that demands only a small increase in friction, weight, and cost. To accomplish this, a pushrod was anchored inside the bladder of a contractile PAM to one of the end fittings (air-thru). The pushrod extends through the inside of the actuator and passes through a rod seal and out of the other end fitting of the PAM, which is known as the fixed end fitting (closed). This design mimics that which was proposed in Ref. [29].

Upon pressurization of the actuator, the PAM portion of the apparatus inflates and expands radially, causing relative contractile motion between the two end fittings. Because the pushrod is of fixed length, this contraction within the PAM causes extension of the pushrod outside of the actuator through the fixed end fitting. As such, when the actuator is supported by the fixed end fitting and the external end of the pushrod, pressurization of the PAM results in an increase in the distance between the mounting points, and a compressive force is generated between them. Although the PAM itself still contracts by an amount ΔL, the pushrod and the specific mounting arrangement change this relative contraction into an overall ΔL extension. This is demonstrated in Fig. 5.1, which shows a schematic of the push-PAM in its resting and actuated states (top and bottom, respectively), with the mounting points marked with a red dashed line. Through this configuration, the push-PAM device allows for the superior force generation capabilities of a contractile PAM to be utilized in an extension/compression scheme.

Given that the push-PAM design makes explicit use of a traditional contractile PAM, the manufacturing techniques used for these two types of actuators are nearly identical. For this study, both the contractile PAM and the push-PAM utilized a latex bladder, a Kevlar braided sleeve, and swaged aluminum end fittings [7]. For the contractile PAM, a thru-hole in one of the end fittings served as a conduit for the air during pressurization. For the push-PAM, the air inlet was incorporated into the free end of the PAM utilizing a series of holes around the attachment point of the pushrod. This configuration allowed for a high mass flow rate of air into the PAM while also simplifying the design of the fixed end fitting, which attached to the rod

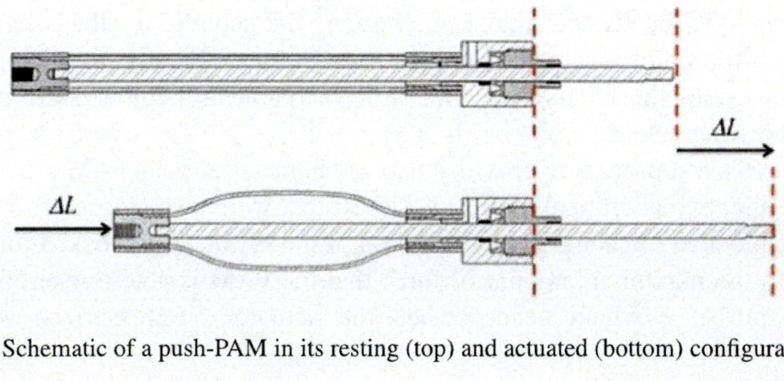

Fig. 5.1 Schematic of a push-PAM in its resting (top) and actuated (bottom) configurations

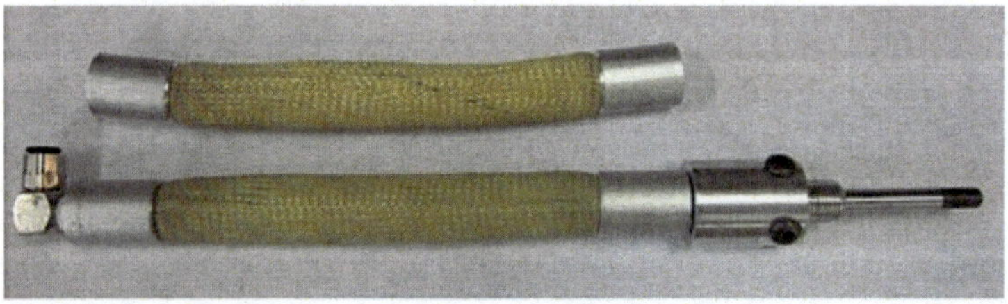

Fig. 5.2 Actuators characterized in this study (traditional contractile PAM on top, novel push-PAM for extensile motion on bottom)

housing. In order to provide a good baseline for comparison during experimental characterization, both the contractile PAM and the push-PAM were designed and manufactured to have matching resting configurations (i.e., braid angle, length, and diameter). The physical actuators used in this study are pictured in Fig. 5.2 to show the similarities between the PAMs used for each type of device (contractile PAM on top, push-PAM on bottom).

5.3 Experimental Characterization

5.3.1 Experimental Procedure

In order to characterize and compare the performance of the contractile PAM and the push-PAM, the actuators underwent quasi-static testing using a 97.9 kN (22 kip) MTS servo-hydraulic test machine, which provided both displacement and force measurements during testing. The PAMs were pressurized with a Husky Pro 227 L (60 gal) air compressor. Meanwhile, a Wilkerson Dial-Air pressure regulator with an SSI Technologies, Inc. digital pressure gauge was used to precisely monitor the pressure supplied to the PAMs.

The overall goal of the experimental characterization was to determine the evolution of force with displacement for a range of internal pressures for both types of PAMs. This was achieved by first inflating each PAM to a specific pressure, ranging from 138 to 621 kPa (20–90 psi) in 69 kPa (10 psi) increments, with both attachment points fixed by the MTS grips. After holding momentarily at this condition, one of the MTS heads was unlocked, allowing the actuator to slowly contract (or extend) to the point where it ceased to exert a force. Finally, following another momentary hold, the actuator was then slowly stretched (or compressed) back to its nominal resting length.

While this test procedure provided data to characterize each PAM's entire actuator load line, two additional points of interest that were also gleaned from this test were the blocked force and free contraction of the PAMs. The blocked force corresponds to the maximum amount of force that the PAM is able to exert at a given internal pressure, which occurs when the actuator is pressurized with both attachment points held fixed. Conversely, free contraction denotes the maximum stroke of the PAM for a given pressure, measured as the amount of contraction at which the PAM no longer produces any force. Although this terminology may seem misleading for the push-PAM because the full apparatus extends, rather than contracts, the conventional PAM nomenclature ("free contraction") has also been adopted to define the maximum stroke of the push-PAM because the device is still driven by the contractile performance of the PAM.

5.3.2 *Discussion of Experimental Results*

The results of the experimental characterization tests are shown in Fig. 5.3 for the contractile PAM and the push-PAM described above, demonstrating each PAM's evolution of force with displacement at a series of discrete internal pressures. For clarity, only the results of the tests at 207, 414, and 621 kPa (30, 60, and 90 psi) are shown, with the contractile PAM results displayed with a red solid line and the push-PAM results shown with a black dashed line. While the full experimental results are shown in Fig. 5.3a, the average actuator force at each pressure (F_{AVG}) is shown in Fig. 5.3b to demonstrate the nature of the actuation without hysteresis effects, which will be important for the modeling work discussed in later sections. In both plots, the horizontal axis is a non-dimensional measure of PAM contraction given by

$$\text{Contraction} = \Delta L / L_0. \tag{5.1}$$

In this equation, L_0 is the resting length of the PAM (otherwise known as the active length), and ΔL is the amount by which the device has contracted (or extended) as described above. For both the contractile PAM and the push-PAM, ΔL is taken to be

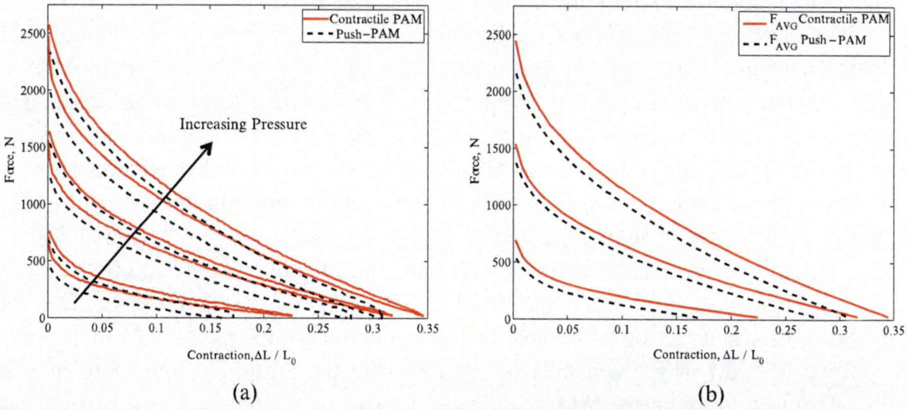

Fig. 5.3 Experimentally characterized actuator load lines at operating pressures of 207, 414, and 621 kPa. (**a**) Full experimental results. (**b**) Average force within hysteresis loop

positive for the desired motion of the actuator. In other words, for the contractile PAM, ΔL is positive when the actuator contracts, and for the push-PAM, ΔL is positive when the pushrod extends as a result of the PAM contraction (see Fig. 5.1).

The data in Fig. 5.3 clearly show that the contractile PAM and push-PAM demonstrate very similar actuation profiles, which is significant in asserting the practicality and usefulness of the push-PAM actuator. As can be seen in these figures, the forces generated by the PAMs are comparable at each actuation pressure; however, the push-PAM is not quite able to achieve as much free contraction as the contractile PAM. The slight performance discrepancies between the two types of PAMs are most likely due to increased friction within the push-PAM as a whole, as a result of pushrod motion within the rod seal. Also, small differences in the fabrication of the two PAMs themselves may contribute to the slight disparity between the actuator load lines.

Examining both sets of actuation profiles in Fig. 5.3, the forces generated by the PAMs decrease with increasing displacement at each given pressure. However, the local slopes of the curves vary depending on the contraction of the actuator. Below about $\Delta L/L_0 = 0.05$, the force drops relatively quickly with increasing contraction, indicating that the PAMs are relatively stiff in this region. This is especially true for the contractile PAM, whereas the push-PAM yields a slightly more gradual decrease in force at low displacements. Again, this discrepancy is likely attributable to additional friction within the push-PAM that mitigates the initial jump in PAM force as the bladder inflates. Above $\Delta L/L_0 = 0.05$, the stiffness of the PAMs decreases, and the forces diminish at a slower rate along a nearly linear path. For the contractile PAM, the nature of the actuation diagram appears to change again at contractions above $\Delta L/L_0 = 0.15$, with the stiffness of the PAM continuing to attenuate as the slope decreases further in this region. At these higher contractions (above $\Delta L/L_0 = 0.05$), it is possible that the latex bladder is being strained beyond the linear portion of a standard stress-strain curve and has entered a softening region in which additional strain causes a nearly negligible increase (or even decrease) in stress. Regardless of the exact characteristics of the PAM stiffness, however, it is clear that the PAM bladders are demonstrating nonlinear elastic behavior.

It is also important to note the aforementioned hysteresis in the actuator load lines (see Fig. 5.3a), with the lower line corresponding to the path from blocked force to free contraction and the upper line corresponding to the return path. This hysteresis results primarily from friction between the components of the PAM. Typically, this refers to friction between the braid and the bladder during inflation and deflation, as well as friction between the individual fibers of the braid as the braid angle changes. For the push-PAM, however, it is likely that friction between the pushrod and the rod seal also adds to the observed hysteretic behavior. As can be seen in Fig. 5.3a, the push-PAM demonstrates more hysteresis (i.e., the distance between the top and bottom curves is larger for the push-PAM than for the contractile PAM at a given pressure), bolstering the argument that there may be more friction in the push-PAM.

5.4 Modeling

Having proven the feasibility of creating a PAM that can extend an integrated push-rod, and having verified that such a push-PAM is capable of replicating large forces and displacements that are comparable to a contractile PAM, the next phase of this study was to demonstrate that a model is able to accurately predict the quasi-static output of this new actuator. Because the push-PAM utilizes the same basic construction as the contractile PAM, a key objective of this work was to validate that both types of actuators may be modeled with the same set of equations, further confirming the ease with which the push-PAM may be used as an alternative to the standard contractile PAM. As with all PAMs, the geometry of the braided sleeve plays a major role in determining the response of the actuator. From the braid angle α (taken from the radial plane of the actuator to the braid fibers), the current PAM length L (given by $L_0 - \Delta L$), and the bladder outer diameter D, the following braid parameters can be calculated through geometric constraints:

$$B \frac{L}{\sin \alpha} \tag{5.2}$$

and

$$N = \frac{B \cos \alpha}{\pi D} \tag{5.3}$$

where B is the length of one fiber and N is the number of turns the fiber makes around the bladder. For ease of measurement, the braid angle, length, and bladder outer diameter used to calculate B and N were measured while the PAM was in its resting state. Although the two PAMs used in this study were intended to be identical, there were slight differences between them as a result of manufacturing. For the contractile PAM, the initial braid angle was 74°, the resting length was 12.8 cm (5.05 in), and the initial bladder outer diameter was 1.59 cm (0.625 in). For the push-PAM, the initial braid angle was 73°, the resting length was 13.1 cm (5.16 in), and the initial bladder outer diameter was 1.59 cm (0.625 in).

5.4.1 *Gaylord Force Model*

Using these basic geometric constraints, a number of models have been developed to predict the quasi-static PAM actuation behavior, with most models focusing on either energy balance or force balance methods of derivation [1, 2, 8, 12, 19–24]. These methods all agree upon the same initial forcing term, known as the Gaylord force,

$$F_G = \frac{P}{4N^2\pi}\left(3L^2 - B^2\right) \tag{5.4}$$

where F_G is the force generated by the PAM upon pressurization and P is the actuation pressure. While this model provides a reasonable estimate for blocked force, it fails to capture the nonlinearity of the actuation diagram and over-predicts the PAM performance. In order to better represent the behavior of the actuator, several correction terms have been added to the Gaylord force to account for factors such as elasticity in the braid and bladder, non-cylindrical tip shapes, and friction between the actuator components [2, 4, 5, 20–25].

5.4.2 Force Balance Model

Of these various models, the force balance model first proposed by Ferraresi et al. [22] and later refined by Kothera et al. [23] and Hocking and Wereley [24] provided an excellent estimation of the output of the PAMs used in this study. In this prior work, the force produced by the actuator was predicted based on force balance methods and derived using the free-body diagram shown in Fig. 5.4. From this diagram, the equilibrium equations in the x- and z-directions are, respectively,

$$\mathrm{PRL} = \sigma_x tL + NT\cos\alpha \tag{5.5}$$

and

$$F + \pi R^2 P = \sigma_z A_B + T\sin\alpha \tag{5.6}$$

where R is the instantaneous bladder outer radius, σ is the stress in the bladder (in the x- or z-direction as denoted by the subscript), t is the instantaneous thickness of the bladder, T is the tension in the braided sleeve fiber, F is the total longitudinal force exerted by the PAM, and A_B is the bladder cross-sectional area.

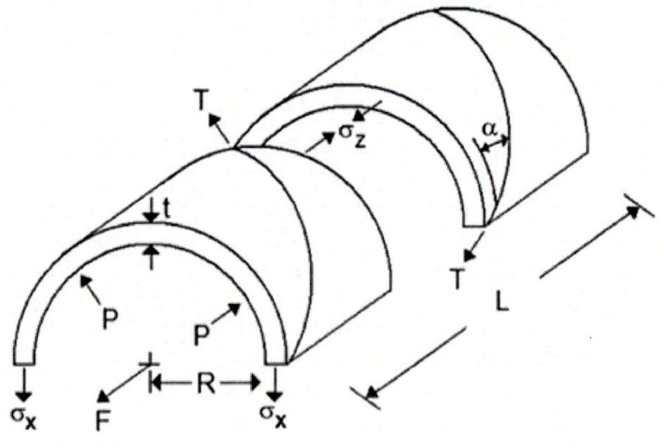

Fig. 5.4 Free-body diagram used in the force balance model derivation [24]

Assuming a constant bladder volume (V_B), the bladder cross-sectional area can be written as $A_B = V_B/L$. Furthermore, the instantaneous bladder thickness can be calculated as

$$t = R - \sqrt{R^2 - \frac{V_B}{\pi L}}. \tag{5.7}$$

Making use of trigonometric identities and the geometric constraints provided in Eqs. (5.2) and (5.3), the instantaneous bladder outer radius may be written as

$$R = \frac{\sqrt{B^2 - L^2}}{2\pi N} \tag{5.8}$$

With this expression for R, as well as the geometric constraints, it is possible to solve Eqs. (5.5) and (5.6) for T, set them equal to each other, and then group like terms such that the force generated by the PAM is given by

$$F = \frac{P}{4N^2\pi}\left(3L^2 - B^2\right) + \sigma_z \frac{V_B}{L} - \frac{\sigma_x t L^2}{2\pi N^2 R} \tag{5.9}$$

Observe that the first term is now equivalent to the Gaylord force term (F_G) from Eq. (5.4) and the remaining terms are corrections that account for elasticity in the bladder.

Former derivations of this model used a linear elastic model for the bladder [22, 23]. However, this material model fails to capture the nonlinear response of the actuator and does nothing to predict the softening in the bladder at high radial strains that was demonstrated during the experimental characterization of both the contractile PAM and the push-PAM. In order to account for this behavior, the nonlinear elastic model proposed by Hocking and Wereley [24] was used in the present study. In this nonlinear material model, the stress in both the x- and z-directions is given by.

$$\sigma = \sum_{k=1}^{M} E_k \varepsilon^k \tag{5.10}$$

where ε is the strain in the bladder and the modulus values (E_k) are design variables that are identified from empirical data at each operating pressure in order to minimize the error between the model and the experimental results. The summation upper limit (M) defines the order of the material model and is chosen such that the model is robust enough to capture the full behavior of the actuator. The strains are defined as in the original force balance model developed by Ferraresi et al. [22] as

$$\varepsilon_x = \frac{\Delta R}{R_0} = \frac{R - R_0}{R_0} = \frac{R}{R_0} - 1 \tag{5.11}$$

and

$$\varepsilon_z = \frac{\Delta L}{L_0} = \frac{L - L_0}{L_0} = \frac{L}{L_0} - 1 \tag{5.12}$$

where ΔR is the change in bladder outer radius and R_0 is the initial bladder outer radius. Substituting this stress vs. strain constitutive material relationship into Eq. (5.9), the force in the PAM is

$$F = \frac{P}{4N^2\pi}\left(3L^2 - B^2\right) + \frac{V_B}{L}\sum_{k=1}^{M} E_k\left(\frac{L}{L_0} - 1\right)^k - \frac{tL^2}{2\pi N^2 R}\sum_{k=1}^{M} E_k\left(\frac{R}{R_0} - 1\right)^k \tag{5.13}$$

As previously mentioned, the first term in this model is the basic Gaylord force term (F_G), while the second two terms are correction terms identified through the use of the nonlinear elastic model in conjunction with the force balance derivation. Labeling these additional two correction terms, collectively, as F_{NLFB}, the overall PAM force presented in Eq. (5.13) can be written as $F = F_G + F_{NLFB}$.

As shown in Ref. [24], note that by setting $M = 1$ (i.e., using a linear elastic constitutive model) in Eq. (5.13), this model simplifies to the force balance model derived in Refs. [22, 23]:

$$F = \frac{P}{4N^2\pi}\left(3L^2 - B^2\right) + E_1 V_B\left(\frac{1}{L_0} - \frac{1}{L}\right) - \frac{E_1 tL^2}{2\pi N^2}\left(\frac{1}{R_0} - \frac{1}{R}\right) \tag{5.14}$$

where E_1 is simply the elastic modulus of the latex bladder, assumed constant over all operating pressures. In this case, the two additional force correction terms added to the Gaylord force are collectively referred to as F_{LFB} such that the overall PAM force in Eq. (5.14) is given by $F = F_G + F_{LFB}$. As demonstrated by Hocking and Wereley [24] and further validated by the current study, however, prudent selection of $M > 1$ in Eq. (5.13) allows for a much more accurate prediction of PAM behavior through the use of the nonlinear elastic material model.

5.4.3 Friction Force Model

In order to fully capture the nonlinear quasi-static actuation behavior of both types of PAMs, an additional force correction term is required to account for the friction between PAM components such that the hysteresis loop demonstrated in the experimental characterization can be accurately modeled. Previous researchers have investigated several methods to account for this behavior, including the use of a fixed friction force [2], a friction force calculated from a static friction model for interactions between the braid fibers at a micromechanical level [8, 20, 25], and a friction force directly proportional to the PAM force [4, 24]. A more comprehensive review of friction and hysteresis modeling within PAM research is offered in Ref. [30].

Given the simplicity and accuracy of the friction model suggested by Woods et al. [4], a similar term has been proposed for use with the nonlinear elastic force balance model, as was suggested by Hocking and Wereley [24] in their initial derivation of the model. For this correction term, the friction force (F_F) opposes the PAM motion such that the friction reduces the overall PAM force during contraction (when the actuator is considered to be moving with positive velocity, v) and increases overall PAM force during extension (when velocity is considered negative). In this model, the magnitude of the friction force is proportional to the magnitude of the PAM force and is thus calculated as

$$F_F = -k_F\left(F_G + F_{NLFB}\right)\operatorname{sgn}(v) \tag{5.15}$$

where k_F is an empirically fit friction factor. For the PAMs used in this study, k_F decreased linearly with actuation pressure. For the contractile PAM, k_F ranged from 0.14 at 207 kPa down to 0.06 at 621 kPa. For the push-PAM, however, more hysteresis was visible in the actuation diagram (most likely due to additional friction in the rod seal), so the friction factor was slightly higher, ranging from 0.16 at 207 kPa to 0.08 at 621 kPa. Combining the friction force from Eq. (5.15) with the nonlinear elastic force balance model given in Eq. (5.13), the full PAM actuation force is now given by $F = F_G + F_{NLFB} + F_F$.

5.5 Pressure Dead-Band

Similar to PAM characteristics reported by previous researchers [2, 12, 21, 24, 26], the actuators used in this study required a small amount of initial pressurization for the bladder to make substantial contact with the surrounding braid and initiate inflation of the PAM, resulting in an actuation dead-band at low operating pressures. To capture this physical phenomenon and its effects on the PAM force generated at a given actuation pressure, a pressure offset term was included in the analysis to account for the initial pressure required for the PAM to begin to actuate. In order to model the pressure dead-band, the actuation pressure P is reduced as

$$P' = P - P_c \tag{5.16}$$

where P' is the corrected actuation pressure to be used in place of P in the PAM force model and P_c is an empirically obtained term corresponding to the pressure at which the actuator initiates contraction. To validate the presence of this dead-band and identify the correct values for P_c for the PAMs used in this study, the pressure-averaged blocked force (i.e., the maximum force exerted by the PAM, neglecting hysteresis effects, at each actuation pressure) was utilized. It is shown plotted against actuation pressure in Fig. 5.5 for both types of actuators for the entire range of pressures tested. As is evident from the figure (and the model), there is a strong linear correlation between blocked force and actuation pressure. With the visual aid of the linear trend lines that have been plotted with the data, it is also clear that neither PAM is likely to generate a blocked force until a minimum actuation pressure threshold has been reached, demonstrating the existence of the dead-band [31]. This minimum pressure is given by the x-intercept of the linear trend line for each data set and is used as the pressure offset, P_c. For the contractile PAM and push-PAM used in this study, P_c was determined from the linear trend lines to be 46.4 kPa (6.7 psi) and 58.1 kPa (8.4 psi), respectively. The slight discrepancy between these offsets is likely due to the aforementioned differences in braid angle and resting length between the contractile PAM and the push-PAM.

Incorporating this model improvement such that the previously described forcing terms in Eqs. (5.13) and (5.15) are a function of the corrected pressure, where applicable, the full PAM force model is given by

$$F = F_G\left(P'\right) + F_{NLFB} + F_F\left(P'\right). \tag{5.17}$$

Unless explicitly stated otherwise, the model results presented in the following section are a function of the corrected pressure (P') in place of the measured value (P) where it appears in the original model expressions previously described in Eqs. (5.13) and (5.15).

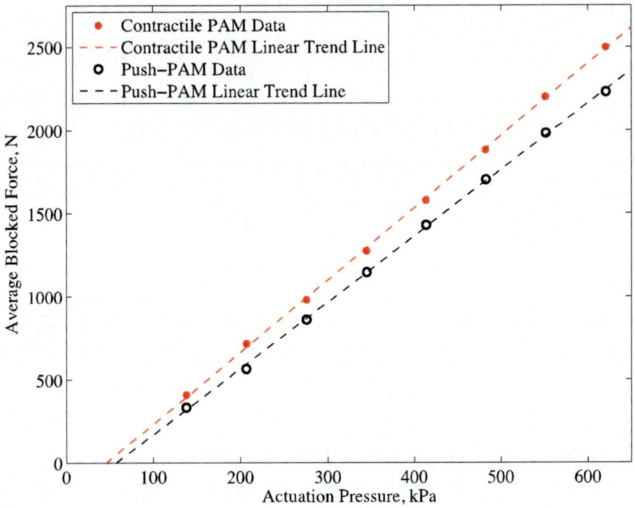

Fig. 5.5 Blocked force as a function of actuation pressure with linear trend lines to demonstrate the presence of a pressure dead-band

5.5.1 Discussion of Modeling Results

As mentioned in Sect. 5.4.2, the proposed force balance model uses the nonlinear material elastic model given in Eq. (5.10) with empirically identified modulus values (E_k). In order to achieve accurate results, the built-in MATLAB function *fmincon* was used to determine the optimum values of E_k at each operating pressure such that the error between the model output without friction force ($F_G + F_{NLFB}$) and the average force within the hysteresis loop (F_{AVG} as depicted in Fig. 5.3b) was a minimum. This process was carried out for increasingly higher-order material models (i.e., increasing M) until the model was able to accurately capture the PAM behavior with less than 0.5% error for all operating pressures, neglecting hysteresis effects. In this optimization, the objective function that was minimized by the MATLAB solver is a normalized least squares regression error given by

$$\text{Error} = \sum_i \left(\frac{\left(F_G + F_{NLFB}\right)_i - \left(F_{AVG}\right)_i}{\left(F_{AVG}\right)_i} \right)^2 \tag{5.18}$$

where $(F_G + F_{NLFB})_i$ is the ith data point predicted by the model described in Eq. (5.13) and $(F_{AVG})_i$ is the ith data point given by the average between the upper and lower hysteresis curves within the actual experimental data set (see Fig. 5.3b).

The results of the proposed nonlinear elastic force balance model are shown with respect to F_{AVG} in Fig. 5.6 for the 207, 414, and 621 kPa tests, demonstrating the effect of increasing the summation upper limit (M) on the accuracy of the model for both the contractile PAM (top row) and the push-PAM (bottom row). Figure 5.6a, d also shows the results of the non-optimized linear force balance model (proposed by previous researchers and presented again in Eq. (5.14)) [22, 23]. From Fig. 5.6a, d, it is apparent that a linear elastic material model (M = 1) is unable to capture the response of either type of PAM. Using the non-optimized linear force balance model (in which E_1 is constant across all pressures), the model accurately predicts blocked force, but greatly over-predicts the PAM force capabilities during contraction, with increased error at higher actuation pressures. While the optimized linear elastic model (in which E_1 is pressure-dependent and has been optimized over the linear elastic portion of PAM contraction from $0 < \Delta L/L_0 < 0.05$) yields much more accurate results, the model tends to under-predict PAM performance at high contractions and high pressures, especially for the contractile PAM, which has slightly better free contraction characteristics. Given the errors associated with the linear elastic model, the accuracy of using a quadratic elastic model ($M = 2$) was explored next, with the output of the model shown with respect to F_{AVG} in Fig. 5.6b, e. In these figures, the results of E_k optimized over both the linear elastic portion of PAM contraction ($0 < \Delta L/L_0 < 0.05$) and over the entire contractile range are shown. For both optimization ranges, it is evident that the quadratic elastic model still lacks the robustness required to capture the PAM behavior throughout actuation. For both the contractile PAM and the push-PAM, the quadratic elastic model causes the force balance model to slightly under-predict the force at low contractions and over-predict the force at

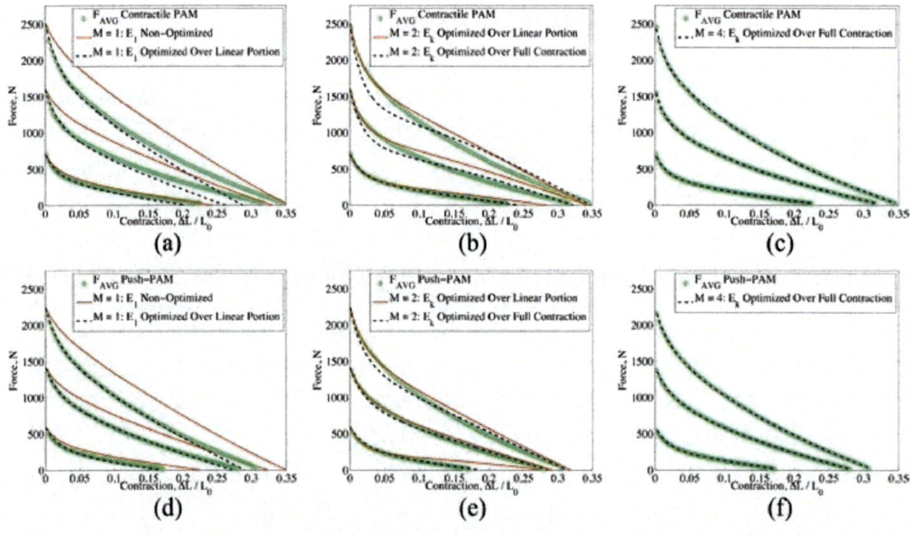

Fig. 5.6 Force balance model results compared with averaged experimental data for the contractile PAM (**a–c**) and the push-PAM (**d–f**). (**a/d**) Linear elastic model results ($M = 1$) for E_1 non-optimized and E_1 optimized over the linear elastic portion of PAM contraction. (**b/e**) Quadratic elastic model results ($M = 2$) for E_k optimized over the linear elastic portion of PAM contraction and E_k optimized over the full PAM contractile range. (**c/f**) Quartic elastic model results ($M = 4$) for E_k optimized over the full PAM contractile range

high contractions. Continuing in this manner, the order of the material model from Eq. (5.10) was increased to $M = 3$ (results not shown) and ultimately to $M = 4$, with the output from the quartic elastic model ($M = 4$) finally producing satisfactorily accurate correlation (i.e., average error less than 0.5%) between the model and the pressure-averaged experimental data for the full actuation range of both the contractile PAM and push-PAM, as shown in Fig. 5.6c, f. Not only does this result demonstrate the excellent modeling capabilities of the nonlinear elastic force balance analysis, but it also confirms that the modeling approaches used to predict traditional PAM behavior may be applied to the push-PAM apparatus with comparable accuracy.

As a complement to the results shown in Fig. 5.6, it is also interesting to consider the physical nature of the material models proposed through both the non-optimized linear elastic model and the optimized elastic models for the contractile PAM and the push-PAM. To that end, Fig. 5.7 shows the bladder stress-strain curves associated with the linear elastic material models ($M = 1$, both non-optimized and optimized) and the quartic material model ($M = 4$) that was ultimately shown to deliver accurate force predictions. For clarity, only the models for the tests at an actuation pressure of 414 kPa are shown in the figure. Here, the stress-strain relationship has been calculated as in Eq. (5.10) using the corresponding values of E_k for each model. This relationship has been calculated up to a maximum strain of 1.67, which corresponds to the highest radial strain experienced during testing at this pressure as defined in Eq. (5.11) and calculated using instantaneous bladder radius values given

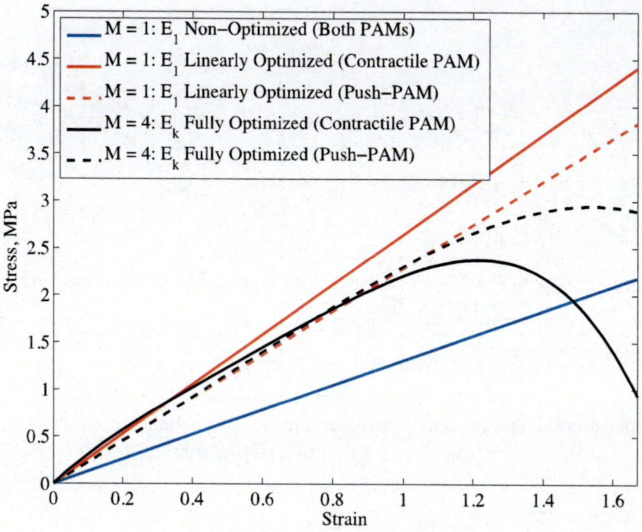

Fig. 5.7 Stress-strain curves associated with a non-optimized linear elastic model and optimized elastic models for $M = 1$ and $M = 4$ for both the contractile PAM and the push-PAM

in Eq. (5.8). As implied by a linear elastic model, the results for $M = 1$ with both non-optimized E_1 and optimized E_1 suggest that stress in the bladder varies linearly with strain over the full range of motion of the PAM, which, based on actual PAM behavior and the nature of elastomeric materials, is physically unlikely. Using the results of the optimization with the quartic elastic model (E_k with $M = 4$), however, the stress-strain curve indicates that the PAM bladders soften with increased strain, which is physically justifiable [32] and yields excellent correlation with experimental results. Furthermore, these results indicate that the contractile PAM bladder undergoes more softening than that of the push-PAM, which was suspected based on the nature of the actuator load lines. Regarding the model associated with the contractile PAM, however, it is clear that such a stress-strain curve yields impractical results at high strains, so additional research into hyper-elastic material models may be required to fully characterize the nonlinear elastic nature of these PAM bladders.

As previously mentioned, two main parameters of interest are the blocked force and free contraction of the PAMs at each internal pressure. These values are represented in Figs. 5.3 and 5.6 as the points at which the actuator load lines reach each axis (blocked force on the vertical axis and free contraction on the horizontal axis). It is evident in these figures that both blocked force and free contraction increase with pressure. For clarity, however, the strict evolution of each parameter with internal pressure is shown separately in Fig. 5.8 for the entire range of actuation pressures tested. Also included in Fig. 5.8 are the model predictions for these parameters using the nonlinear elastic force balance model with $M = 4$. Regarding the experimental data alone, it is clear that the contractile PAM is able to achieve a slightly higher blocked force and free contraction than the push-PAM at each actuation

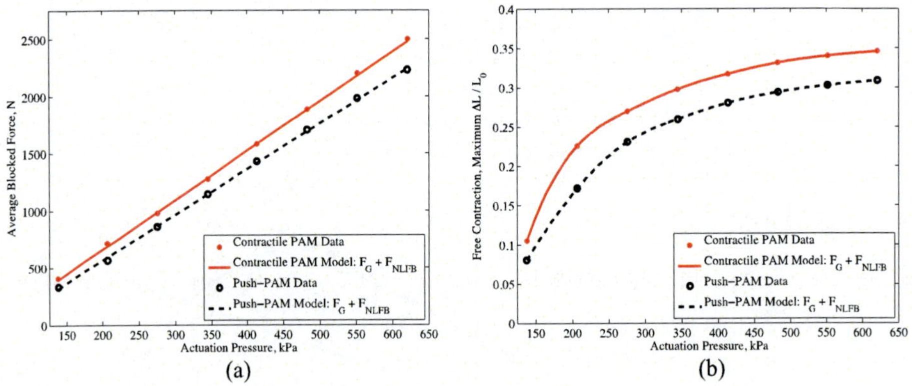

Fig. 5.8 Experimental results and nonlinear elastic force balance model predictions for (**a**) blocked force and (**b**) free contraction as a function of pressure

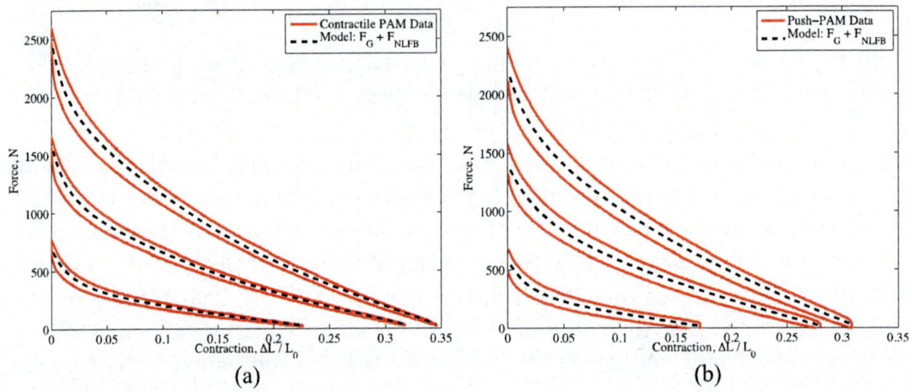

Fig. 5.9 Results of the nonlinear elastic force balance model (using $M = 4$) compared to full experimental results for (**a**) the contractile PAM and (**b**) the push-PAM

pressure. However, it is important to acknowledge that the push-PAM yields comparable results and some of this disparity is due to manufacturing differences between the contractile PAM and the push-PAM (e.g., active length, braid angle, pressure offset). Studying the model predictions for blocked force and free contraction, the overwhelming accuracy of the nonlinear elastic force balance model (with a pressure dead-band correction term) is confirmed, and the hypothesis that the contractile PAM and the push-PAM can be effectively modeled with the same set of equations is validated.

To further demonstrate the accuracy of the nonlinear elastic force balance model with $M = 4$, the results of this model ($F_G + F_{NLFB}$) with E_k empirically optimized for each actuation pressure are shown with respect to the to the full experimental results for the 207, 414, and 621 kPa tests in Fig. 5.9. For clarity, the results from the contractile PAM and the push-PAM have been plotted in two separate figures (Fig. 5.9a, b, respectively). It is evident from these figures that the PAM behavior predicted by the nonlinear elastic force balance model given by Eq. (5.13) yields excellent agreement with the experimental data (neglecting hysteresis), including the rapidly changing stiffness of the PAMs below $\Delta L/L_0 = 0.05$. As such, these results help to pave the way forward for both the traditional contractile PAM and the novel push-PAM to be better utilized in controlled actuation schemes.

Although the nonlinear elastic force balance model was originally developed by Hocking and Wereley [24] to accurately capture the actuation characteristics of small-scale PAMs (with diameters on the order of millimeters), the results shown in the present work indicate that this model is appropriate and useful for larger-scale PAMs as well. To illustrate the impact of the nonlinear elastic force balance model on the accuracy of force predictions for larger PAMs, Fig. 5.10 shows the progression in model capabilities from the basic Gaylord force model to the nonlinear elastic force balance model with a pressure offset term (using only the 414 kPa case for the sake of clarity). While Fig. 5.10a, c show the actual model output compared to the experimental results for the contractile PAM and the push-PAM (respectively), Fig. 5.10b, d show the associated error between the model prediction (F_{MODEL}) and the experimental data for each type of PAM (contractile PAM results in the top row,

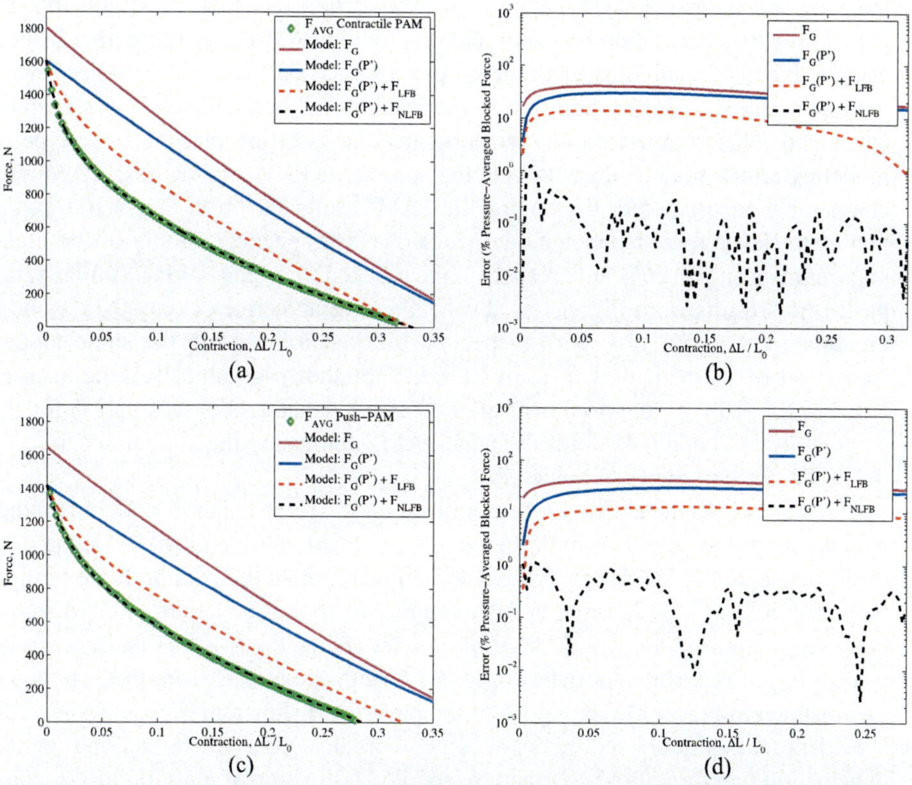

Fig. 5.10 Progressive accuracy of modeling techniques for the contractile PAM (**a**, **b**) and the push-PAM (**c**, **d**) at 414 kPa. (**a/c**) Model output compared to experimental data. (**b/d**) Associated model error (% of pressure-averaged block force)

push-PAM results in the bottom row). As the models shown here neglected hysteresis, the experimental data used to calculate the error is the average force (F_{AVG}) as used above. The error at each contraction value has been calculated as

$$\text{Error}_i = \left| \frac{(F_{\text{MODEL}})_i - (F_{\text{AVG}})_i}{F_{\text{B}}} \right| \tag{5.19}$$

where F_{B} is the pressure-averaged blocked force for each type of actuator. As is evident in Fig. 5.10, the Gaylord force alone (F_{G}) provides a poor estimate of each PAM's blocked force capabilities and uniformly over-predicts the PAM performance as the actuators contract. Upon including the pressure correction to account for the actuation dead-band (see $F_{\text{G}}(\text{P}')$), the model yields a much more accurate prediction of blocked force, but still fails to capture the appropriate loss in PAM force with increasing contraction. Using the basic linear elastic force balance model used by Ferraresi et al. [22] and again by Kothera et al. [23], that is, $F_{\text{G}}(P') + F_{\text{LFB}}$, the model is able to more closely forecast the PAM force, although it still tends to over-predict the PAM output during contraction for both the contractile PAM and the push-PAM. Finally, the nonlinear elastic force balance model proposed by Hocking and Wereley [24], $F_{\text{G}}(P') + F_{\text{NLFB}}$, shown here using $M = 4$, is able to fully capture the PAM actuation behavior and predict the PAM force (neglecting hysteresis) with considerably less error, demonstrating that this model does indeed work well with larger PAMs, in addition to the smaller PAMs for which it was initially developed. This comparison also corroborates the assertion that the same type of modeling efforts may be used on both the contractile PAM and the push-PAM with comparable accuracy. For the contractile PAM results shown in Fig. 5.10a, b, the nonlinear elastic force balance model yielded an average error of only 0.12% and a maximum error of 1.26% with respect to the blocked force at 414 kPa. Similarly, the push-PAM results from Fig. 5.10c, d reflect an average error of only 0.31% and a maximum error of 1.13% with respect to blocked force using the same model. Across all of the actuation pressures tested (not shown graphically), the average error associated with the nonlinear elastic force balance model was 0.22% for the contractile PAM and 0.34% for the push-PAM, confirming the similar accuracy of the model for both types of actuators.

Having verified the ability of the nonlinear elastic force balance model to predict the average force output of both the contractile PAM and the push-PAM, the next phase of the model validation process was to incorporate the friction force term as proposed in Eq. (5.15) in order to fully capture the hysteretic nature of PAM actuation. The output of the full PAM force model ($F_{\text{G}} + F_{\text{NLFB}} + F_{\text{F}}$) is shown with respect to the experimental data in Fig. 5.11, with the results from the contractile PAM shown in Fig. 5.11a and those of the push-PAM shown in Fig. 5.11b. Here, it is evident that the proposed model accurately predicts the PAM force as a function of actuation pressure, PAM contraction, and PAM direction of motion, fully encompassing all aspects of the actuation behavior for both types of actuators.

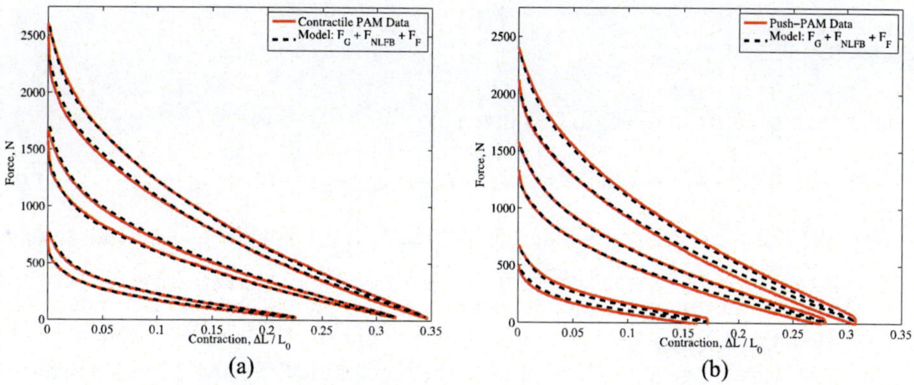

Fig. 5.11 Full PAM force model ($F_G + F_{NLFB} + F_F$) compared to experimental data for (**a**) the contractile PAM and (**b**) the push-PAM

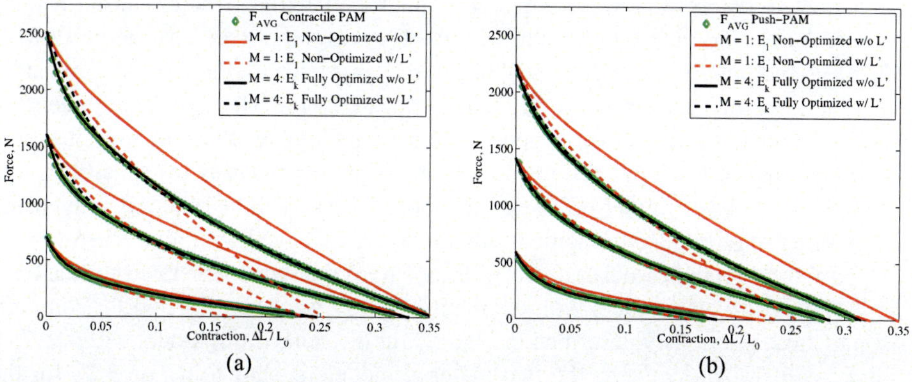

Fig. 5.12 Force balance model results with and without the corrected length term (L') compared to experimental data for (**a**) the contractile PAM and (**b**) the push-PAM

5.5.2 Impact of Length Correction Term on Modeling Efforts

Although prior work on PAMs supports the inclusion of a length correction term to account for the non-cylindrical tip shape of the actuator when it is inflated [4, 12, 20, 21, 23], this study ultimately concluded that such a term was unnecessary when using the nonlinear elastic force balance model. Figure 5.12 shows the modeling results for both the contractile PAM (Fig. 5.12a) and the push-PAM (Fig. 5.12b) using the linear elastic force balance model (E_1 non-optimized) and the nonlinear elastic force balance model ($M = 4$), both with and without a corrected length term. While Tondu [30] offers a review of several methods used to account for the non-cylindrical tip shape, the correction offered by Woods et al. [4] was used to generate these results. For this correction, the tip shape was modeled as a 90° circular arc, which effectively reduced the length of the actuator at each end by

$$\Delta l = \left(\frac{\pi}{2} - 1\right)\left(R - R_0\right) \tag{5.20}$$

such that the corrected length was given by

$$L' = L - 2\Delta l \tag{5.21}$$

where L' was the corrected actuator length that was used in place of the true actuator length (L) in the PAM force models.

As shown in Fig. 5.12, the inclusion of the length correction term reduces the predicted PAM force for the non-optimized linear elastic force balance model ($M = 1$). This change allows the model to better capture the PAM actuation behavior at low contractions, but ultimately causes the model to vastly under-predict the PAM force at higher contractions. For the nonlinear elastic force balance model, on the other hand, including the length correction term produced a less substantial change with respect to the same model without the length correction term. In some cases, the two nonlinear models nearly overlay each other (see contractile PAM at 207 kPa and push-PAM at 207 and 414 kPa). This is because the error minimization routine identified the optimum modulus values (E_k) for both types of models, so both are able to achieve close correlation to experimental results. However, it should be noted that at low contractions and high actuation pressures (especially for the contractile PAM), a higher-order elastic model ($M > 4$) will be required when using L' in order to reproduce the same accuracy as was achieved without the use of L'.

Furthermore, the associated stress-strain curves that were generated for the optimized quartic elastic model with the inclusion of L' were investigated for comparison to those previously described in Fig. 5.7 for the same order model without L'. A direct comparison between the two material models is shown in Fig. 5.13 for the 414 kPa test of both the contractile PAM and the push-PAM, with the maximum strain plotted once again corresponding to the maximum radial strain encountered during testing. As shown in this figure, the stress-strain relationship associated with the nonlinear elastic model using L' suggests that the stress in the bladder is negative as a result of the strains encountered during testing, which is not physically meaningful. Given this result, as well as the small decrease in model accuracy associated with the use of L', this term can be neglected when using the nonlinear elastic force balance model.

5.6 Conclusions

This study proposed a novel push-PAM actuator that utilized the advantageous performance characteristics of a traditional contractile PAM, but converted the normal direction of force and motion such that the resulting push-PAM achieved excellent compressive force generation and extensile motion with just a small increase in friction, weight, and cost. This motion conversion was accomplished by a pushrod that

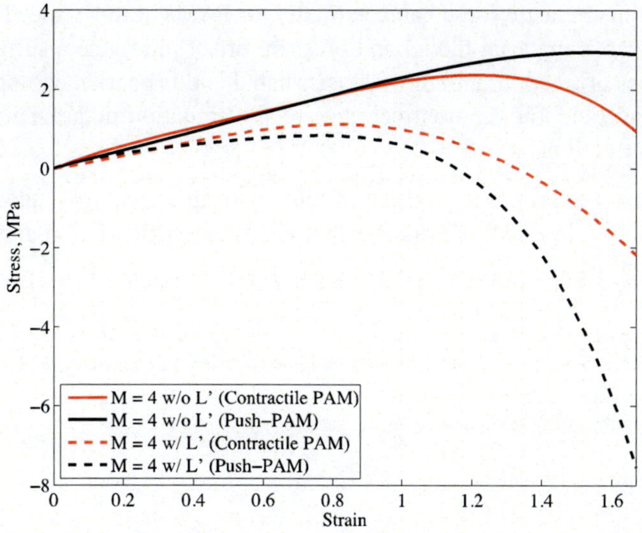

Fig. 5.13 Stress-strain curves associated with an optimized nonlinear elastic model ($M = 4$), with and without L', for the contractile and push-PAMs

was attached to the inside of one of the contractile PAM's end fittings and extended through the actuator body and out of the other end fitting. Upon pressurization, the flexible body of the actuator inflated and expanded radially, causing relative contractile motion between the two end fittings, which ultimately extended the pushrod such that the push-PAM actuator demonstrated extensile motion and compressive force generation. The key results of this study are the following:

1. The experimental characterization of the push-PAM highlighted the quasi-static behavior of the actuator and verified that the push-PAM was capable of producing forces and displacements comparable to a contractile PAM. For both types of actuators, three key nonlinear phenomena were observed during testing: nonlinear PAM stiffness, hysteresis of the force vs. displacement response, and a pressure dead-band.
2. In terms of modeling the behavior of the push-PAM, it was shown that the same analysis methods may be applied to both the push-PAM and the contractile PAM with similar accuracy. In order to account for the nonlinear behavior of the actuators, both PAMs were ultimately modeled using a refined force balance model. This model included additional terms to accurately capture the nonlinear stress vs. strain relationship in the PAM bladder, friction effects that contributed to hysteresis, and the actuation dead-band at low operating pressures, offering excellent agreement with experimental results for both the contractile PAM and the push-PAM. Additionally, the accuracy of these quasi-static modeling efforts validated the applicability of the nonlinear elastic force balance model proposed by Hocking and Wereley [24] for use with larger PAM actuators.
3. Further model analysis revealed that the inclusion of a length correction term to account for the non-cylindrical tip shape of the PAM is unnecessary when using

the nonlinear elastic force balance model for PAMs at this scale. The utilization of the quartic material model and pressure offset introduce a sufficient number of degrees of freedom into the analysis such that this phenomenological model is able to account for the nonlinearities of the actuation diagram without the tip shape correction.

In conclusion, the proposed push-PAM offers a simple and highly effective method to use a contractile PAM in a manner that yields extensile motion and compressive force generation.

Acknowledgments E. Hocking was supported by a DoD, Air Force Office of Scientific Research, National Defense Science and Engineering Graduate (NDSEG) Fellowship, 32 CFR 168a.

References

1. D. Caldwell, N. Tsagarakis, G.A. Medrano-Cerda, bio-mimetic actuators: Polymeric pseudo muscular actuators and pneumatic muscle actuators for biological emulation. Mechatronics **10**(4–5), 499–530 (2000). https://doi.org/10.1016/S0957-4158(99)00071-9
2. C. Chou, B. Hannaford, Static and dynamic characteristics of McKibben pneumatic artificial muscles, in *Proc. of ICRA*, vol. 1, (1994), pp. 281–286
3. R.M. Robinson, C.S. Kothera, B.K.S. Woods, R.D. Vocke III, N.M. Wereley, High specific power actuators for robotic manipulators. J. Intell. Mater. Syst. Struct. **22**(13), 1501–1511 (2011). https://doi.org/10.1177/1045389X11417653
4. B.K.S. Woods, C.S. Kothera, N.M. Wereley, Wind tunnel testing of a helicopter rotor trailing edge flap actuated via pneumatic artificial muscles. J. Intell. Mater. Syst. Struct. **22**(13), 1513–1528 (2011). https://doi.org/10.1177/1045389X11424216
5. R.D. Vocke III, C.S. Kothera, A. Chaudhuri, B.K.S. Woods, N.M. Wereley, Design and testing of a high-specific work actuator using miniature pneumatic artificial muscles. J. Intell. Mater. Syst. Struct. **23**(3), 365–378 (2012). https://doi.org/10.1177/1045389X11431743
6. B.K.S. Woods, Pneumatic artificial muscle driven trailing edge flaps for active rotors, Ph.D. Thesis, University of Maryland, College Park, 2012
7. B.K.S. Woods, M.F. Gentry, C.S. Kothera, N.M. Wereley, Fatigue life testing of swaged pneumatic artificial muscles as actuators for aerospace applications. J. Intell. Mater. Syst. Struct. **23**(3), 327–343 (2011). https://doi.org/10.1177/1045389X11433495
8. H.F. Schulte, D.F. Adamski, J.R. Pearson, Characteristics of the braided fluid actuator, Report at Department of Physical Medicine and Rehabilitation, the University of Michigan, Ann Arbor, MI, 1961, pp. 105–120.
9. K. Bharadwaj, K.W. Hollander, C.A. Mathis, T.G. Sugar, Spring over muscle (SOM) actuator for rehabilitation devices, in *Proc. of the 26th Annual Int. Conf. of the IEEE EMBS*, vol. 1, (2004), pp. 2726–2729
10. M.K. Philen, On the applicability of fluidic flexible matrix composite variable impedance materials for prosthetic and orthotic devices. Smart. Mater. Struct. **18**, 104023 (10 pp) (2009). https://doi.org/10.1088/0964-1726/18/10/104023
11. M.C. Birch et al., A miniature hybrid robot propelled by legs, in *Proc. of the 2001 IELE/RSJ Int. Conf. on Intelligent Robots and Systems*, vol. 2, (2001), pp. 845–851
12. M. De Volder, A.J.M. Moers, D. Reynaerts, Fabrication and control of miniature McKibben actuators. Sens. Actuators A **166**(1), 111–116 (2011). https://doi.org/10.1016/j.sna.2011.01.002

13. Y. Chen, W. Yin, Y. Liu, J. Leng, Structural design and analysis of morphing skin embedded with pneumatic muscle fibers. Smart Mater. Struct. **20**, 085033 (8 pp) (2011). https://doi.org/10.1088/0964-1726/20/8/085033
14. L.D. Peel, J. Mejia, B. Narvaez, K. Thompson, M. Lingala, Development of a simple morphing wing using elastomeric composites as skins and actuators. ASME J. Mech. Des., 131, 091003 (8 pp) (2009). https://doi.org/10.1115/1.3159043
15. Z. Zhang, M.K. Philen, Review: pressurized artificial muscles. J. Intell. Mater. Syst. Struct. **23**(3), 255–268 (2011). https://doi.org/10.1177/1045389X11420592
16. M.B. Pritts, C.D. Rahn, Design of an artificial muscle continuum robot, in *Proc. IEEE Int. Conf. on Robotics and Automation*, vol. 5, (2004), pp. 4742–4746
17. D. Trivedi, C.D. Rahn, Soft robotic manipulators: Design, analysis, and control (Chap. 7), in *Plants and Mechanical Motion: A Synthetic Approach to Nastic Materials and Structures*, ed. by N. M. Wereley, J. M. Sater, (DEStech Publications, Inc., Lancaster, 2012), pp. 141–165
18. W. Liu, C.D. Rahn, Fiber-reinforced membrane models of McKibben actuators. ASME J. Appl. Mech. **70**(6), 853–859 (2003). https://doi.org/10.1115/1.1630812
19. R. Gaylord, Fluid actuated motor system and stroking device, U.S. Patent No. 2,844,126, 1958.
20. B. Tondu, P. Lopez, Modeling and control of McKibben artificial muscle robot actuators. IEEE Control. Syst. Mag. **20**(2), 15–38 (2000). https://doi.org/10.1109/37.833638
21. N. Tsagarakis, D. Caldwell, Improved modeling and assessment of pneumatic muscle actuators, in *Proc. of ICRA*, vol. 4, (2000), pp. 3641–3646
22. C. Ferraresi, W. Franco, A.M. Bertetto, Flexible pneumatic actuators: A comparison between the McKibben and the straight fibres muscles. J. Robot. Mechatron. **13**(1), 56–63 (2001)
23. C.S. Kothera, M. Jangid, J. Sirohi, N.M. Wereley, Experimental characterization and static modeling of McKibben actuators. ASME J. Mech. Des., 131, 091010 (10 pp) (2009). https://doi.org/10.1115/1.3158982
24. E.G. Hocking, N.M. Wereley, Analysis of nonlinear elastic behavior in miniature pneumatic artificial muscles. Smart Mater. Struct. **22**(1), 014016 (2013). https://doi.org/10.1088/0964-1716/22/1/014016
25. S. Davis, D. Caldwell, Braid effects on contractile range and friction modeling in pneumatic muscle actuators. Int. J. Robot. Res. **25**(4), 359–369 (2006)
26. B. Solano, C. Rotinat-Libersa, Compact and lightweight hydraulic actuation system for high performance millimeter scale robotic applications: Modeling and experiments. J. Intell. Mater. Syst. Struct. **22**(13), 1479–1487 (2011). https://doi.org/10.1177/1045389X11418860
27. M.K. Philen, Y. Shan, P. Prakash, K.W. Wang, C.D. Rahn, A.L. Zydney, C.E. Bakis, Fibrillar network adaptive structure with ion-transport actuation. J. Intell. Mater. Syst. Struct. **18**(4), 323–334 (2007). https://doi.org/10.1177/1045389X06066097
28. B. Zhu, C.D. Rahn, C.E. Bakis, Actuation of fluidic flexible matrix composites in structural media. J. Intell. Mater. Syst. Struct. **23**(3), 269–278 (2011). https://doi.org/10.1177/1045389X11428676
29. B.K.S. Woods, N.M. Wereley, K.S. Kothera, S.M. Boyer, Extensile fluidic muscle actuator, U.S Patent Appl. No. 12/955,242, 2011
30. B. Tondu, Modelling of the McKibben artificial muscle: A review. J. Intell. Mater. Syst. Struct. **23**(3), 225–253 (2012). https://doi.org/10.1177/1045389X11435435
31. A. Gelb, W.E. Vander Velde, *Multiple-input Describing Functions and Nonlinear System Design* (McGraw-Hill, New York, 1968), pp. 4–5, Chap. 1
32. J.T. Bauman, *Fatigue, Stress, and Strain of Rubber Components – A Guide for Design Engineers* (Hanser Publishers, Munich, 2008), pp. 9–41, Chaps. 2–3

Erica G. Hocking is a Graduate Research Assistant at the University of Maryland Department of Aerospace Engineering.

Benjamin K. S. Woods is a Senior Lecturer in Aerospace Structures at the University of Bristol, with expertise in morphing and lightweight composite structures. He earned his Ph.D., M.S., and B.S. degrees in Aerospace Engineering from the University of Maryland. He moved to the UK in 2012 to take up a post-doctoral research post at Swansea University, before joining the University of Bristol in 2015. He is currently an EPSRC Early Career Fellow and is leading the Horizon 2020-funded SABRE program. He holds 7 US patents and has authored 28 journal publications, 2 book chapters, and 57 conference papers.

Norman M. Wereley is the Department Chair and Minta Martin Professor of Aerospace Engineering at the University of Maryland. Dr. Wereley's research interests are in dynamics and control of smart structures applied to helicopters and other aerospace and automotive systems, with emphasis on active and passive vibration isolation, shock mitigation (especially occupant protection systems), and actuation systems. A key focus in his research is the application of magnetorheological fluids in various occupant protection applications with adaptive suspension systems: vibration isolation and crashworthy helicopter crew seats, sliding crashworthy automotive seats, blast-mitigating crew seats for ground vehicles, and shock-mitigating crew seats for high-speed water craft. In each of these applications, Prof. Wereley was the lead technical expert on magnetorheological energy-absorbing systems. Dr. Wereley has published over 230 journal articles, 18 book chapters, and over 275 conference articles. Dr. Wereley is an inventor on 20 patents and several patents pending. Dr. Wereley was the editor of *Plants and Mechanical Motion – A Synthetic Approach to Plants and Mechanical Motion* (2012) and *Magnetorheology: Advances and Applications* (2013) from *The Royal Society of Chemistry*. Dr. Wereley is Editor of the *Journal of Intelligent Material Systems and Structures* and associate editor of *Smart Materials and Structures* and *Journal of the American Helicopter Society*. He recently served as Chair (2012–2013) of the SPIE Symposium on Smart Structures/NDE. Dr. Wereley is a Fellow of the American Helicopter Society (AHS), the American Institute of Aeronautics and Astronautics (AIAA), the American Society of Mechanical Engineers (ASME), the Institute of Physics (IOP), and the International Society for Optics and Photonics (SPIE). He is also a Senior Member of the Institute of Electrical and Electronics Engineers (IEEE). Dr. Wereley is the recipient of several awards including AIAA National Capital Section Engineer of the Year (2009), AIAA Sustained Service Award (2011), the AHS Harry T. Jensen Award in Crashworthiness (2011), and the ASME Adaptive Structures and Materials Systems Best Paper Award in Structural Dynamics and Control (2004, 2012). Dr. Wereley is the recipient of the ASME Adaptive Structures and Material Systems Prize (2012), the SPIE Smart Structures and Materials Lifetime Achievement Award (2013), and the SPIE Smart Structures and Materials Product Implementation Award (2013). He has a B.S. in Mechanical Engineering from McGill University and M.S. and Ph.D. from the Massachusetts Institute of Technology.

Contact: Prof. Norman M. Wereley, Minta Martin Professor and Department Chair, Director, Smart Structures Laboratory, Director, Composites Research Laboratory

Department of Aerospace Engineering, University of Maryland, College Park MD 20742

Tel: 301-405-1927; wereley@umd.edu; http://aero.umd.edu/faculty/wereley

Curt S. Kothera is a Senior Research Engineer at InnoVital Systems, Inc.

Chapter 6
Soft Pneumatic Actuators: Modeling, Control, and Application

Erik Howard Skorina and Cagdas D. Onal

Abstract Soft robots, robots that are constructed out of soft materials or using compliant actuation methods, can operate safely in complex environments without fear of damaging their surroundings or themselves. However, the soft materials and structures can be imprecise and difficult to control. This thesis seeks to remedy this problem for reverse pneumatic artificial muscle (rPAM) actuators, which can apply force by extending when pressurized. First, we discuss the modeling of simple linear rPAMs as well as the motion control of a linear rPAM-driven 1-degree-of-freedom (DoF) revolute joint. Control is done through an iterative sliding mode controller with and without a feedforward term. Next, we adapt this control approach to a soft planar bending segment and use a model reference adaptive controller to compensate for the variability of soft material mechanical properties. From there, we expanded the iterative sliding mode controller and used it to control 2-DoF joint modules using three linear rPAM actuators. We subsequently replaced this universal joint with the human wrist and developed a system to provide haptic feedback to the users, improving their performance computerized line-following task. We then developed a model to predict the behavior of bending actuators, both under load and while pressurized internally, which we used to perform inverse kinematic path following. These techniques represent a meaningful advancement in understanding and improving soft actuators, allowing them to move with speed and precision while resisting external forces.

Keywords Soft robotics · Reverse pneumatic artificial muscle · Haptic feedback · Modeling

E. H. Skorina · C. D. Onal (✉)
WPI Soft Robotics Lab, Robotics Engineering Department, Worcester Polytechnic Institute, Worcester, MA, USA
e-mail: ehskorina@wpi.edu; cdonal@wpi.edu

© Springer Nature Switzerland AG 2022

L. Rasmussen (ed.), *Smart Materials*,
https://doi.org/10.1007/978-3-030-70514-5_6

Advances in soft robotics presented in this chapter help to make progress toward a future vision of soft mobile manipulators that can operate in three dimensions and interact with users, objects, and terrains gracefully and safely

6.1 Introduction and Background

Due to their weight and rigidity, robots operated by traditional motors can be dangerous to humans, limiting their ability to operate efficiently in human-trafficked environments. Soft actuators can absorb energy to enable safe and compliant physical interaction with the environment in a way that is similar to biological muscles, allowing for a bio-inspired approach to robotics and actuation.

This section revolves around a soft linear actuation concept inspired by biological anatomy we call the *reverse pneumatic artificial muscle* (rPAM). This actuator is called the rPAM because it operates on similar principles to the traditional PAM (also known as the McKibben actuator [1]), only with a reversed direction of actuation (similar to the work of [2]). Our approach utilizes these fiber-reinforced elastomer tubes either in a pre-strained state when driving a rigid kinematic skeleton or as part of a soft bending continuum manipulator (a segment that bends gradually and not at along a single axis).

This section will start with a series of literature reviews for the parts of this section covering soft pneumatic actuators, bending actuator control, 2-degree-of-freedom (DoF) soft-actuated modules, wrist device, and soft bending static modeling.

6.1.1 *Soft Pneumatic Actuators*

The rPAM is an example of a fiber-reinforced elastomeric enclosure (FREE). The first example of this is the McKibben muscle which contracts under pressurization due to the geometry of an external braided mesh [1, 3]. McKibben actuators use fluidic power to offer large forces and fast responses in a way that is analogous to biological muscles [4] but may suffer from significant radial expansion as well as modeling and control difficulties. The authors of [5] showed that a large deformation membrane model with two families of inextensible fibers accurately predicts the static response of McKibben actuators. McKibben actuators can be connected together in either series or parallel to achieve more complex movements [6]. Vacuum is usually not used in soft pneumatic systems, as negative pressure will often cause the soft actuator walls to collapse in undesirable ways. However, negative pressure was used in [7] to increase chamber venting speed and thus improve actuator response time.

A different type of FREE was invented in [2] and recently studied by Bishop-Moser, Krishnan, and Kota [8, 9]. This soft actuator consists of a hollow cylinder of silicone rubber with fibers wrapped around the outside. The relative inextensibility of the external fibers with respect to the silicone creates geometric constraints on the deformation of the actuator when pressurized. Depending on the exact geometry of the fiber reinforcements, they can cause the actuator to undergo axial expansion, contraction, bending, or twisting when pressurized. Bishop-Moser, Krishnan, and Kota [8, 9] explore the mobility characteristics of FREEs with double-helix fiber configurations, both symmetrical and asymmetrical. This work has also been extended to FREEs with double helical fibers and an additional thread in order to achieve the mobility needed for snake-like soft robots [10]. These differ from McKibben actuators in the nature of the fiber reinforcement, which is a braided mesh for McKibben actuators but a simpler composition of helical fibers for this newer work. For linear muscle-like soft actuation and the ability to exert larger forces, we consider a single cylindrical pressure chamber reinforced radially by fiber wound in two symmetrical small-angle helices to approximate a series of circles along the length, as an extensile version of McKibben actuators. Thus, hoop stresses in the composite are opposed by the inextensible thread, while the axial stresses lead to deformation. The resulting actuators are inexpensive, easier, and faster to fabricate and can provide large forces due to a larger range of input pressures (than fluidic elastomer actuators used in our earlier work [11–16] and other similar articles in the recent literature [17–20]). Soft linear actuation is not very useful for extension motion outputs, as it can easily buckle under a payload. Our solution to this challenge is to operate the fiber-reinforced soft actuators *in reverse* to achieve rPAMs, by not applying vacuum, but ensuring that the actuator always encounters tensile stresses. We achieve this by pre-straining the actuator and releasing the corresponding stresses through pressurization. Figure 6.5 shows a picture of the proposed soft linear actuators.

A modeling approach to these actuators has previously been developed using conservation of energy and the geometrical constraints created by fiber enclosures [9, 21, 22]. In addition [23], expanded upon this model to consider the distortion effects at the termination nodes and the radial pressure loss due to rubber elasticity. However, these analytical models do not cover all of the forces because the material properties of the elastomeric substrate (typically silicone rubber) contribute to the force balance. The authors of [24] use a nonlinear Mooney-Rivlin term to consider the nonlinear elasticity of the bladder in order to reduce the modeling error. However, the Mooney-Rivlin model has limitations for large deformations of rubbers [25]. Therefore, a more accurate model needs to incorporate the properties of silicone rubber to obtain the stored strain energy and to accurately calculate the forces and torques generated by the actuator.

6.1.2 Bending Actuator Control

One important issue for soft actuator control is the significant variation exhibited in soft materials. The same material can exhibit different static and dynamic properties due to uncertainties in fabrication [26], leading to unpredictability in soft robotic behavior. In addition, the dynamics of soft pneumatic actuators is nonlinear and involves inherent time delays as pressurized air flows from the source to the expansion chamber.

There is also the difficulty of controlling the pressure inside soft actuators. Air flow can be controlled using analog pneumatic cylinders [27], which are effective but bulky and expensive, prohibiting their use in a mobile robot. As an alternative, [28, 29] control solenoid valves using a pulse width modulation (PWM) signal, while [30, 31] use two solenoid valves to allow the system to hold a constant pressure inside the soft actuator. These methods are less capable than the pneumatic cylinder method, but are more conducive to mobile robots. The primary control methods for binary solenoid valves are sliding mode or bang-bang control [13, 28–31].

In this section, we investigate a solution to some of these system-level challenges in soft robotics. When a soft robotic system is scaled up with multiple segments, it is important for the usability of the system that all the segments function similarly. To this end, we formulate a model reference adaptive controller [32] (MRAC) to soft bending actuators. A model-based adaptive controller was previously used to control a humanoid arm driven by McKibben actuators [33]. However, this approach has not previously been taken to control recent fluidic elastomer actuators (FEAs), which are the focus of this section.

6.1.3 2-Degree-of-Freedom Soft-Actuated Modules

Recent robotic research has seen the emergence of soft robotic arms [16, 33–41]. Compared with the rigid manipulators, these soft arms are low cost and more appropriate for human-robot interaction. Soft robotic systems commonly utilize pneumatic actuators [16, 38–41]. However, motion control of pneumatic soft actuators is challenging and remains an open topic of research. First, theoretical modeling of a soft-actuated robot is not trivial, because the continuous shape of the soft actuator influenced by gravity is difficult to predict. Second, the dynamic behavior of these actuators includes a nonlinear time delay as pressurized air is introduced and released through valve commands. The control schemes of [16, 40, 41] are used on 2-degree-of-freedom (DoF) soft manipulators, manipulators that can bend in two directions. In general, these papers directly control the input flow rates or air pressures, the infrastructure for which is often both bulky and expensive.

6.1.4 Wrist Device

Haptic feedback devices have been used to convey subtle informational cues to users. These come in two basic categories: tactile and kinesthetic. Tactile haptic feedback uses purely sensory cues, such as vibrations, to inform the user of events or provide the illusion of forces, as in [42]. This has been used for a range of activities, such as gait training [43] and vision aid [44].

Kinesthetic haptic feedback, which this section focuses on, utilizes real forces in order provide feedback to the user. The simplest example of this is the force feedback-enabled joystick. [45] used a force feedback joystick to teleoperate a robotic arm and showed that the physical feedback provided a significant improvement to the performance of the user. Another example of a force feedback joystick was developed by [46] for teleoperating robots in space. This system was developed to use the rapid and intuitive flow of information from the force feedback joystick to the operator to compensate for the communication lag in long-distance robot teleoperation. These joystick systems can provide strong, high-bandwidth feedback, but require the user to manipulate an external device. Other applications of rigid, kinesthetic feedback include robotic surgery as discussed by [47], where haptic feedback reduced errors by a factor of 3. In addition, work has been done by [48] using kinesthetic feedback for rehabilitation.

Similar electric motor techniques can be directly applied to the user's joints. Margineanu et al. [49] created a 5-DoF haptic arm exoskeleton for use in space telerobotics. This system was effective, but was heavy and bulky. The authors of [50] developed an exoskeletal glove which used rigid links protruding above the hand to apply forces to the user's fingertips. This method is bulky and can easily become a hindrance when not used in a lab setting. The authors of [51] developed a similar glove, but used magnetorheological fluids to convey variable stiffness information to the user. The authors of [52] developed a similar device, but one that used pneumatic pistons to apply the forces. An example of a haptic wrist actuator was presented by [53]. This device used piezoelectric actuators to provide relatively compliant (and magnetic resonance imaging-compatible) forces, but required a complicated mechanism, making it suited for use in controlled environments. A hybrid technique combining properties of tactile and kinesthetic was used by [54]. This work uses a motor to physically stretch the skin of the user to simulate force effects, representing a skin-based haptic feedback that nevertheless involves real forces. These haptic technologies rely on the user manipulating an external device or being constrained by an inflexible joint. However, there is a significant discrepancy in this approach, since human bodies are soft and flexible. Thus, in order to provide a seamless experience for haptic interactions, we posit that it would be appropriate for haptic devices to be soft and flexible to match the mechanical behavior of the human body. Soft feedback can be applied to the user in many ways, including using soft pneumatic actuators. One example of a soft pneumatic actuator is the McKibben muscle (or the pneumatic artificial muscle) as discussed by [55]. This actuator type consists of a rubber tube wrapped in a mesh which causes it to contract when pressurized. McKibben muscles were used by [56] as part of a soft glove for haptic feedback for a piano-playing VR experience. The soft actuators provided safe forces to the user's fingers, while the compliant nature of the glove made it comfortable and adaptable for a wide range of hand sizes. The authors of [57] compared several haptic feedback devices used in active-prosthetic grasping. They found that pneumatic feedback was more reliable and easier to interpret than vibro-tactile haptic or visual feedback. Soft techniques can also be used for purely tactile feedback. One example is the work of [58], where the authors used a series of small electroactive polymer nodules to apply stimulation to the user without additional electromechanical transmissions.

A different type of soft actuator used in soft robotics is the PneuNet. Presented by [59], this actuator type consists of a series of connected chambers within a soft manifold. When pressurized, the chambers expand, causing the PneuNet to deform in a desired direction, usually out-of-plane bending. This bending can follow a similar path to the curling of fingers and has been used in a soft pneumatic glove by [60]. This system was intended for hand rehabilitation and only had a response time of around 2 s, not fast enough for real-time haptic feedback.

An example of a soft wrist device was discussed by [61]. This system uses pneumatic actuators similar to our rPAMs, but with an inextensible constraint layer that causes them to bend when pressurized. This work focused on EMG-enabled rehabilitation and only performed simple motions, with no precise control or angle feedback. In addition, the nature of the bending actuators used to apply forces meant that this system could only apply forces in a single plane, limiting its usability for haptic feedback. Similarly, a soft wrist device was created by [62] using both extensile and contractile McKibben muscles to actuate the user's wrist. Mounted on a glove (further analyzed in [63]), these actuators can apply high forces, although the authors did not investigate the control of the wrist.

6.1.5 *Soft Bending Static Modeling*

Continuum manipulators tend to fall into two categories: pneumatically driven and cable driven. Cable-driven continuum manipulators, such as in [64–66], can combine complaint structures with electric motors. Forward and inverse kinematics can be performed in relation to cable lengths [67], assuming limited external forces. One group who did incorporate external forces into a dynamic was [68], where the authors used numerical solutions to Cosserat rod theory to perform sliding mode control, though they did not perform physical experiments.

There are difficulties with this approach, such as the physical challenges of dealing with cables, and many groups use pneumatically driven continuum actuators, such as the OctoArm [69]. Cosserat rod theory was adapted to this platform as well in [70]. The authors highlighted the advantages in accuracy over using constant-curvature segments, though they did not discuss the speed of calculations. Neural networks were used in [71] to develop an inverse kinematic (IK) approximation to be used as a feedforward term in low-level control. A separate group [72] incorporated fixed universal and prismatic joints in their manipulator, which aided kinematics but reduced the system flexibility.

Another approach that is less common is the discretized constant curvature method, which uses simple constant-curvature beam bending. This approach was described by [73], though the authors did not do any verification or inverse kinematics. The authors of [74] used a discretized model with inverse kinematics and path following, but ignored the effects of external forces.

6.1.6 Contributions

The contributions of this section are as follows, with more specific contributions listed within the sections of this chapter themselves:

- The modeling and control of rPAM actuated rigid kinematic skeletons. We used valve pulse width modulation and iterative sliding mode control to achieve fast, precise signal tracking for both revolute and universal joints with different actuator configurations (Sects. 6.2 and 6.4).
- The application of model reference adaptive control to soft robotics, ensuring that constant-curvature bending actuators with differing material properties could behave similarly (Sect. 6.3).
- The application of soft rPAM actuators and our valve-PWM methodology to wrist haptics, creating the first soft haptic wrist system capable of real-time feedback (Sect. 6.5).
- The creation and verification of a discretized curvature for soft pneumatic continuum bending manipulators. This model can operate in three dimensions, takes into account external tip forces, and incorporates the rPAM chamber deformation (Sect. 6.6).

6.2 Soft Actuator Modeling and Control

While having a soft exterior can cushion a robot in the event of a collision, this can only do so much. To further increase the safety of robots, it is important to have actuation methods that are compliant. In this section, we consider an inexpensive and reliable soft linear actuator, called the reverse pneumatic artificial muscle (rPAM), which consists of a silicone rubber tube that is radially constrained by symmetrical double-helix threading. When the pressure is added to the inside of the tube, the threading prevents it from expanding radially and instead causes it to extend. This is in contrast to the pneumatic artificial muscle, which contracts under pressure [1].

The contributions of this section include:

- A computationally efficient and modular analytical model of rPAMs that incorporates the nonlinear material properties of silicone rubber under large strains
- A finite element analysis model of rPAMs to compare with the analytical model
- Analytical modeling, dynamic characterization, and control of a 1-degree-of-freedom (DoF) revolute joint driven antagonistically by a pair of rPAMs

6.2.1 Reverse Pneumatic Artificial Muscle Fabrication, Analytical Modeling, and Verification

We talk about our model describing the linear extension of the reverse pneumatic artificial muscles (rPAMs) under pressure and external axial loads.

6.2.1.1 Actuator Fabrication Process

The rPAM presented in this section was made by molding silicone rubber (Smooth-On Dragon Skin® 10) in a three-dimensional (3D) printed mold. The step-by-step fabrication process (shown in Fig. 6.1) is described below:

Step 1: Insert a carbon fiber or metallic rod of appropriate diameter into the center of the body mold to create the hollow cylindrical core inside the actuator. Introduce silicone rubber into the body mold with the rod inside, the top connector mold, and the bottom mold.

Step 2: After the silicone rubber has cured, remove the rod inside the body mold, and then remove the silicone rubber from the body mold.

Step 3: Tie two symmetrical helices of thread around the cylindrical silicone rubber body guided by the grooves. Apply a thin layer of uncured silicone rubber on the threads to immobilize the thread around the actuator.

Step 4: Embed the pneumatic fitting connector into the top mold before the silicone rubber had fully cured. Once cured, remove the top and bottom silicone rubber pieces from their respective molds. Bond these to the silicone rubber body using a thin layer of uncured silicone rubber.

6.2.1.2 Analytical Model

Our soft actuator analytical model combines a constraint model and the material model. The constraint model was based on geometrical relationships of the helical threads, and the material model was based on the data from stretch and compression tests described by the Ogden model. Table 6.1 and Fig. 6.2 display the parameters used in the analytical model. For simplicity, we make the following standard assumptions about the system:

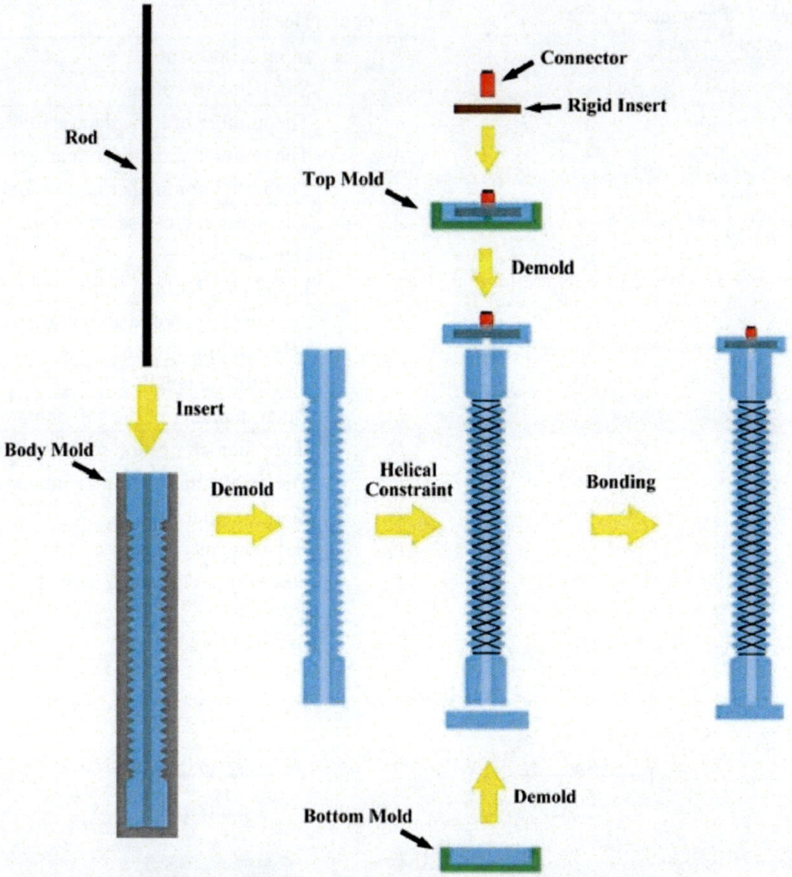

Fig. 6.1 The fabrication process of the proposed rPAM model

Assumption 1: There is no shift of or friction from the thread when actuated. This is ensured by the fact that the threads are slotted into grooves and then glued in place.
Assumption 2: The actuator remains cylindrical.
Assumption 3: The thread is inextensible.
Assumption 4: The silicone rubber material is incompressible.
Assumption 5: The analytical model will be quasi-static, describing the position of the actuator at steady state. All experimental data is taken at steady state as well.

The analytical model of rPAMs is based on force balance in steady state as:

$$F_{ext} = F_{cons} + F_{int}, \tag{6.1}$$

where F_{ext} is the external axial load on the actuator in tension, F_{cons} is the helical constraint force, and F_{int} is the force due to internal material stresses calculated based on the Ogden model.

Table 6.1 Parameters of the soft actuator model

Symbol	Description
P	Input air pressure
V	Soft actuator volume
n	The number of helical turns of the thread
A	The nominal cross-sectional area
A_o	The initial nominal cross-sectional area
A_{in}	The internal cross-sectional area
θ	Helix angle
L_o	Initial length of the soft actuator
L	Length of the soft actuator after deformation
b	Total thread length
D_{out}	Outer diameter of the soft actuator
D_{in}	Inner diameter of the soft actuator
σ	The axial stress within the material
λ	The principal stretch
F_{cons}	Helical constraint force
F_{int}	Force due to internal material deformation
F_{ext}	External force

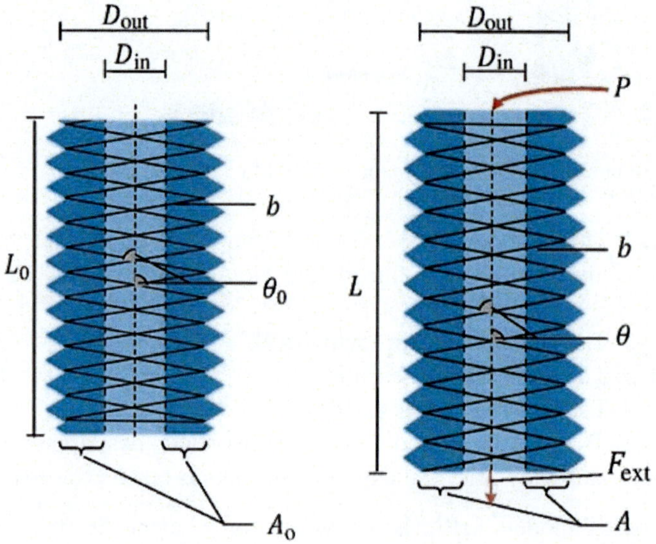

Fig. 6.2 Geometric parameters of the rPAM before (left) and after deformation (right)

To develop the model for the helical constraint force, we consider calculating an effective internal area based on the McKibben actuator analytical model [9, 21, 75], also taking the effect of wall thickness into account. As the pneumatic actuator is a cylinder, its external volume can be calculated as follows:

$$\frac{V_{\text{out}} = \pi D_{\text{out}}^2 L}{4} \tag{6.2}$$

where $\pi D_{\text{out}}^2 / 4$ is the cross-sectional area of the actuator and L is its length, as discussed in Table 6.1. From here, we can use trigonometric relations to calculate V_{out} in terms of b, θ, and n. From Fig. 6.2, we can deduce that $D_{\text{out}} = b \sin(\theta)/\pi n$ and $L = b \cos(\theta)$.

Thus, substituting in, we get:

$$\frac{V_{\text{out}} = b^3 \sin^2\theta \cos\theta}{4\pi n^2} \tag{6.3}$$

Input pressure acts on the internal cross-sectional area. Thus, it is necessary for us to calculate the internal cross-sectional area. First, it can be put as:

$$A_{\text{in}} = A_{\text{out}} - A \tag{6.4}$$

where A_{out} is cross section of the entire actuator and A is the cross-sectional area of the actuator wall. We can substitute in using the work in [75] to get:

$$\begin{aligned} A_{\text{in}} &= \frac{dV_{\text{out}}}{dL} - \frac{A_{\text{o}}}{\lambda} \\ &= \frac{b^2 - 3L^2}{4\pi n^2} - \frac{A_{\text{o}}}{\lambda} \end{aligned} \tag{6.5}$$

where the initial cross-sectional are area $A_{\text{o}} = \pi\left(D_{\text{out}}^2 - D_{\text{in}}^2\right)/4$ and the principal stretch $\lambda = L/L_{\text{o}} = A_{\text{o}}/A$ for incompressible material [76]. Thus, the helical constraint force can be calculated as:

$$F_{\text{cons}} = -PA_{\text{in}} = -P\left(\frac{b^2 - 3L^2}{4\pi n^2}\right) - \frac{A_{\text{o}}}{\lambda} \tag{6.6}$$

The next step is to determine the internal material force F_{int}. To model the internal forces within the actuator, we turned to Ogden hyperelastic solid model [77], which is a powerful tool to analytically describe the deformation of a broad range of elastomeric materials assuming that the nonlinear stress-strain relationship in the material can be described using a strain energy density function [78–80].

Using this model, the axial stress [79] follows the form:

$$\sigma = \sum_{i=1}^{3} \left(\frac{2\mu_i}{\alpha_i} \right) \left(\lambda^{\alpha_i} - \lambda^{-\alpha_{i/2}} \right) \tag{6.7}$$

for three Ogden elements, where μ_i and α_i are material constants obtained by experimental tensile testing data. Consequently, the internal forces due to material deformation can be written as:

$$\begin{aligned} F_{\text{int}} &= \sigma A \\ &= \sum_{i=1}^{3} \left(\frac{2\mu_i A_o}{\alpha_i} \right) \left(\lambda^{\alpha_i - 1} - \lambda^{-\alpha_{i/2} - 1} \right) \end{aligned} \tag{6.8}$$

To calculate the Ogden parameters for our silicone rubber formulation, we performed tension [81] and compression tests [82]. We combined the resulting data and used it to calculate parameters for Eq. (6.7). Figure 6.3 shows the nonlinear fit curve, and Table 6.2 shows the calculated Ogden parameters.

Together, these components create our analytical model, which can calculate the force output of the actuator at a given length. For static behavior, such as that tested, we can also solve the equation for the actuator length that will result in an equilibrium.

6.2.1.3 Numerical Model

Finite element analysis (FEA) is a common tool for modeling soft actuator behavior because of its ability to deal with geometric complexities. To this end, we modeled rPAM behavior using Abaqus CAE to provide a numerical solution to support and

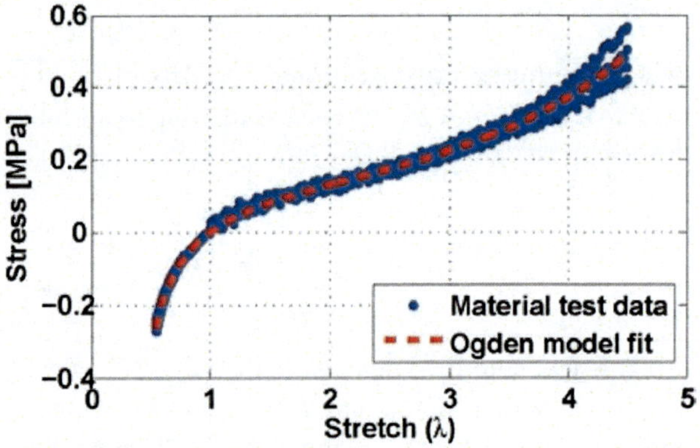

Fig. 6.3 Static stress-strain response of Dragon Skin 10 silicone rubber. Blue dots are experimental data points from multiple compressive and tensile material loading tests, while the red dashed line is the Ogden model curve fit. The mean error for this Ogden fit is 1.1 ± 15 kPa

Table 6.2 Parameter values of the Ogden model

μ_1	μ_2	μ_3	α_1	α_2	α_3
80031.09	1049.90	32.89	1.62	5.18	−13.42

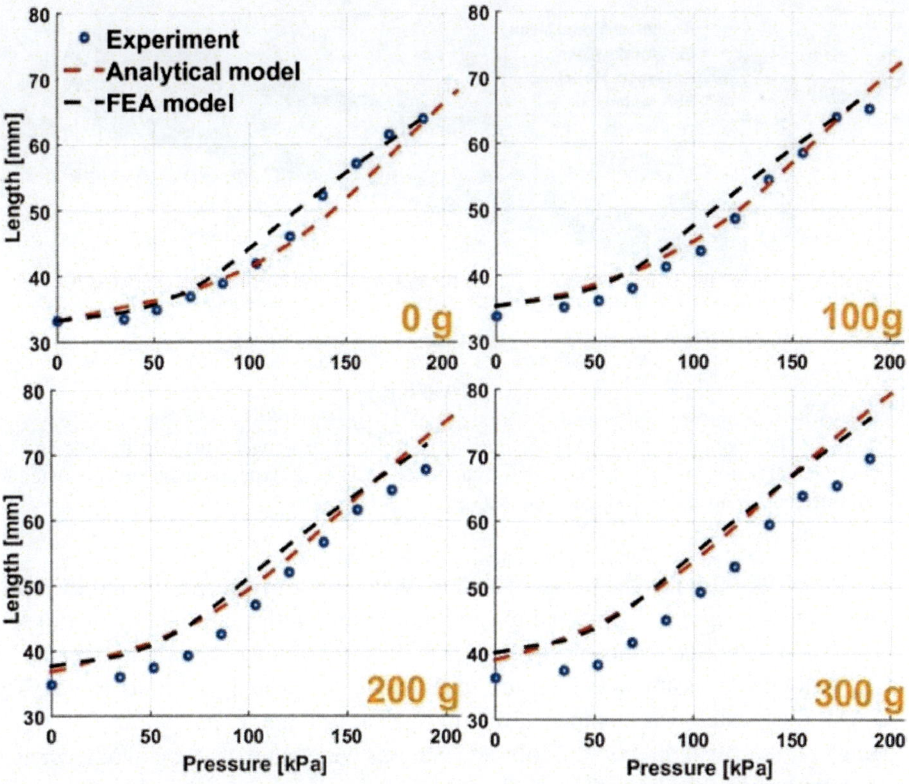

Fig. 6.4 Experimental (blue dots), analytical (red dashed curve), and numerical (black dashed curve) results for four different external loads of 0, 100, 200, and 300 g under input pressures of 0–190 kPa

compare with our analytical one. Our model uses axis of symmetry to approximate the full 3D object. In order to factor in the limitations due to helical threading, we set the outside edge of the actuator to be constrained in the horizontal direction. Thus, it maintains a constant radius of 7 mm while allowing the actuator to extend. The lower end is fixed to represent the actuator mount, while the upper end was capped with more of the silicone rubber and left free. Pressure input is applied directly to the inside surfaces of the actuator as a constant uniform pressure, which simulates steady state. The numerical simulation approximates the effect of the helix shown in Fig. 6.4 with a radial constraint because an accurate representation of the complete helical constraint would require a computationally expensive non-linear geometric relation.

6.2.1.4 Actuator Static Deformation Response Experimental Setup

Experiments were performed to evaluate the accuracy of the developed analytical and numerical models. The test setup consisted of the rPAM, a pressurized air source, solenoid valves, a valve control system, and an object tracking system. An

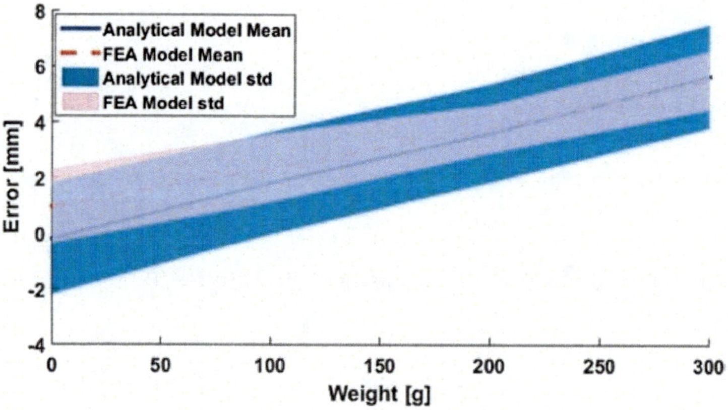

Fig. 6.5 The mean and standard deviation (std) for the error between the two models and the experimental results of static actuator deformation (see Fig. 6.4). Experiments were conducted for four different payloads: 0, 100, 200, and 300 g

air nozzle was used as the pressure source, which was driven and filtered by a precise pressure regulator. For testing, pressure inputs within a range of 0–196 kPa (0–20 psi) with increments of 17 kPa (2.5 psi) were given via the pressure regulator. The regulated pressure was then fed to the actuator through solenoid valves. The rPAM to be tested was clamped at the tip from a rigid holder and allowed to freely hang in the vertical direction. Calibrated weights were hung off the end of the actuator from an attached hook (Fig. 6.5).

A distinctly colored dot was marked at the tip of the soft actuator, tracked by an external camera throughout the experiments. A line of known length was marked next to the actuator as a reference for data analysis in an object tracking software [83]. The deformation of the actuator under different pressures was measured exactly by the tracking system, giving us precise static deformation data. The solenoid valves receive pressure inputs from the pressure regulator. A control system was designed to turn the solenoid valves on and off at specified periods controlled by an Arduino Uno, in serial communication with a workstation running MATLAB. L298N full-bridge drivers were connected to the controller along with a power supply. As all experiments were performed under static conditions after the actuator had reached steady state, the communication delays were not a concern, and the data from the tracking system was calculated offline.

6.2.1.5 Actuator Deformation Results

Table 6.3 shows the experimental conditions and the parameter values of the rPAM. The initial total length of the actuator is 50 mm which consists of primarily the helix-constrained section with a few millimeters of non-helix-constrained section at either end. In order to match the models, we only considered the middle helix constraint part. The experiment involved exerting a range of pressures on the actuator while hanging different weights on it. Figure 6.4 shows a comparison of the

Table 6.3 Experimental parameters of the rPAM model

P	Input pressure	0–196 kPa
n	Number of turns of the thread	8
L_o	Initial length of the helix section	33.23 mm
b	The total length of the thread	250 mm
D_{out}	The nominal outer diameter	11.54 mm
D_{in}	The inner diameter	5.08 mm
F_{ext}	The external weight	0–300 g

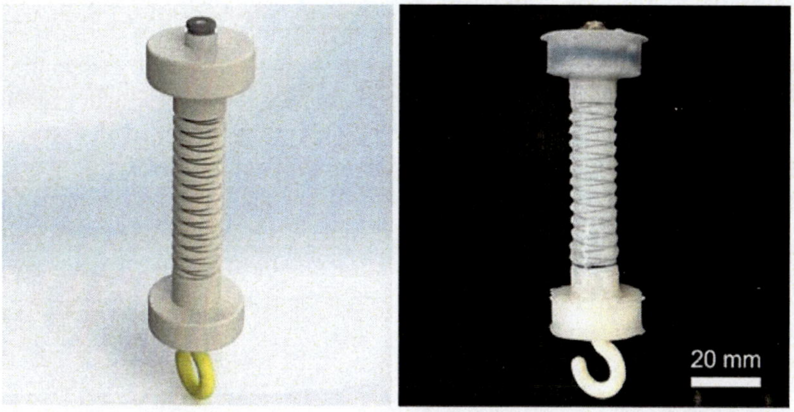

Fig. 6.6 A CAD model (left) and a physical prototype (right) of the proposed soft linear actuator, called reverse pneumatic artificial muscle (rPAM), which offers convenient physical and fluidic connectors to operate rigid kinematic linkages

static deflection results for each load case. From this figure, we can conclude that both models predict similar actuator behavior and are the most accurate without any external forces. We also plotted the mean and standard deviation of the errors for each model over all four trials, shown in Fig. 6.6.

The error between the two models and the experimental results get progressively larger with increasing payload. This is a result of inadequate modeling of the helical constraint force. The FEA model does not incorporate the helical constraint, which plays a larger role the greater the actuator deformation. On the other hand, the analytical model does not consider the effect of additional deformation on the helical constraint force. From Eq. (6.6), when $P = 0$, then $F_{cons} = 0$. However, an external load will stretch the actuator and deform the helix, which will change the constraint force. The FEA model has a smaller error standard deviation, meaning that the shape of its behavior more closely matches that of the actual system, but with a consistent offset. The analytical model displays a smaller mean error at lower payloads, meaning it is on average more accurate, while the FEA model is more precise. Both models perform similarly over the range of weights and pressures tested. Thus, we do not need to worry about adapting the FEA model for online use. Instead, we can use the analytic model, which is much better suited to our applications of computationally efficient control.

6.2.2 rPAM-Driven Revolute Joint

To verify our modeling efforts and develop motion control algorithms for soft actuation in a bench-top setting, we designed a simple 1-DoF revolute joint setup, as described previously in [84]. This setup allows us to gauge the rPAM's usability for driving arbitrary kinematic chains. The increased stroke length of rPAMs over regular PAMs as well as their increased efficiency allows the joint to be driven to much larger angles by smaller pressures. As the proposed rPAMs extend when pressurized, we use two actuators operating antagonistically to drive our joint, as shown in Fig. 6.7. While this joint could also be driven by a single rPAM and a spring, this would limit the workspace and make the behavior of the joint non-symmetrical. The joint is designed so that the actuators are both stretched nominally when the joint is at the neutral point. Without this pre-strain keeping the rPAMs under tension, the joint would have to compress one of the rPAMs every time it rotated, forcing that muscle to buckle outward and reducing the mobility and usability of the joint. The original length of the threaded component of the actuators is approximately 50 mm, which is stretched to 75 mm at the neutral angle of the joint, for a pre-strain of 50%. In order to perform motion control using a reliable feedback signal, an optical encoder (CUI Inc. AMT 203) was mounted on the joint axis.

6.2.2.1 Analytical Model

In order to calculate the geometric relationship between the joint and the two rPAMs, we assumed that each of the actuators would remain straight during actuation and apply force directly from the lower mount to the upper mount. This assumption holds up to mid-range pressures, but the actuators were observed to bow under higher pressure differentials and larger joint angles. In the experimental setup, each

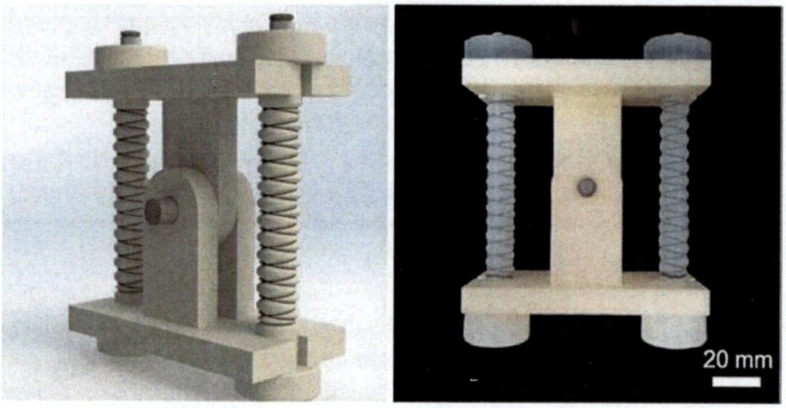

Fig. 6.7 A CAD model (left) and a physical prototype (right) of the 1-DoF revolute joint operated by the proposed rPAMs in antagonism

actuator had an independent pressure regulator. We adjusted each regulator from 41 to 96 kPa in increments of 7 kPa (1 psi) and recorded the resulting angle for each combination. This range was chosen to maintain the assumption that the rPAMs did not bend or buckle.

Our analytical model utilized the 1-DoF revolute kinematic model shown in Fig. 6.8. When the joint reaches steady state, the two external moments exerted by actuators A and B should be equal:

$$\begin{aligned} F_{\text{exta}} L_{\text{ma}} &= F_{\text{extb}} L_{\text{mb}}, \\ \left(F_{\text{inta}} + F_{\text{consa}}\right) L_{\text{ma}} &= \left(F_{\text{intb}} + F_{\text{consb}}\right) L_{\text{mb}}, \end{aligned} \tag{6.9}$$

where L_{ma} and L_{mb} are the moment arms of the actuators A and B, L_a and L_b are the actuator lengths, and L_1 and L_2 are internal dimensions of the setup. L_a and L_b can be determined as a function of the joint angle α and can be used to calculate L_{ma} and L_{mb} as follows:

$$\begin{aligned} L_{\text{ma}} &= \sqrt{L_1^2 + L_2^2 - L_a^2/4}, \\ L_{\text{mb}} &= \sqrt{L_1^2 + L_2^2 - L_b^2/4}. \end{aligned} \lim_{x\to\infty} \tag{6.10}$$

From Eq. (6.9), the joint rotation angle can be determined with respect to input pressures and the proposed analytical model for rPAMs.

Figure 6.9 displays both the analytical and experimental responses of the joint to combinations of pressure inputs for both rPAMs between 41 and 96 kPa (6–14 psi). This figure indicates that the steady-state joint angle is a function of the pressure difference between the two rPAMs. It can be seen that the simulated and experimental results match closely, with the experimental results showing slight variations

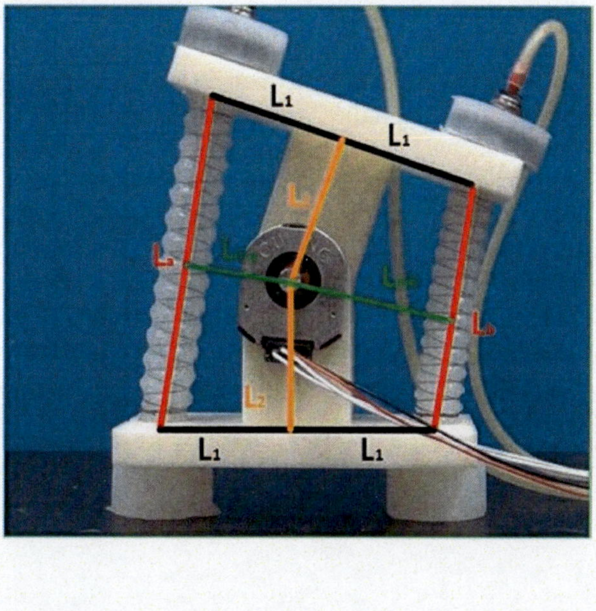

Fig. 6.8 The geometric model of the 1-DoF joint setup. The black and brown lines represent the rigid joint links, while the red lines represent the rPAMs. The green lines are the calculated moment arms for each soft actuator. It should be noted that the actuator is vertically symmetric

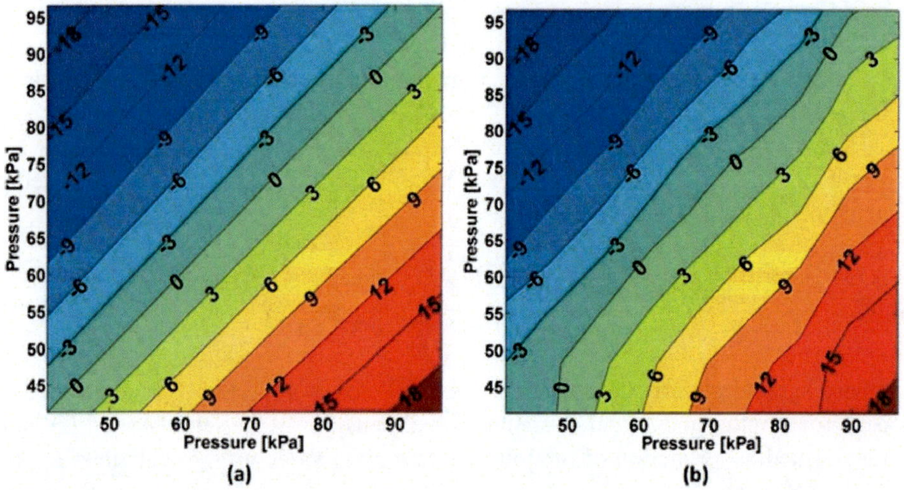

Fig. 6.9 Comparison of model predictions and experimental results for a range of input pressures ranging from 41 to 96 kPa for both rPAMs. Contour plots of the resulting steady-state joint angle in degrees are displayed for simulation (**a**) and experimental (**b**) results. Joint angle values are stepped at 3°, annotated on the curves, and indicated as color coding from blue to red. The mean error between the model and the experimental results is 0.27 ± 1.1°

inherent to the physical world. This could be a result of slight bending in the actuators when pressurized, from shifts in actuator mounts, or from inconsistencies between the two actuators.

6.2.3 Control Strategy

Each of the two rPAMs driving the joint was connected to a 138 kPa (20 psi) pressurized air line and controlled by a separate solenoid valve. To approximate pressure inputs between 0 and 138 kPa, we operated the valves with a 30 Hz PWM signal. This resulted in a 1-DoF system being controlled by two independent inputs. Arguably, this additional control authority may be utilized to provide a common pressure value within both actuators, to control the stiffness of the overall system. In this work, to simplify joint control, we created a purely antagonistic scheme, where the binary valves are always driven in opposition to each other (i.e., *when one valve is being pressurized, the opposite valve is being vented*). Thus, to control the joint angle, we specify a single number between 0 and 100 as the PWM duty cycle of one of the valves corresponding to the positive rotation of the joint and used as the control input in the rest of this section.

6.2.3.1 Sliding Mode Controller Design Based on Lumped System Dynamics

Using the electrical circuit equivalence of pneumatic systems from [13], we approximated the dynamic behavior of the 1-DoF joint as the following lumped second-order dynamic equation:

$$\alpha = C_1 e^{(-t/\tau 1)} + C_2 e^{(-t/\tau 2)} + C_0 \quad (6.11)$$

where C_1 and C_2 are constant coefficients, τ_1 and τ_2 are the time constants, and C_0 is the steady-state static angle, with a one-to-one relation to the PWM duty cycle, indicating the angle the joint will converge after the dynamic terms dissipate.

The dynamic response of the actuator (Eq. 6.11) can then be represented by the following second-order system equation:

$$\hat{a}\ddot{\alpha} + \hat{b}\dot{\alpha} + \alpha = \hat{c}u(t) \quad (6.12)$$

where α is the rotation angle with respect to time, $u(t)$ is the duty cycle of the PWM signal to the input valves, and $\hat{a} \in (a_o - \Delta a, a_o + \Delta a)$, $\hat{b} \in (b_o - \Delta b, b_o + \Delta b)$ and $\hat{c} \in (c_o - \Delta c, c_o + \Delta c)$ are the bounded uncertainty parameters. a_0, b_0, and c_o and Δa, Δb, and Δc are the mean and standard deviation values for a, b, and c.

Figure 6.10 shows the variation of parameters $\hat{a}$, $\hat{b}$, and $\hat{c}$ for duty cycles ranging from 35% to 65% under 138 kPa (20 psi) pressure input.

Equation (6.12) can be rewritten in state-space form as:

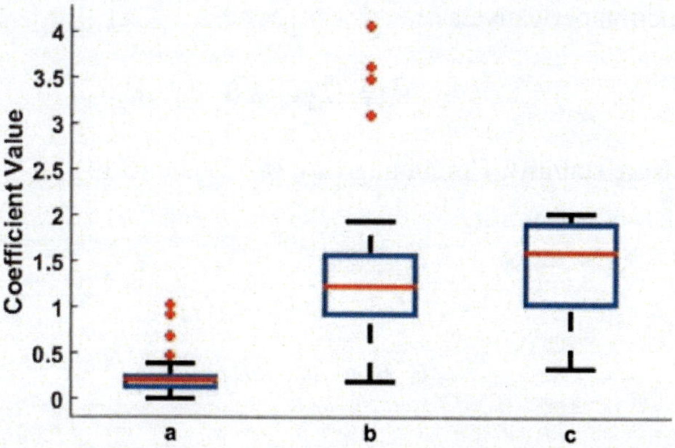

Fig. 6.10 A box plot representation of the dynamic coefficients for joint response to duty cycles of 35–65% under 138 kPa (20 psi). The red line is the median, the blue box is the 25th and 75th percentile, the black lines are the bounds of the non-outlying data, and the red circles are the outliers. The 25th and 75th percentiles of $\hat{a}$, $\hat{b}$, and $\hat{c}$ are (0.4123, 0.3240) rad/s^2, (0.9859, 1.7316) rad/s, and (1.2577, 1.637) rad/PWM. The median of the $\hat{a}$, $\hat{b}$, and $\hat{c}$ is 0.2006 rad/s^2, 1.2319 rad/s, and 1.6022 rad/PWM

$$\dot{\boldsymbol{x}} = \boldsymbol{A}\boldsymbol{x} + \boldsymbol{B}u(t) \tag{6.13}$$

where $\mathrm{x} = [\alpha, \dot{\alpha}]^{\mathrm{T}}, \mathrm{A} = \begin{bmatrix} 0 & 1 \\ -\frac{1}{\hat{a}} & -\frac{\hat{b}}{\hat{a}} \end{bmatrix}$, and $\mathrm{B} = [0, \frac{\hat{c}}{\hat{a}}]$.

Thus, we can design an iterative sliding mode controller for motion control based on our dynamic model and bounded uncertainty parameters. For a given reference $\boldsymbol{x}^{\mathrm{ref}}$, the position error is given as:

$$e_x = \boldsymbol{G}_x\left(\boldsymbol{x}^{\mathrm{ref}} - \boldsymbol{x}\right), \tag{6.14}$$

where $\boldsymbol{G}_x = [C, 0]$.

We define a non-negative Lyapunov function candidate and its derivative as follows:

$$V_x = \frac{e_x^2}{2} \geq 0 \tag{6.15}$$

$$\dot{V}_x = e_x \dot{e}_x, \tag{6.16}$$

and select a desired dynamic error manifold as:

$$\dot{e}_x + D_x e_x = 0 \tag{6.17}$$

where D_x is a positive constant. Plugging (6.17) into (6.16), we obtain a non-positive Lyapunov function derivative as:

$$\dot{V}_x = -D_x e_x^2 \leq 0 \tag{6.18}$$

which will ensure stability. Combining Eqs. (6.13) and (6.14),

$$\begin{aligned} \dot{e}_x &= \boldsymbol{G}_x\left(\dot{\boldsymbol{x}}^{\mathrm{ref}} - \dot{\boldsymbol{x}}\right) \\ &= \boldsymbol{G}_x\left(\dot{\boldsymbol{x}}^{\mathrm{ref}} - \boldsymbol{A}\boldsymbol{x} - \boldsymbol{B}u(t)\right) \\ &= \boldsymbol{G}_x\boldsymbol{B}\left(u_{\mathrm{eq}}(t) - u(t)\right), \end{aligned} \tag{6.19}$$

$$\boldsymbol{G}_x\boldsymbol{B}\left(u_{\mathrm{eq}}(t) - u(t)\right) + D_x e_x = 0, \tag{6.20}$$

where $u_{\mathrm{eq}}(t)$ is the continuous equivalent control input, which is difficult to calculate [85].

Rearranging Eq. (6.19) reveals:

$$u_{\mathrm{eq}}(t) = u(t) + \left(\boldsymbol{G}_x\boldsymbol{B}\right)^{-1} \dot{e}_x. \tag{6.21}$$

Approximating $u_{eq}(t)$ with $u_{eq}(t - \Delta t)$, where Δt is the time step, yields the iterative sliding mode control law:

$$u(t) = u(t - \Delta t) + (\boldsymbol{G}_x \boldsymbol{B})^{-1} (\dot{e}_x + D_x \boldsymbol{e}_x) \tag{6.22}$$

For simplicity, we define $K = (\boldsymbol{G}_x\boldsymbol{B})^{-1}$ as a scalar positive tuning factor as:

$$u(t) = u(t - \Delta t) + K(\dot{e}_x + D_x e_x). \tag{6.23}$$

6.2.3.2 Feedforward Controller Design

An improvement to this controller can be developed by taking advantage of the fact that a given control input to our system converges to a single angle at steady state. We characterized this static response by recording the resulting steady-state joint angles from a range of duty cycle control inputs. Next, we mapped the solenoid valve duty cycle with respect to the resulting pressure, which we measured using a pressure sensor.

Figure 6.11a shows the mean and standard deviation of the resulting pressure within the rPAM for a range of PWM duty cycle inputs. After characterizing this relation, we plugged it into (6.9) to determine the relation between the steady-state angle of the joint and solenoid valve duty cycle. Figure 6.11b displays this mapping function from the model prediction (red solid line) and our experimental data (blue dotted line). The result shows that the amplitudes of the two mapping functions are close, but the experimental data displays an offset, indicating a slight bias to one direction, which is likely caused by manual fabrication differences between the two rPAMs. We fit a third-order polynomial to the experimental data and incorporated it into the sliding mode controller as a feedforward term (SM + FF) as follows:

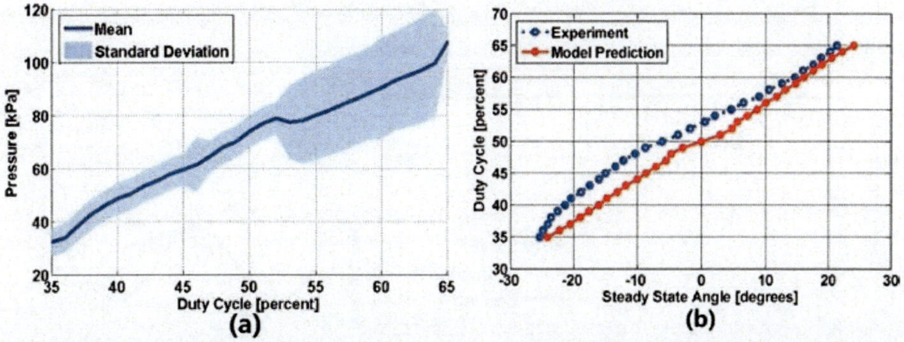

Fig. 6.11 (**a**) The relation between the duty cycle of the solenoid valve and the resulting pressure, with an input pressure of 137 kPa (20 psi) and a pulse width modulation frequency of 30 Hz. (**b**) Steady-state angle response at various duty cycles and the corresponding mapping function from the experiment (blue dotted line) and the model prediction (red solid line)

$$u_h(t) = \mathrm{Map}(\alpha^{\mathrm{ref}}) + u(t) \tag{6.24}$$

In this equation, $u_h(t)$ is the SM + FF output signal, $u(t)$ is the controller signal from (6.23), and Map(α^{ref}) is the mapped PWM duty cycle, which results in the reference angle in steady state. Substituting in, we obtain:

$$u_h(t) = u(t - \Delta t) + K(\dot{e}_x + D_x e_x) + \mathrm{Map}(\alpha^{\mathrm{ref}}) \tag{6.25}$$

It should be noted that $u(t)$ is still updated according to the iterative feedback law in Eq. (6.23), which prevents the output of the mapping function from compounding and skewing the control input.

6.2.3.3 Joint Control Results

To verify and compare the performance of our proposed feedback motion control approaches, we performed a number of experimental studies using our 1-DoF system. First, we tested the ability of our controllers to follow step reference signals, representative results of which are shown in Figs. 6.12a and 6.13. As shown in Fig. 6.12a, the sliding mode and SM + FF controllers perform equally well, with a reaching time of approximately 0.8 s. The two data sets start at two slightly different angles (SM + FF at −1.5° and sliding mode at 1.5°, approximately). This is a result of friction in the joint and shifting of the actuators, which cause the neutral position of the joint to change slightly between uses. We have observed that this shift in the starting position does not affect the long-term joint trajectory. The control coefficients used for all experimental results were tuned through preliminary experiments to be $K = 30$ and $D_x = 0.0033$.

We next tested our controllers with more complicated signals: offset sinusoidal waves. Using a range of reference frequencies between 0.2 and 1 Hz, we observed the performance of our controllers in a dynamic context. Experimental results of

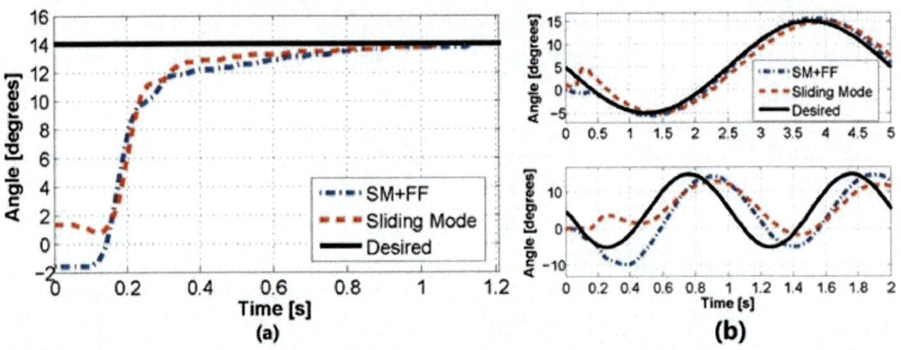

Fig. 6.12 (**a**) A comparison of the two control algorithms on a step response. Small variations in the starting point of each trial are a result of frictional effects. (**b**) A comparison of the two control algorithms tracking 0.2 Hz (top) and 1 Hz (bottom) sinusoidal waves

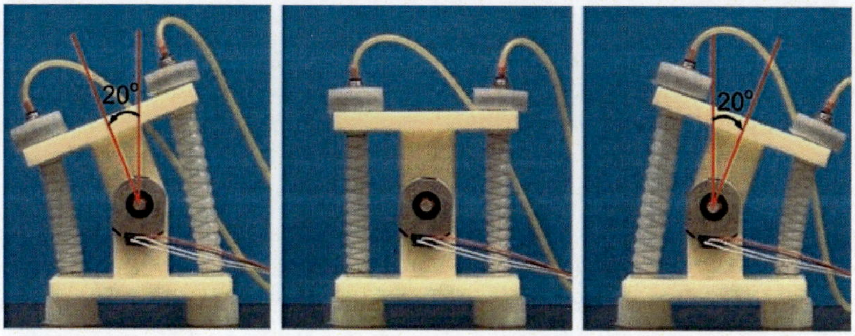

Fig. 6.13 The feedforward assisted sliding mode controller offers precise position control under step reference signals in both directions

Fig. 6.14 A series of snapshots taken from a 0.5 Hz sinusoidal wave tracking experiment using the SM + FF controller

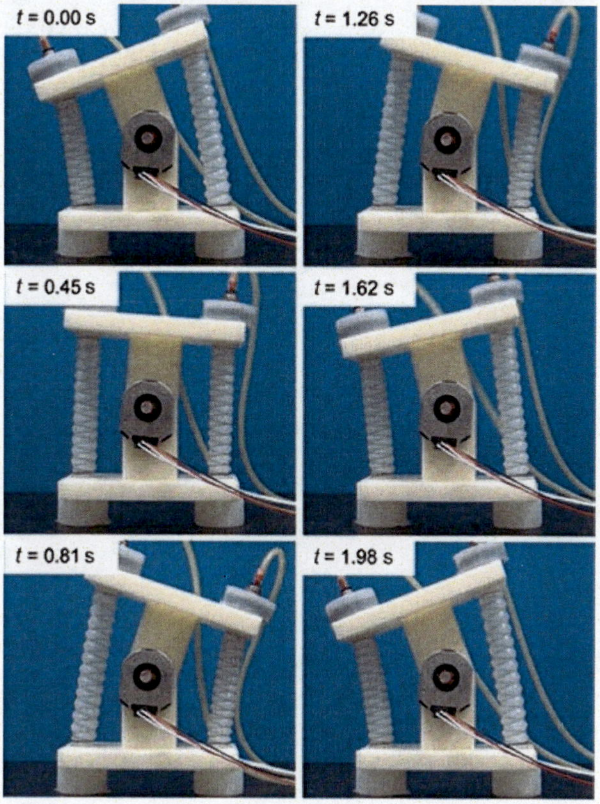

these trials can be seen in Figs. 6.12b and 6.14. These figures indicate that the sliding mode controller follows a sinusoidal trajectory with a constant time delay with respect to the reference signal. At lower frequencies, it follows the input signal with a slight lag, while at higher frequencies, it lags behind with a reduced amplitude due to the slow dynamic characteristics of the soft actuators.

The SM + FF controller follows a more unique path. At lower frequencies, it catches up with the input signal before each peak and overshoots slightly. Though its max amplitude is higher than the reference signal, the SM + FF response has minimal phase shift at lower frequencies, providing a level of compensation for slow actuator dynamics. At higher frequencies, the SM + FF controller can no longer catch up with the reference signal, but still appears to follow better than the sliding mode controller.

To confirm this observation on a wider scale, we investigated the controller responses over a series of dynamic tracking trials, each with a different frequency. Figure 6.15 displays a plot of the closed-loop frequency response (amplitude and phase delay) for the two controllers over the given range of sinusoidal frequencies, from 0.1 to 1 Hz. It is clear from this data set that the SM + FF controller maintains larger amplitudes and smaller phase lags for all frequencies tested.

We also performed the 0.2 Hz sinusoidal tracking experiment with a 200 g weight pulling perpendicular to the joint in the positive angular direction (for a total torque of approximately 0.1 N-m) to quantify the performance of the SM + FF controller when the mapping function is disturbed through external loading. The results of this

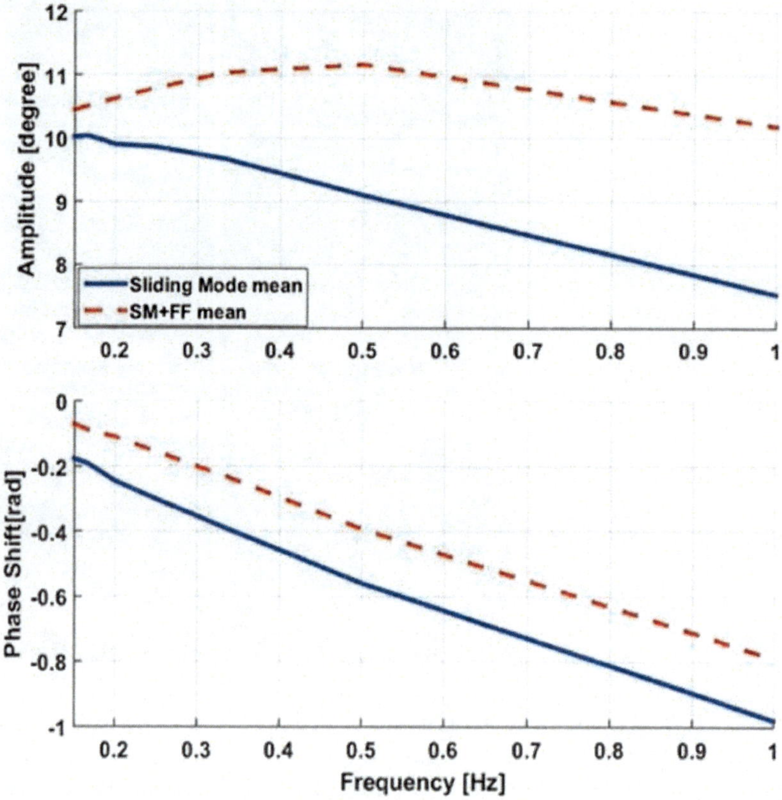

Fig. 6.15 Amplitude (top) and the corresponding phase shift values (bottom) of tracking sinusoidal waveforms of 10-degree amplitude over a range of frequencies. The standard deviations for these experiments were small enough (around 0.04°) not to warrant plotting, highlighting the repeatability of our joint behavior

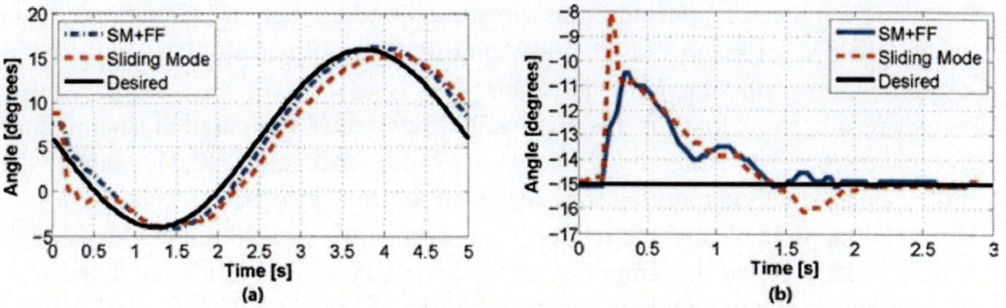

Fig. 6.16 (**a**) A comparison of the two control algorithms following a sine wave with an external torque of 0.1 N-m acting in the positive direction. (**b**) Experimental comparison of the two control algorithms when a 200 g weight is added as a sudden disturbance after the step function has been reached

experiment can be seen in Fig. 6.16a, which indicates that, even with the unloaded mapping function as the feedforward term, the SM + FF controller still outperforms the sliding mode feedback controller. In other words, the feedforward component of the SM + FF controller allows for an improved system response, even when the feedforward mapping function is incorrect for the tested loading case. Thus, the SM + FF controller is robust under changes to the system.

Finally, we explored the ability of the controllers to respond to a sudden disturbance, performed similar to the previous experiment. For these experiments, we utilized a step reference signal and added the 200 g payload (applied perpendicular to the joint in the positive angular direction, as before) shortly after the set point was reached. The results, shown in Fig. 6.16b, indicate that the two controllers respond similarly, with a small improvement from the SM + FF controller.

6.2.4 Conclusion

This section represents a seamless progression from fabrication and physics-based analytical modeling to dynamic motion control of a soft pneumatic actuator we call the *reverse pneumatic artificial muscle (rPAM)*. We describe the design and fabrication of the rPAM, which is made of molded silicone rubber helically wrapped in inextensible thread. We describe numerical and analytical models of static actuator behavior and study their accuracy over a range of pressures and with external payloads up to 300 g. The models function effectively for all pressures without any external payload, but exhibit increasing error as payload increases, resulting from imprecise modeling of the constraint thread under load.

Additionally, we used a pair of rPAMs in antagonism to drive a 1-DoF revolute joint. The analytical actuator model is further extended to this joint, and its accuracy is verified through experimental studies. We also derived an iterative sliding mode feedback motion controller for this system. This controller was then combined with a static mapping function to provide feedforward assistance to the sliding mode controller (SM + FF). The SM + FF controller was notably more effective in following a dynamic sinusoidal trajectory even under external loading and disturbance. The SM + FF and sliding mode controllers performed similarly in following a step function without external payload.

Our results show that for this application, numerical computation such as finite element method was generally not necessary for describing the rPAM behavior. The finite element method required computational time and effort and resulted in a similar prediction to the analytical model. This is probably a result of complexity of the helical constraint and the simplicity of the geometry which further emphasizes the value of our analytical model.

Our analytical model represents a simple and computationally efficient method of predicting soft actuator response, ideal for implementation on a microcontroller. The proposed model is modular and can easily be divided into discrete components. The material and constraint forces can be individually modified for greater fidelity or to utilize alternative material models without affecting the rest of the overall model. Additional force terms may be added as necessary (e.g., to incorporate empirical correction forces [84] or additional constraint relations).

Performance of the proposed joint was studied experimentally and analytically. It can be seen from Fig. 6.9 that multiple pressure input combinations will result in identical joint rotation angles. The implementation of the SM + FF controller is simple and requires minor computational complexity, since the mapping function is based on the characterization of the static response of the rPAM-driven joint under varying PWM duty cycles. For this small cost, it gains significant accuracy in dynamic signal tracking. The only shortcoming of this approach is the requirement of the initial calibration step for each new rPAM and that potential shifts in the muscle attachment point may slightly change the static relation. The former is a relatively simple process, which can be completely automated. The latter may reduce the effectiveness of the SM + FF controller. However, it can be as seen from external loading experiments that the SM + FF maintains an improvement over the original sliding mode controller under experimental variations.

Our current work includes refining our actuator model and the development of more complex kinematic chains driven by multiple rPAM actuators. We are working on expanding our models predicting the behavior of more general arrangements of rPAMs, allowing them to be used to drive a robotic arm [86]. We are also investigating improvements in control algorithms to drive these kinematic chains to desired states. Future work includes further refining the proposed SM + FF controller by combining the calibrated system dynamics with the static mapping function for a more reliable dynamic feedforward term. We also seek to improve actuator dynamics in order to achieve a higher performance to better perform biological tasks.

6.3 Soft Actuator Bending Control

While using the linear rPAM actuators to control a rigid kinematic skeleton allows for compliant actuation, we would like to investigate the usability of a more fundamentally soft system. To this end, we removed the rigid kinematic skeleton and mounted the two rPAMs adjacent to each other as part of a single soft unit. When a single rPAM is pressurized, it still extends, but the asymmetrical nature of the force causes the entire segment to bend away from the extending rPAM chamber. This can be increased by adding an inextensible constraint layer. We demonstrated previously that the iterative sliding mode controller worked for this type of bending actuator, but we also observed that the physical properties and subsequent dynamic response of soft actuators made of the same material varied. To counteract this, we developed a model reference adaptive controller to force different actuators to behave according to the same dynamic model.

The contributions of the work are as follows:

- The development of an efficient model reference adaptive controller for soft bending actuators, demonstrating its usability in modifying actuator behavior to follow a desired dynamic model
- The formulation of an inverse dynamic feedforward controller for soft bending actuators to track desired continuous signals

6.3.1 Actuator Design and Experimental Setup

In our previous work [31], we designed a bidirectional soft bending actuation module with integrated proprioceptive curvature sensing. This module, as well as a representation of the curvature sensor, can be seen in Fig. 6.17. Our custom magnetic curvature sensor utilizes a magnet and a Hall effect sensor each mounted on a flexible circuit board [87]. The Hall effect sensor can measure changes in the magnetic field as the position of the magnet changes due to the deformation of the substrate, enabling accurate measurement of the curvature of the system. Our soft bending actuator is comprised of two soft linear muscles [88] and an inextensible constraint layer in between. The individual linear actuators are made of tubes of silicone rubber with engineered cross sections wrapped in inextensible thread, which causes

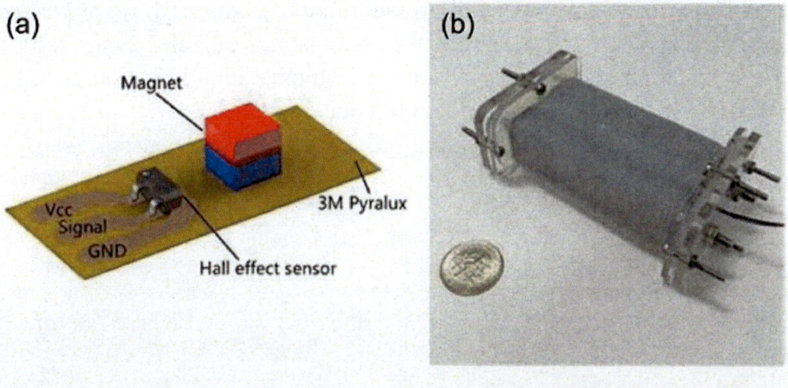

Fig. 6.17 Our flexible magnetic curvature sensor (**a**) embedded in our 1-degree-of-freedom bidirectional bending actuator segment (**b**)

them to extend with minimal radial deformation when pressurized. The constraint layer, with a custom integrated curvature sensor, inhibits this linear extension, resulting in the bending of the entire soft module away from the pressurized actuator. Caps are attached to both ends of the actuator to seal the chambers and allow for modular connections with other segments. The caps are made of two acrylic boards sandwiching the rim of the silicone tube to prevent leaking.

The actuator is driven by two 3–2 (3-port, 2-state) binary solenoid valves, each connecting one pressure chamber to a common 8 psi (55 kPa) pressure source. This pressure value was chosen because it is the highest that the actuators can withstand continuously. The valves can either inflate or deflate a given actuator chamber. We control the pressure in each chamber using a 60 Hz PWM of the valves. We set the valves to operate in complete antagonism, so when one chamber is inflating, the other is always deflating. Thus, we can reduce the number of required inputs to one, corresponding to the single (active) degree of freedom (DOF) of the bending actuator. We have previously observed that the valves begin to saturate at duty cycles below 20% or above 80%. Thus, we constrained the duty cycles sent to the system to stay between these two values.

Valve PWM was used to ensure that pneumatic system had a response time that was fast enough to match. We experienced valve delays of 0.1 s when we employed an electronic pressure regulator, which was unsuitable for the proposed controller.

6.3.2 System Identification

We can treat the dynamic response of the actuator as a generic second-order system:

$$\dddot{x} + a_1 x + a_2 \dot{x} = bu, \tag{6.26}$$

where x is the bending angle of the actuator, u is the system control input, and a_1, a_2, and b are constant dynamic parameters.

The first step in building a reference model for the soft actuator is understanding its nominal performance under full pressure input. To this end, we characterize the values for these constant parameters. We note that this model does not represent the nonlinear system at every pressure input as we have found previously in [29], but only provides a starting point for the reference model.

The general solution of the second-order system under constant (step) input is given as:

$$x(t) = C_1 e^{(-t/\tau 1)} + C_2 e^{(-t/\tau 2)} + C_0 \tag{6.27}$$

with the boundary conditions $x(t = 0) = 0$ and $\dot{x}(0) = 0$ and where t is time and C_0, C_1, C_2, τ_1, and τ_2 are constants.

The relationship between the coefficients in Eqs. (6.26) and (6.27) can be described as follows:

$$C_0 + C_1 + C_2 = 0$$

$$C_1 / \tau_1 + C_2 / \tau_2 = 0$$

$$a_1 = \frac{1}{\tau_1 \tau_2} \tag{6.28}$$

$$a_2 = \frac{1}{\tau_1} + \frac{1}{\tau_2}$$

$$b = \frac{1}{\tau_1 \tau_2 u} C_0.$$

As this is a constant-input solution, u in Eq. (6.28) is constant. We collected the trajectory data from a single actuator using vision tracking software with an input pressure of 8 psi, the maximum pressure applied to the actuator. We fit (6.27) via least squares to the resulting dynamic trajectory, yielding time constants $\tau_1 = 0.1107$ and $\tau_2 = 0.0021$ as well as coefficients $C_0 = 0.8366$, $C_1 = -0.8531$, and $C_2 = 0.0165$.

To calculate u, we assumed that the steady-state output of the system is a linear function of valve duty cycles between 20% and 80% and that an 80% duty cycle, being fully saturated, is equivalent to 8 psi, the pressure used for the experiment. As 20% would saturate in the opposite direction, and thus be equivalent to 8 psi in the opposite direction, we shifted u for purposes of the dynamic model. Our shifted u varied between −30 and 30 (i.e., $u \in [-30, 30]$). Thus, for system identification in 3.3, we used $u = 30$. Solving these equations, we get $a_1 = 4221.8$, $a_2 = 476.4912$, and $b = 117.74$.

6.3.3 Model Reference Adaptive Controller

For embedded operation, the MIT rule provides a computationally efficient approach to formulate an MRAC for use with our soft actuators, allowing us to standardize their behavior. This rule introduces an adaptive gain, which changes the system input to match the system behavior to a desired model behavior. The equation translating the nominal input to the system input is as follows:

$$u = u_c \theta + 50 \tag{6.29}$$

where u_c is the nominal input (bounded between −30 and 30), u is the duty cycle input to the physical system (bounded between 20 and 80), and θ is the adaptive gain. θ is updated at every experimental time step using the following equation:

$$\theta(n) = \theta(n-1) + \gamma\left(x(n) - x_m(n)\right) x_m(n) \Delta t \tag{6.30}$$

where x is the position (bending angle) of the actuator, x_m is the position of the actuator model, and Δt is the time step. x_m was calculated using a standard constant-acceleration model over each time step as follows:

$$\ddot{x}_m(n) = -a_{1m}x_m(n) - a_{2m}\dot{x}_m(n) + u_c(n)b_m \tag{6.31}$$

$$x_m(n+1) = x_m(n) + \dot{x}_m(n)\Delta t + \frac{1}{2}\ddot{x}\dot{x}_m(n)\Delta t^2 \tag{6.32}$$

$$\dot{x}_m(n+1) = \dot{x}_m(n) + \ddot{x}_m(n)\Delta t \tag{6.33}$$

where a_{1m}, a_{2m}, and b_m are the system model parameters used in the experiment.

Thus, the controller constantly keeps track of the model trajectory and uses it as a reference for the adaptive controller. As we constrained u to remain between 20 and 80, we also saturate u_c to remain within −30 and 30. This prevents the model from being driven in ways the physical system is incapable of, which would hinder consistency between the reference model and the experimental data.

6.3.4 Inverse Dynamic Control

We investigated the use of a feedforward inverse dynamic controller on our system. Using the dynamics of the system model, this controller calculates the control input needed for the system to reach the desired position within a single time step. It does so by first calculating the acceleration required to reach the desired position in a single time step, as shown in the following equation:

$$\ddot{x}_R = \frac{x_d - 2x_m(n) + x_m(n-1)}{\Delta t^2} \tag{6.34}$$

where $\ddot{x}_R$ is the required acceleration and x_d is the desired actuator position.

This equation is used to calculate the required input to the system by solving Eq. (6.26) for u_c, resulting in the following:

$$u_c = \frac{(\dot{x}_R + a_{1m}x_m + a_{2m}\dot{x}_m)}{b_m} \tag{6.35}$$

Under ideal circumstances, this controller, which is entirely open-loop, would allow the system to quickly reach a desired angle. However, since it uses an imperfect model of the system to calculate the required inputs, the feedforward inverse dynamic inputs result in errors in a physical system. We can address this by using the adaptive controller to ensure that the behaviors of the dynamic model and the plant match each other.

6.3.5 *Experimental Results*

For our experiments, we ran our control loop at a constant 1 kHz (thus, $\Delta t = 0.001$). This was set to allow the constant acceleration discretization (Eqs. 6.31, 6.32, and 6.33) to function, as a significantly longer time step would cause the constant-acceleration model to diverge. Because of the nature of the communication link between our computer and the microcontroller, we were unable to relay system information at the control frequency. Thus, data was collected separately at around 40 Hz.

We arbitrarily modified the identified model of the system for use in our experiments. For our reference model, we set $a_{1m} = 4000$ and $a_{2m} = 460$. These changes created additional differences between the nominal and desired behavior of the actuator, allowing us to highlight the strength of our approach. The value of b_m varied between the experiments.

In addition, we implemented an initial delay of 50 ms for the model. This was to match a similar delay observed in the behavior of the actuator. The model acceleration in (6.31) would not be updated until after the first 50 ms of operation, after which it would be updated as normal. This significantly improved the similarity between the plant and the model, as during that period of inaction by the plant θ would increase dramatically, causing significant overshoot at a later point.

6.3.5.1 Model Reference Adaptive Control System Results

Our first experiments involved testing the functionality of the MRAC alone. We applied it under various cases using a sinusoidal input function with amplitude 20 (in modified duty cycle units) as the input u_c. We collected data in each experiment for 30 s and calculated the root mean square (RMS) error between the model and the physical system. We performed this experiment three times for each of three actuators, three signal frequencies (0.5, 1, and 2 Hz), and four γ values (0, 0.005, 0.01, and 0.01). The experiments where $\gamma = 0$ represented a control group, where the adaptive controller does not modify the plant behavior. For this experiment, we used $b_m = 80$, instead of the value 117.74 characterized above in Sect. 6.3.2. The reduced b_m value helped the continuous dynamic trajectories of the model be achievable by all of the actuators tested, regardless of any inconsistencies in their fabrication. We had previously observed that higher values of b_m may result in a divergence of θ for actuators that physically cannot provide the desired bending moment under maximum pressure.

Figure 6.18 shows three examples of actuator trajectories with and without adaptation, one for each of the frequencies tested for $\gamma = 0.005$. The MRAC has the best results at 0.5 Hz, where it can almost perfectly match the model performance apart from oscillations resulting from valve PWM and sensor noise. It still must adapt within a single cycle, and the adaptive trajectory falls behind and then catches up to the model on the falling component of each period. The advantages are less apparent for 1 Hz, where the adaptive trajectory cannot compensate fast enough to match both the peaks and valleys of the trajectory, as the nominal motion of the actuator is

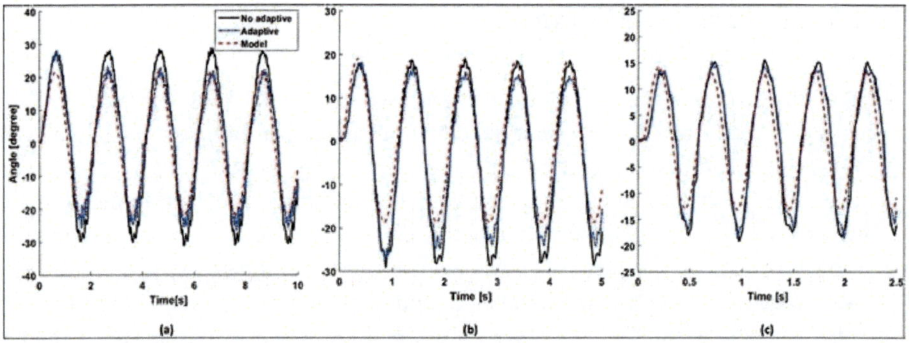

Fig. 6.18 Three sample trajectories of the system with and without the adaptive control compared to the trajectory of the reference model. (**a**) Actuator 2, 0.5 Hz, $\gamma = 0.005$. (**b**) Actuator 3, 1 Hz, $\gamma = 0.005$. (**c**) Actuator 2, 2 Hz, $\gamma = 0.005$

skewed in the negative direction. This is likely the result of differences in the material properties or dimensions between each chamber of the actuator, which were fabricated separately. The same is true for the 2 Hz signal, where there are only slight benefits to the adaptive control.

We calculated the mean and standard deviation of the RMS errors between the model and actual trajectories for all experiments, as illustrated in Fig. 6.19. From this figure, we can see that the addition of the proposed adaptive controller resulted in an improvement for at least one value of γ for all experiments except Actuator 3 (Fig. 6.19c) at 2 Hz. Overall, $\gamma = 0.005$ provided the most consistent improvement over the actuators and frequencies, though in two cases higher γ values provided additional improvement (Actuator 1 at 2 Hz and Actuator 3 at 0.5 Hz).

We isolated an illustrative example of the behavior of the adaptive gain θ during actuator operation. We used Actuator 3 at 0.5 Hz using $\gamma = 0.02$, and the results of this can be seen in Fig. 6.20. The adaptive gain varies wildly within each sinusoidal cycle, reaching 1.2, while the actuator is in the positive half of its cycle and dropping to 0.3 during the negative half. On the rising part of the signal, a nonlinearity in the actuator causes it to lag behind the model, driving the adaptive gain up to compensate. This causes it to slightly overshoot, and the gain drops rapidly to allow the actuator to catch up to the model as it drops.

6.3.5.2 Inverse Dynamic Controller Results

We tested our inverse dynamic feedforward controller against step functions of various bending angles [30°, 20°, 10°, −10°, −20°, and −30°] using each of the three actuators. To show the advantages of the adaptive controller, we performed this experiment with and without adaptive control enabled ($\gamma = 0.02$). The higher γ was chosen because, for step references, rapid convergence is desired. We chose $b_m = 100$ as it better matched the identified system model and the application meant we would not risk a divergence of θ. In addition, we observed that using a Δt of 0.001 in Eq.

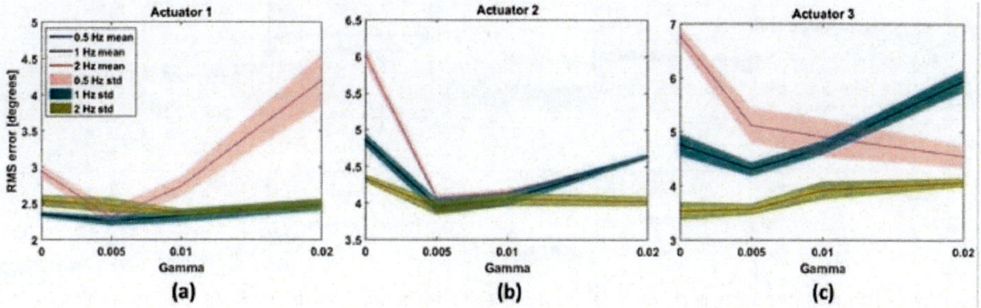

Fig. 6.19 The mean and standard deviation of the RMS error between the plant and the reference model. (**a–c**) Represent different actuators, highlighting the performance of the MRAC controller on physical systems with slightly different properties

Fig. 6.20 An example of the behavior of the adaptive gain Θ in (6.3) during the operation of the Actuator 3. For this plot, the frequency was 0.5 Hz and $\gamma = 0.02$

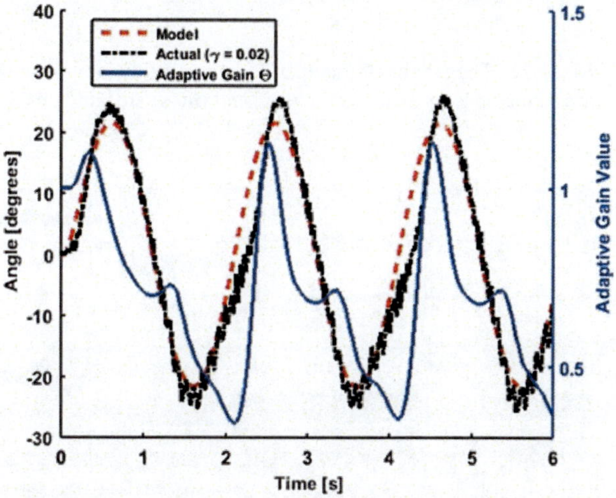

(6.34) would result in too much input saturation. To prevent this, we used $\Delta t = 0.01$ and replaced $x_m(n-1)$ with $x_m(n-10)$, making for a less exacting trajectory, inherently reducing the effect of any noise in sensor measurements. The results of this experiment without adaptation can be seen in Fig. 6.21, while the results with adaptation can be seen in Fig. 6.22.

The inverse dynamic controller can quickly reach a desired angle, even when operating without any feedback. However, it does result in some steady-state errors resulting from the differences between the reference model and the physical system. The addition of the MRAC allowed the system to maintain its fast approach while eliminating the steady-state error. We can also observe that the model and the plant behave similarly during the approach to the desired angle, though the plant consistently overshoots before reacquiring the target angle.

One thing we can observe from the non-adaptive trajectories is that the plant has a tendency to drift slowly even after the model input has stabilized. This represents time-variant dynamic behavior which has not been captured by our model. Fortunately, the adaptive controller is able to compensate for it. In addition, we can

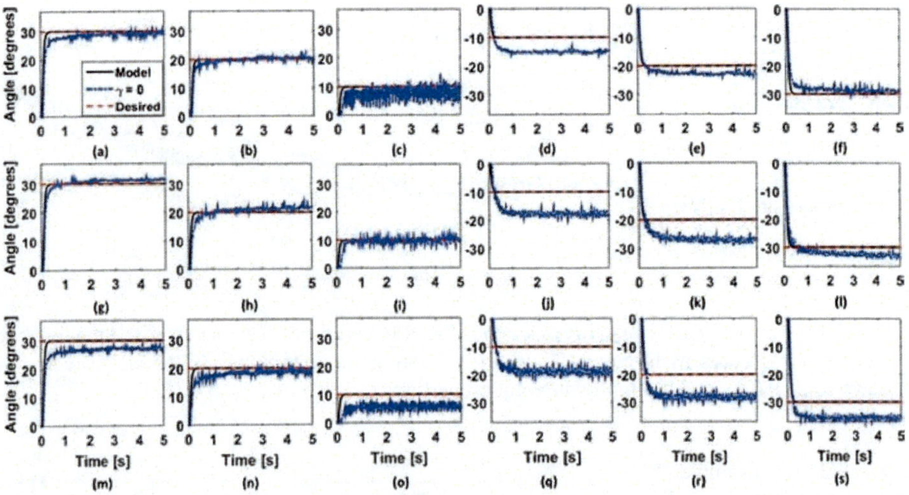

Fig. 6.21 The results of the open-loop inverse dynamic controller following step functions. (**a–f**) are Actuator 1, (**g–i**) are Actuator 2, and (**m–s**) are Actuator 3

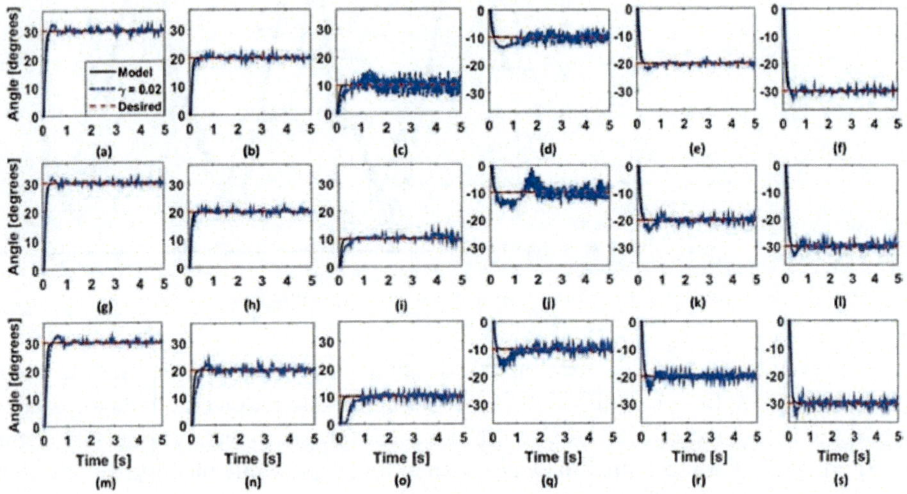

Fig. 6.22 The results of the inverse dynamic controller augmented by the MRAC ($\gamma = 0.02$) while following step functions. (**a–f**) are Actuator 1, (**g–i**) are Actuator 2, and (**m–s**) are Actuator 3

observe that the steady-state error for the system varies widely with respect to the desired angle. The model actually gets more accurate at higher angles, and the errors are significantly different between the positive and negative sides. This is because the characterization experiment used a constant input in the positive direction that resulted in a large angle, so the model is more accurate in these conditions.

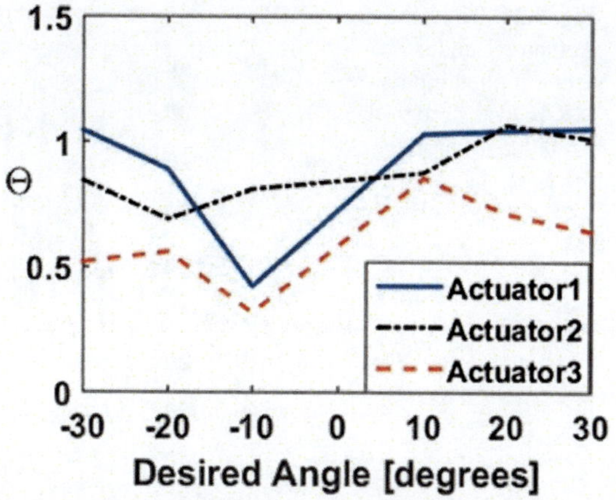

Fig. 6.23 The steady-state results of θ for the adaptive control augmented inverse dynamic control

We also recorded the steady-state value of the adaptive gain θ for each experiment in Fig. 6.22. This was done by taking the mean of the last 20 data points for each experiment after convergence is observed, and its results can be seen in Fig. 6.23. As every experiment converged to the angle specified by the model, these differences in θ provide a visual representation of the nonlinearities in the actuator dynamics, as well as the variation between the behavior of different soft bending actuators.

6.3.5.3 Unstructured Signal Tracking

Our final experiment was to investigate the performance of the adaptive controller in tracking unstructured continuous references. We connected a user input knob to the embedded controller and mapped its positions to system inputs between $-30°$ and $30°$. A user manipulated the knob, and the system responded to the input, while the MRAC worked to try and keep the system in sync with the model. Results of these trials can be seen in Fig. 6.24 for $\gamma = 0.005$ and Fig. 6.25 for $\gamma = 0.02$.

Results show that the model and the system behave similarly, even under the unstructured inputs. We can see the effect of the lower γ, as in Fig. 6.24 the system does not respond fast enough to reach some of the peaks the model does. In addition, we can observe that the adaptive controller has difficulty matching the system to the model when the model is around 0. This is because of the x_m term in the adaptive update equation in Eq. (6.30) indicates that when the model output is near zero, θ will be updated slowly. Our control method requires the valves to be constantly operating, with an equal 50% duty cycle for each valve supposedly corresponding to an angle of 0. Unfortunately, differences between our actuators can cause this state to have a slight bias (for this actuator, in the negative direction). This is something that the MIT rule adaptive MRAC is unable to compensate.

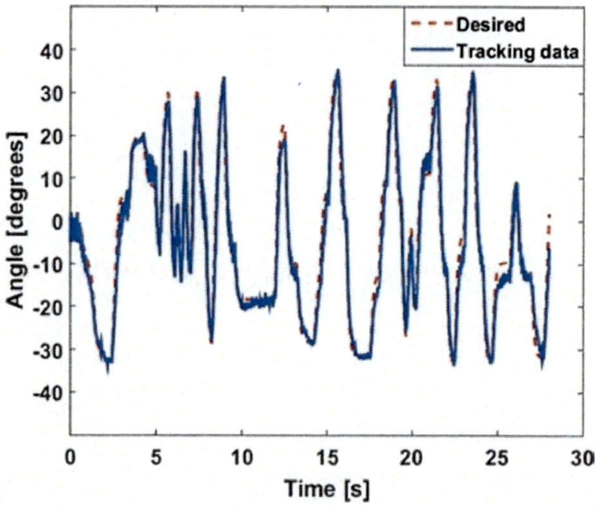

Fig. 6.24 The behavior of Actuator 1 under user-supplied inputs. MRAC was used with $\gamma = 0.005$

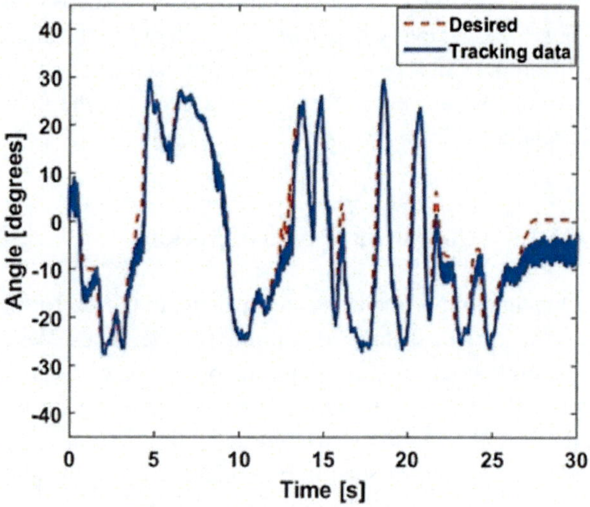

Fig. 6.25 The behavior of Actuator 1 under user-supplied inputs. MRAC was used with $\gamma = 0.02$

6.4 Degree-of-Freedom Soft-Actuated Skeleton Module

Up to this point, we have dealt with soft systems that can only operate in planar situations. The world is unfortunately not planar, and the revolute joint and the bending actuator are both bulky enough that mounting them in series, perpendicular to each other, is an impractical method for creating soft actuation in three dimensions. In addition, the pneumatic infrastructure required to support two rPAMs per DoF can easily become burdensome. Thus, we created rigid 2-DoF universal joint modules actuated by three rPAM actuators. We adapted the iterative sliding mode controller to one of these modules and then used it to control two of them mounted in series.

The contributions of this section include:

- The improvement of our rPAM actuators to increase their reliability
- Development of a soft-actuated 2-DoF universal joint module and associated controller to increase control pneumatic efficiency
- The cascading of these modules to create a 4-DoF soft-actuated manipulator with associated motion controllers

6.4.1 Design and Fabrication

We developed a new fabrication method for the rPAM with increased reliability and resistance to bursting under unexpectedly high pressure. In this section, we discuss this new fabrication method, as well as their use in a soft-actuated universal joint module.

6.4.1.1 Fabrication of the Soft Actuator

The reverse pneumatic artificial muscles (rPAMs) used in this section are fabricated by molding silicone rubber (Smooth-On Dragon Skin 10) in a 3D printed mold. The step-by-step fabrication process (shown in Fig. 6.1) is described below:

Step 1: Insert a carbon fiber or metallic rod of appropriate diameter into the center of the body mold to create the hollow cylindrical core inside the actuator. Introduce silicone rubber into the body mold.

Step 2: After silicone rubber has cured, remove the rod inside the body mold, and then remove the silicone from the body mold. Tie two symmetrical helices of thread around the cylindrical silicone rubber body guided by the grooves. Apply an outer layer of uncured silicone rubber around the threads to permanently bond the thread to the actuator.

Step 3: Sandwich each of the ends of the actuator between two layers of acrylic with an integrated vent screw to serve as pneumatic fitting and tighten to form the end caps.

This fabrication process is a simplification of our previous approach, requiring fewer fabrication steps and less molding. In addition, the acrylic and vent screw provide a more secure and repeatable interface for the input air line, which was previously the most common point of failure. These actuators can reliably withstand actuation pressures of 30 psi. In addition, the screws on either end of the actuator allow it to be easily mounted to external rigid structures.

6.4.1.2 Module Design

To demonstrate the usability of this type of actuation, we designed a universal joint to be driven by three rPAMs. This design methodology is scalable, to be used for controlling robotic limbs in three-dimensional space. In addition, the coupled use of three actuators for 2 degrees of freedom (compared to two antagonistic actuators for each DoF) reduces the electronic and pneumatic infrastructure. The final prototype can be seen in Fig. 6.7. The joint was 3D printed out of PLA with steel axles. The two axles of the universal joint are offset slightly and mounted with bearings to reduce friction. In addition, we used miniature potentiometers on each of the axles to provide reliable feedback for closed-loop control.

The module is designed to be longer than the nominal length of the actuators, forcing them to be pre-strained during assembly. This prevents the actuators from getting in the way of joint movement when pressure is released. Without a pre-strain that keeps the soft actuators under tension, the joint will collide with an actuator during operation, forcing it to compress and buckle outward, thus reducing the mobility of the joint. The amount of pre-strain is not trivial to determine a priori. A trade-off seems to exist between the motion range and reliability of the modules, as a function of pre-strain. A large pre-strain value causes the joint to be unstable under increasing equal pressure inputs. Instead of staying perfectly vertical, the module would toggle between extreme angles. A moderate pre-strain eliminates this problem.

6.4.2 Soft-Actuated 2-DoF Module

In this section, we discuss the development of a kinematic model for the soft-actuated universal joint and its usage in designing a controller.

6.4.2.1 Kinematic Model and Control

For any robotic manipulator, it is useful to have a kinematic model. In the proposed module, rPAMs are mounted at radially symmetric increments of 120° from each end, starting at [40 mm 0 0]. A graphical representation of the module kinematics is shown in Fig. 6.26. The general parameters we used can be seen in Table 6.4. This model was used to develop the joint control algorithm described in detail below.

Our previous work [29] resulted in an iterative sliding mode control law shown in Eq. (6.36), which we used to control a 1-DoF joint.

$$u(t)=u(t-\Delta t)+K(\dot{e}_x+De_x), \tag{6.36}$$

where u is the control input; e_x and $\dot{e}_x$ are the error and its time derivative, respectively; and K and D are the control coefficients.

Expanding this control law to a 2-DoF system is not straightforward, as multiple actuators must interact to move the joint to a desired state. To address this, we used the assumption that changing the control input to a given actuator (representing the PWM valve duty cycle) will cause that actuator to extend or contract accordingly.

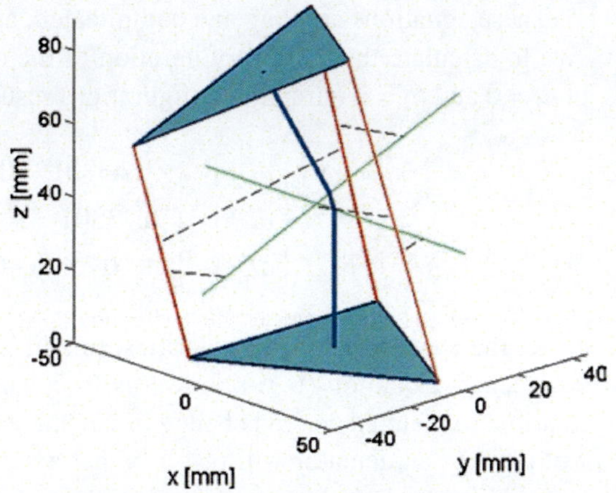

Fig. 6.26 Kinematic model of the 2-DoF module. Blue lines represent the links, red lines represent the actuators, green lines represent the axes of rotation, and the dashed black lines represent the moment arms for the torque generated by each actuator

Table 6.4 Kinematic parameters of the 2-DoF module model, describing the translations and rotations to calculate the end-effector position

Change	Quantity
Z-translation	37.5 mm
X-rotation	θ_1
Z-translation	5 mm
Y-rotation	θ_2
Z-translation	37.5 mm

Using the kinematic model, we can determine the level of extension or contraction required from each actuator in order to move the module toward the desired angle combination. Using the forward kinematics of the 2-DoF module, the following equation can be written for the length of a given rPAM:

$$L_i = f_i\left(\theta_1, \theta_2\right), \tag{6.37}$$

where L_i is the length of actuator i.

From here, we can calculate a Jacobian, taking partial derivatives with respect to each joint angle.

$$\begin{bmatrix} dL_a \\ dL_b \\ dL_c \end{bmatrix} = \begin{bmatrix} \frac{dL_a}{d\theta_1} & \frac{dL_a}{d\theta_2} \\ \frac{dL_b}{d\theta_1} & \frac{dL_b}{d\theta_2} \\ \frac{dL_c}{d\theta_1} & \frac{dL_c}{d\theta_2} \end{bmatrix} \begin{bmatrix} d\theta_1 \\ d\theta_2 \end{bmatrix} \tag{6.38}$$

These equations are long and complicated, and thus it is computationally expensive to calculate them at every iteration of the control loop. Linearizing Eq. (6.38) at $\theta_1 = 0$ and $\theta_2 = 0$ (straight configuration) results in the following:

$$\begin{bmatrix} dL_a \\ dL_b \\ dL_c \end{bmatrix} = \begin{bmatrix} -0.43301 & 0.25 \\ 0.43301 & 0.25 \\ 0 & -0.50 \end{bmatrix} \begin{bmatrix} d\theta_1 \\ d\theta_2 \end{bmatrix} \tag{6.39}$$

As the system has no singularities, this linearization would work for the entire workspace. Equation (6.39) allows us to calculate the change in actuator lengths required to match a desired change in the angles. Assuming that an increase in control input to an actuator will result in that actuator increasing in length (or at least exerting force along its axis), we can combine Eqs. (6.39) and (6.36), yielding the control law:

$$\begin{bmatrix} u_a(t) \\ u_b(t) \\ u_c(t) \end{bmatrix} = \begin{bmatrix} u_a(t-\Delta t) \\ u_b(t-\Delta t) \\ u_c(t-\Delta t) \end{bmatrix} + \begin{bmatrix} -0.43301 & 0.25 \\ 0.43301 & 0.25 \\ 0 & -0.50 \end{bmatrix} + \begin{bmatrix} K\left(\dot{e}_{\theta 1} + De_{\theta 1}\right) \\ K\left(\dot{e}_{\theta 2} + De_{\theta 2}\right) \end{bmatrix} \tag{6.40}$$

6.4.2.2 Experimental Results

To demonstrate the effectiveness of this controller, we applied it to various trajectories. Figure 6.27 is a representative example of the controller trying to reach an angle combination. In this trial, θ_2 reached the desired angle of 20° after less than 0.25 s, while θ_1 reached the same target after 0.5 s. Not only was the target reached in less than a second, but the controller was capable of holding that position within ±1°.

In addition, we have the controller follow a series of circular paths. These paths were generated by giving offset sinusoidal waves to each angle. An example of the controller following such a trajectory with period 8.5 s (0.1176 Hz) and 15-degree amplitude can be seen in Fig. 6.28. The controller follows the given trajectories with relative accuracy, while occasional oscillations are observed. This is likely a result of latent dynamic effects from the PWM-operated valves. In addition, θ_1 overshoots on both sides just after the wave switches direction.

We performed this test over a range of reference periods to determine the amplitude and phase lag of the closed-loop system. We performed this experiment three times and calculated the mean and standard deviation. A plot of these results can be seen in Fig. 6.29. The phase lag for θ_2 was consistently lower, though both increased linearly with respect to frequency. θ_1 had a slightly larger amplitude than the desired 15° at lower frequencies, while θ_2 had the desired amplitude. Both amplitudes slightly decreased as the frequency increased, as expected, but remained close to the desired value for the tested frequency range.

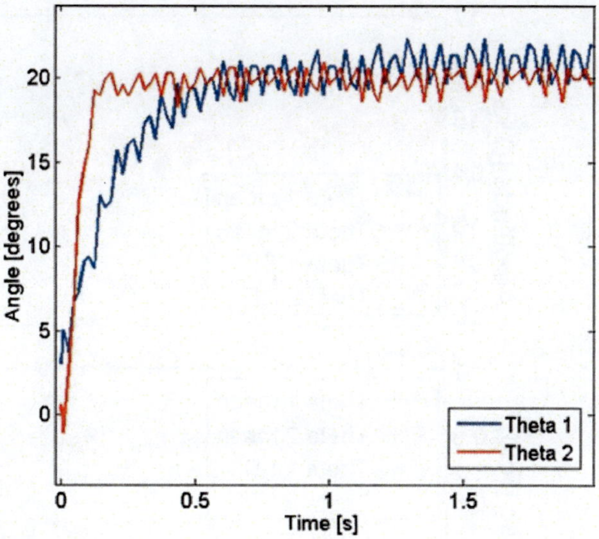

Fig. 6.27 The step response behavior of the controller with the 2-DoF joint. The desired angles are 20°

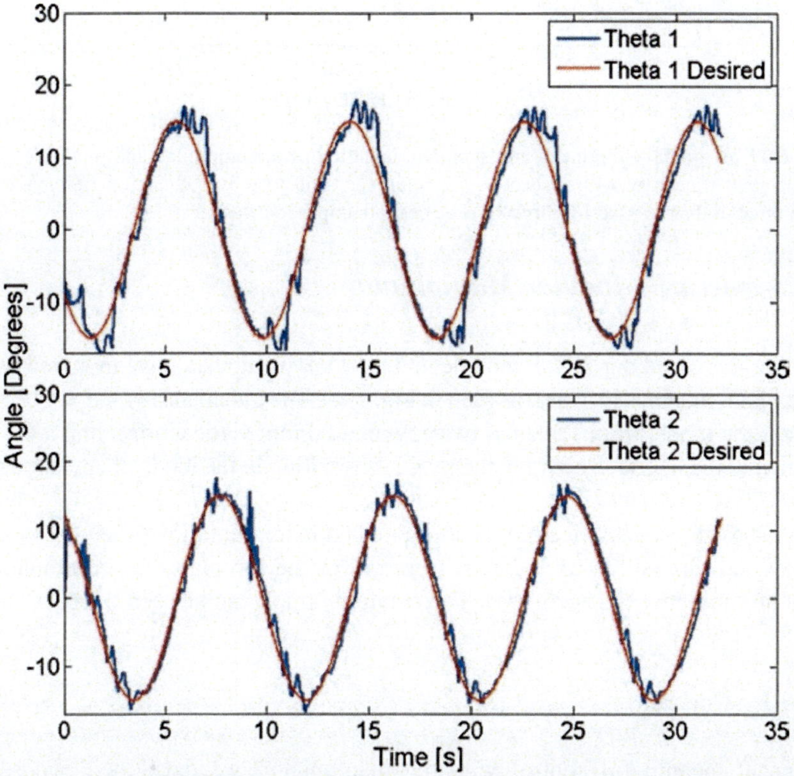

Fig. 6.28 The trajectory following results for the 2-DoF joint following a circular path. The period is 8.5 s and the amplitude is 15°

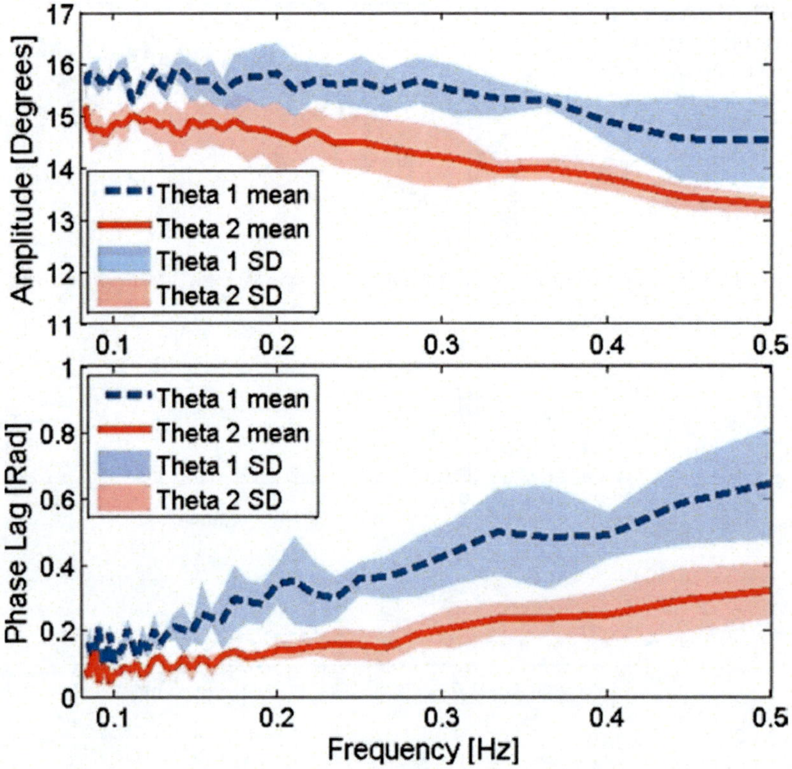

Fig. 6.29 A graph of the mean and standard deviation of the amplitude and phase delay of the 2-DoF joint trajectory with respect to the reference trajectory over changing frequencies. Data from three trials was used for this plot. The desired amplitude was 15°

6.4.3 4-DoF Modular Manipulator

In order to expand our 2-DoF module into a 4-DoF manipulator, we mounted two universal joint modules in series, as seen in Fig. 6.30. The kinematics of the 4-DoF system utilizes the model from Table 6.4 twice, with a 30 mm vertical offset and a 60-degree rotation at the interface between the modules resulting in the model that can be seen in Fig. 6.31.

To show the workspace of this soft-actuated manipulator, we used the kinematic model and plotted the tip location (centroid of the top plate) for combinations of joint angles between $[-\pi/6, \pi/6]$. The resulting graph can be seen in Fig. 6.32.

6.4.3.1 Control

To enact meaningful control with the manipulator, we developed two separate motion controllers. The first controller was a direct inverse kinematics (IK)-based controller. Because we were using a 4-DoF manipulator in 6-DoF space, we ignored the z-position and z-rotation of the end effector. For simplicity, we assumed that the

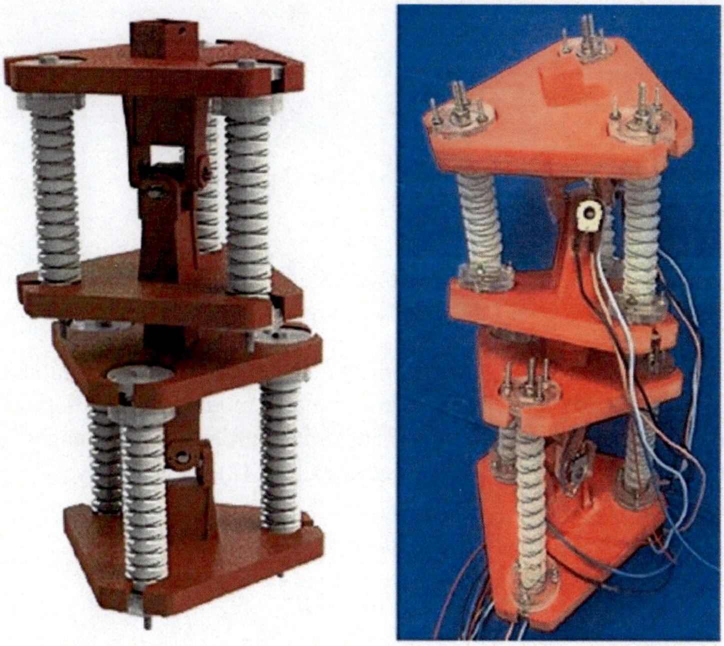

Fig. 6.30 A CAD model (left) and a physical prototype (right) of the 4-DoF soft-actuated modular manipulator

Fig. 6.31 Kinematics of the 4-DoF manipulator. Blue lines represent the joint, itself. Red lines represent the actuators, green lines represent the axes of rotation, and the dashed black lines represent the moment arms for the torque of each actuator around each axis

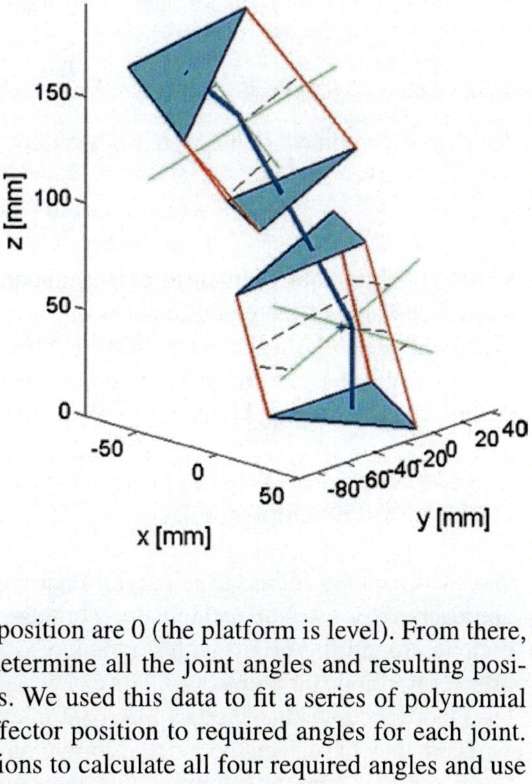

rotational components of the final position are 0 (the platform is level). From there, we used the kinematic model to determine all the joint angles and resulting positions that satisfied those conditions. We used this data to fit a series of polynomial functions relating a desired end-effector position to required angles for each joint. Our IK controller used these functions to calculate all four required angles and use the controller in Eq. (6.40) to reach the target.

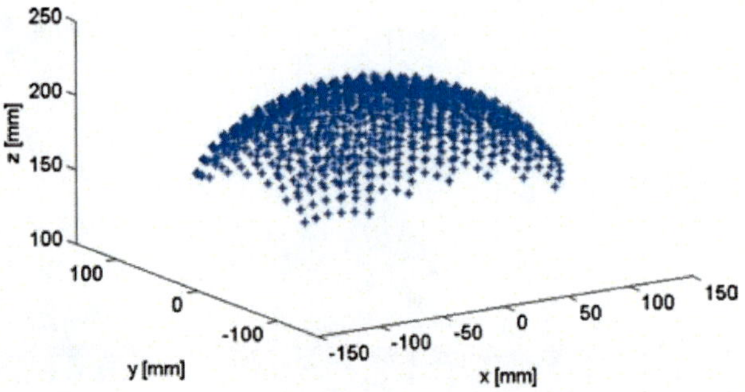

Fig. 6.32 A graph of the workspace of the 4-DoF manipulator when the joints can travel $\pi/6$ radians in each direction

The second controller used a geometric Jacobian approach similar to the one presented in Sect. 6.4.2.1. Using the kinematic model, we developed linearized Jacobian matrix equations relating changes in joint angles to changes in actuator length (Eq. 6.41) and end-effector position (Eq. 6.42), respectively:

$$d\vec{L} = \mathbf{A} d\vec{\theta} \tag{6.41}$$

$$d\vec{x} = d\vec{\theta}\,\boldsymbol{B} \tag{6.42}$$

where $\vec{L}$ is the vector of actuator lengths, $\vec{\theta}$ is the vector of joint angles, $\vec{x}$ is the end-effector position, and **A** and **B** are the linearized Jacobian matrices. Just like Eq. (6.39), **A** and **B** were linearized at $d\vec{\theta} = [0000]^T$. We multiplied Eq. (6.42) by the pseudo-inverse of **B** on the left and substituted into Eq. (6.41) as:

$$d\vec{L} = \boldsymbol{AB} \wedge \{+\}\, d\vec{x} \tag{6.43}$$

which was combined with Eq. (6.36), yielding:

$$\vec{u}(t) = \vec{u}(t - \Delta t) + \boldsymbol{AB} \wedge \{+\} K \left(\dot{e}_{\vec{x}} + De_{\vec{x}}\right) \tag{6.44}$$

where $\vec{u}$ is the control input to each actuator, $\dot{e}_{\vec{x}}$ is the change in error for each controlled value of the end-effector position, $e_{\vec{x}}$ is the end-effector position error, and K and D are the scalar control coefficients.

This controller requires calculating the forward kinematics of the manipulator within the control loop, but this was observed to not reduce performance.

6.4.3.2 Experimental Results

To experimentally compare their performance, we applied the direct IK and Jacobian approaches to the 4-DoF manipulator. For these tests, we had the end effector follow a circle around the origin with a radius of 30 mm at a period of 15 s. The end-effector positions for both controllers can be seen in Fig. 6.33 where the IK controller is more accurate than the Jacobian controller, but exhibits a similar level of noise.

There may have been too much information loss when the control matrices were linearized, reducing the fidelity of this controller. On the other hand, doing entirely

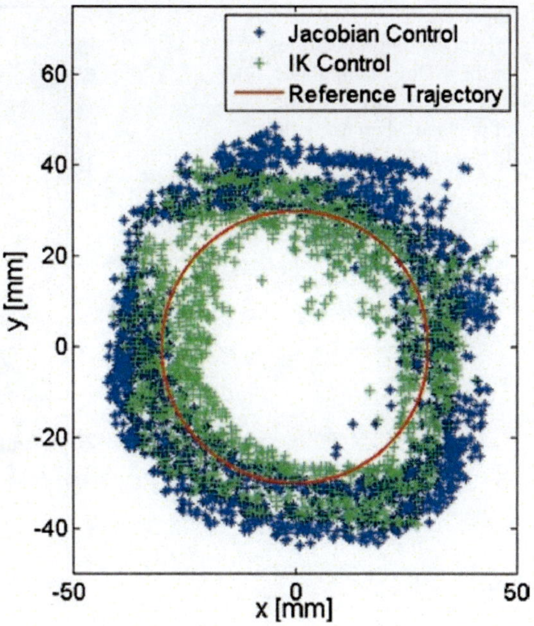

Fig. 6.33 The actual and reference joint trajectories of the 4-DoF manipulator end effector when using the IK versus Jacobian control

Table 6.5 Mean and standard deviation of the radius of the circular end-effector trajectory

	No payload	80 g payload
IK	32 ± 5.8 mm	32 ± 8.4 mm
Jacobian	38 ± 5.2 mm	35 ± 4.4 mm

separate control of the two 2-DoF joints likely introduces additional noise in the IK controller, as both joints separately oscillate around their desired angle combinations. Quantitative analysis of the end-effector trajectories can be seen in Table 6.5, where data is shown in mean ± standard deviation form.

In addition, we tested both controllers while the manipulator was carrying a payload of 80 g, the end-effector data for which can be seen in Fig. 6.34. With 80 g of payload, IK controller still follows the desired trajectory with similar accuracy, but the noise in the data increases. It is likely that the additional mass amplifies the dynamic effects of the six valves operating simultaneously. On the other hand, the noise of the Jacobian controller did not change significantly, though the motion of the end effector remained outside of the desired trajectory.

6.5 2-Degree-of-Freedom Actuator Application

As computers become increasingly prevalent, the ability of humans and computers to communicate becomes more important. While much can be conveyed visually, humans have access to other senses that can be used to communicate information and provide feedback to a human user. One way of this is through kinesthetic haptic

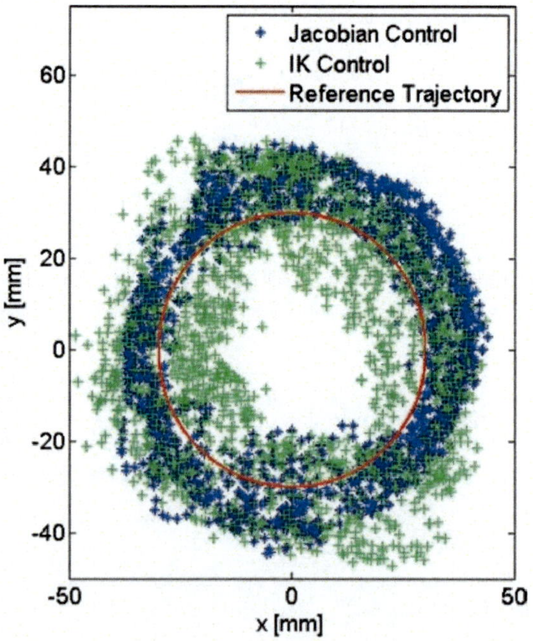

Fig. 6.34 The actual and reference joint trajectories of the 4-DoF manipulator end effector carrying 80 g of payload when using the IK versus Jacobian control

feedback, applying forces to the user's joints. It should be unsurprising that soft pneumatic actuators have advantages that make them suitable for this application. The soft nature of rPAM allows them to easily conform to users of various sizes, while the compliant nature of the actuation ensures the user's safety and comfort. We create a system with four rPAM actuators mounted around the user's wrist (shown in Fig. 6.35), which can apply safe forces to users in order to improve their performance in a line-following task.

The contributions of this section are as follows:

- The first use of rPAM actuators in a haptic system
- A haptic device that is both faster than other soft kinesthetic devices and safer, lower profile, and more flexible than existing rigid kinesthetic devices

6.5.1 Fabrication

The wearable soft haptic wrist device described in this section uses reverse pneumatic artificial muscles (rPAMs) to apply compliant forces to the wearer's wrist. These rPAM actuators consist of a tube of silicone wrapped in a helix of thread. When pressurized, the thread provides a physical constraint that prevents the actuator from expanding into a sphere. Instead, the actuator simply extends, imparting compliant forces as discussed in [29, 89].

The compliant nature of these linear actuators makes applying compressive forces a difficult proposition. When doing so, the actuators have a tendency to

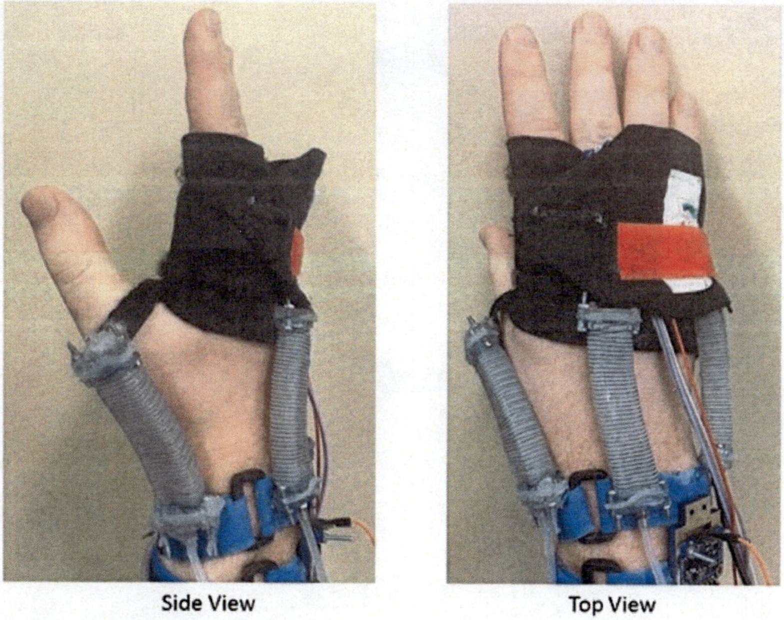

Fig. 6.35 Views of the haptic wrist device worn by a user

buckle. To compensate for this, we have previously mounted these actuators antagonistically around a 3D printed joint in a pre-strained condition. Thus, when pressurized, the actuators relieve the pre-strain without buckling, allowing them to apply their antagonistic forces.

The rPAMs used in this section were fabricated using a multi-step process. First, the hollow core of the actuator was created out of silicone rubber (Dragon Skin 10) in a 3D printed mold. Next, nylon thread is wrapped in a uniform helical pattern around the actuator. The mold includes grooves around the outside, making it easy for the thread to be wrapped uniformly. To hold the thread in place, we applied a thin layer of uncured Dragon Skin 10 over the outside.

To provide a seal for each end of the actuator, we used a technique we developed in [90]. This consists of a pair of acrylic plates sandwiching a flange at each end of the actuator, with the actuator fitting through the inner plate. These two plates were bolted together over the flange to form a tight seal. On one side of each actuator, a vent screw (a machine screw with a hole drilled through) was slotted though the outer plate, allowing pressure to be applied to the actuator chamber.

Securing these actuators onto the forearm and hand was a challenge, especially considering the desired pre-strain in the rPAM actuators. We had to provide solid mounts for the actuators that would both resist actuator forces and stay put as the user moved their arm. After investigating several options, including nitrile gloves, Velcro® straps, and an elbow support, we settled on using different attachment methods at the hand and the wrist. To attach at the hand, we used a Flarico Hand Wrap, which is slotted over the thumb and then wrapped above the thumb and

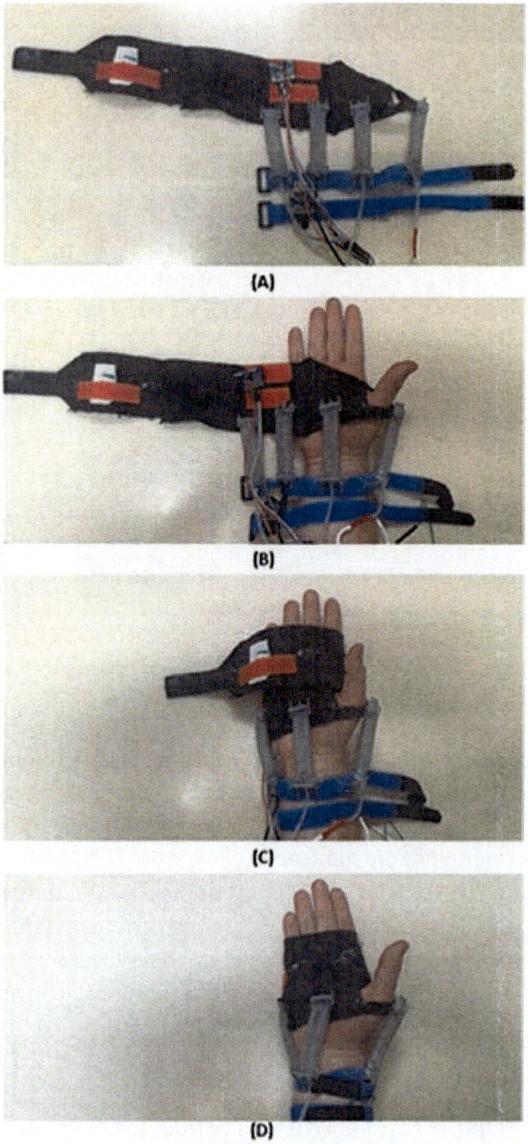

Fig. 6.36 (**a**) The full haptic wrist device, unattached to a user. (**b**) Step 1 in donning the device, put loop over the thumb. (**c**) Next step in donning the device, begin wrapping. (**d**) Fully wrapped, tighten the forearm straps

around the hand. The acrylic endplates of the actuators were attached to the lowest layer of this wrap, so that they would be situated around all sides of the user's wrist when the device is worn. The material of the wrap is stretchy enough to adapt to different hand sizes. The donning process can be seen in Fig. 6.36.

To mount the actuators on the forearm, we used a Bluecell Velcro® strap secured below the user's wrist. Without tightening this Velcro strap to an uncomfortable level, this mount was still subject to sliding, but the bumps below the radius and ulna ensure that the actuators stay attached at a usable position. Again, the acrylic endplates of the actuators were glued to the strap, which allowed for a simple, effective mount.

In order for the system to measure the state of the user's wrist, we need to accurately measure the angle of the wrist in real time. To do this, we mounted a pair of inertial measurement units (IMUs) on the wrist device, one on the forearm and one on the back of the hand. We chose the BNO0055 9-axis absolute IMU for its small size, ease of use, and high reliability. This IMU performs all sensor fusion on-board and only outputs its orientation.

We mounted the IMU on the forearm by bolting it to an acrylic plate and attaching the plate to Velcro strap. In order to ensure that this IMU remained stationary during motion, we added a second Velcro strap farther down the forearm, which was attached to the other side of the IMU attachment plate. The hand IMU was initially attached to the hand wrap using the same acrylic plate with the forearm IMU. However, initial user tests indicated that the fixed position of the IMU on the hand wrap was vulnerable to differences in user's hand size. To remedy this, we replaced the direct connection between the IMU plate and the hand wrap with a Velcro® connection. Thus, the IMU could be placed at a variety of points on the wrap to ensure its location coincided with the back of the user's hand.

6.5.1.1 Physical Capabilities

We created a test setup to verify the forces that the rPAMs can impart on the wrist and ensure that the torques are within a safe range. We 3D printed a replica of a human hand and forearm. This model included a hinge joint at the wrist, which allowed it to pivot forward and backward. This joint also included a potentiometer for reliable and direct angle measurement. We applied known torques around the wrist joint and adjusted the pressure in the corresponding actuator to return the joint to a neutral angle. This allowed us to determine the torque output of rPAM actuators on a user's wrist. The experimental setup can be seen in Fig. 6.37.

We found that the maximum torque the actuators can apply while the hand is maintained at a straight pose is 0.14 Nm at around 8 psi input pressure. If the actuators are given a higher pressure, they will start buckling, and the force applied to the wrist will not increase. This is significantly lower than the maximum wrist torque humans can sustain [91] ensuring that our haptic device will be safe to operate regardless of any malfunctions. Thus, this was the pressure we used for all experiments.

6.5.2 Path-Following Test

In order to test the usability of the haptic feedback provided by our device, we created a simple scenario where the user would use their wrist to control an agent operating in a simple virtual environment. We initially considered having the user control an agent in a maze or other obstacle-filled environment. However, we decided that the obstacles would add an element of problem-solving to any tests that

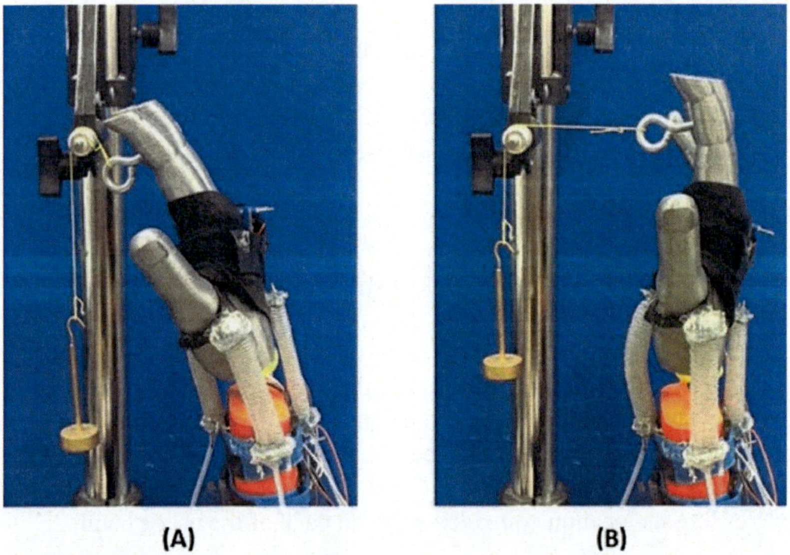

Fig. 6.37 The haptic wrist force experiment. (**a**) Torque applied, no pressure applied; (**b**) torque applied, pressure applied to return the hand to the zero angle. Note: The pressurized actuator is against the palm and is inflated and extended

were performed, adding noise to any direct analysis of the effect of the haptic feedback on user performance.

Instead, we chose to test the benefits of the haptic device on a simple path-following task. This would allow the user to only focus on performing the intricate task as steadily as possible without having to worry about internal path planning. In addition, the nature of following an infinitely thin curve means that there is no performance ceiling, that there will always be room for the user to improve. Thus, there will always be room for a haptic device to assist the user.

The environment consists of a path and an agent. We tested the usability of the system on both a straight line and a sinusoidal curve, both traveling left to right. The agent that the user controls within this environment is represented by a "*" icon. It has a forward velocity, represented by a line branching out from the agent position indicating the direction of travel. The length of this line was adjusted in conjunction with the velocity of the agent, to give some additional visual indication of velocity. A scene from the test environment is shown in Fig. 6.38. The goal of the user was to follow the path as closely as possible while traveling all the way to the right side of the screen. We also instituted a time limit in the event that the user was traveling too slowly, to ensure that experiments were completed in a timely fashion.

6.5.2.1 Virtual Agent Control

At each simulation time step, the system calculates the motion angle and linear velocity of the agent from user input and iterate agent position on the screen. The user controls the agent using the angle of their wrist. In order to do this, it is first

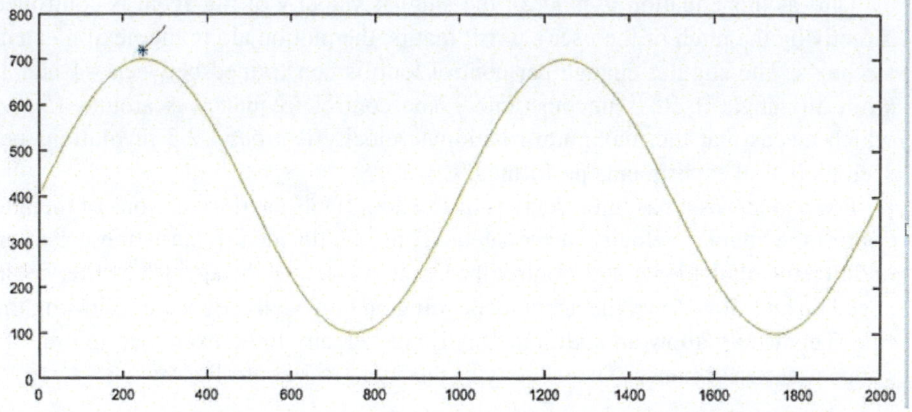

Fig. 6.38 An example state the user would see when performing a line-following experiment

necessary to extract the wrist angle from the readings from the two IMUs. The control code running on the desktop computer receives global orientations of the IMUs in the form of quaternions. From there, we converted them to rotation matrices and used the following equation to find the rotation matrix of the wrist:

$$R_D = R_2 R_1^T \tag{6.45}$$

where $R_D \in \mathbb{R}^{3\times3}$ is the rotation matrix between the two IMUs, $R_2 \in \mathbb{R}^{3\times3}$ is the global rotation matrix to the wrist IMU, and $R_1 \in \mathbb{R}^{3\times3}$ is the global rotation matrix to the hand IMU. We convert this back to $D_r \in \mathbb{R}^{3\times1}$, the raw local Euler angle state vector of the wrist. Because of the inconsistencies between human morphologies and the way the wrist device is wrapped around the hand, different users, or even the same user during different trials, would have different rest angles for their hand. Thus, we implemented a calibration routine every time the user puts on the device. This routine calculated the average D_r over 50 iterations of the user holding their hand loose and steady; this was used to calculate an offset value used during the subsequent experiments as follows:

$$D = D_r - D_o, \tag{6.46}$$

where $D \in \mathbb{R}^{3\times1}$ is the adjusted angle of the wrist, D_r is the measured raw state discussed previously, and $D_o \in \mathbb{R}^{3\times3}$ is the offset calculated during calibration.

From this adjusted value of the state of the wrist, the algorithm calculates the angle of the next state using the following equation:

$$A_i = A_{i-1} - \mathrm{con}\left(D(2)k_a, -1, 1\right), \tag{6.47}$$

where A_i is the angle of the current state, A_{i-1} is the angle at the previous state, k_a is a sensitivity constant, con(.) is a constraint function, and $D \in \mathbb{R}^{3\times1}$ is the Euler angle state vector of the wrist in "yaw-pitch-roll" order.

Thus as this equation uses $D(2)$, the angular velocity of the agent is controlled directly by the pitch of the user's wrist, that is, the motion along the flexion/extension axis. The angular change per control loop is constrained between −1 and 1, while the angle itself is unconstrained. The control frequency is around 15 Hz, which means that the maximum rotational velocity is around 2.5 revolutions per second. For all experiments performed, $k_a = 1$.

The human wrist has three degrees of freedom (DoF), and we use one of them to control the angular velocity of the agent. Thus, we decided to study the use of an additional DoF in the wrist to control the linear velocity of the agent. The user could speed up and slow down the agent, depending on how well they were following the line. This would allow an additional metric of velocity to be examined in order to gauge user performance. The velocity of the agent was controlled by:

$$V_i = \text{con}\left(V_{i-1} + \text{con}\left(D(1)k_v, -0.3, 0.3\right), 15, 2\right) \tag{6.48}$$

where V_i is the linear velocity at the current state, V_{i-1} is the linear velocity at the previous state, and k_v is a sensitivity constant.

As this equation uses $D(1)$, the linear acceleration is controlled directly by the yaw of the wrist, that is, the lateral motion in the radial/ulnar direction. The constraint functions added here constrained both the acceleration of the agent applied by the user and the total velocity. We also observed that users could more easily move their hands in the ulnar direction (toward the little finger) than in the radial direction (toward the thumb). Thus, we used a different value for k_v depending on which direction the user was turning their wrist, as follows:

$$k_a = \begin{cases} 15, & \text{if } D(1) < -0.01 \\ 7, & \text{if } D(1) > 0.08 \\ 0, & \text{otherwise} \end{cases} \tag{6.49}$$

This equation also creates a deadzone, where it would theoretically be easier for the user to steer the agent without accidentally changing its velocity. However, initial experiments indicated that this method of controlling the agent velocity was still difficult for most users. Some users found that the mental coupling between the two axes of rotation in the wrist made it difficult to control one without changing the other. Thus, we looked into different ways of controlling the linear velocity of the agent using the user's body. We settled on controlling the linear velocity of the agent directly using the angle of their forearm. This would decouple the two DoF while using our already-existing hardware. The velocity control equation for this method is written as:

$$V_i = \text{con}\left(E_2(1)k_{v2}, 20, 3\right), \tag{6.50}$$

where $E_2 \in R^{3\times 1}$ is the global Euler angle of IMU2 (the IMU mounted on the forearm) and k_{v2} is a sensitivity constant.

Thus, if the user held their hand and forearm with the little finger downward, they could directly control the velocity of the agent by moving their forearm up and down. For all of these experiments, we used $k_{v2} = 30$. The velocity was constrained between 20 and 3. This comes out to a maximum velocity of one sixth of the horizontal length of the environment per second, and the minimum velocity being one fortieth of the horizontal length of the environment per second. Thus, the user would reach their maximum velocity when the arm was held at around 40° above horizontal where $E_2 \in R^{3\times 1}$ is the global Euler angle of IMU2 (the IMU mounted on the forearm) and k_{v2} is a sensitivity constant.

6.5.2.2 Feedback Control

In order for the system to guide the user to the desired path, we created an algorithm to determine the pressure in each rPAM. To do this, we calculated how far the agent was from the nearest point on the path. For faster computation, this distance was only calculated for the 400 nearest points on the horizontal axis. The distance to each of these points was calculated, and the shortest distance was determined. Once that point was determined, we calculated the angle needed for the agent to reach that point, as well as the angle of the path at that point. The desired angle for the agent was then calculated as a weighted average between these two angles:

$$A_d = \alpha a_1 + (1-\alpha) a_2, \tag{6.51}$$

where A_d is the desired angle, a_1 is the angle for the agent to reach the nearest point on the path, a_2 is the angle of the path at that point, and $\alpha \in [0, 1]$ is the weight. α becomes smaller as the minimum distance to the nearest point on the path becomes smaller and is calculated using the following equation:

$$\alpha = \min\left(\frac{m}{200}, 1\right) \tag{6.52}$$

where m is the minimum distance to the desired path.

Thus, if $m > 200$, $\alpha = 1$, and the desired angle is equal to the angle between the agent and the nearest point on the path. As the virtual agent approaches the path, the desired angle becomes more and more aligned with the direction of the path. Entirely following the desired angle, an agent would asymptotically approach the desired path.

Using the desired angle, we calculate the angle error A_e using the following equation:

$$A_c = \begin{cases} A_d - A_i, & \text{if}\left[A_d - A_i\right] > dz \\ 0 & \text{otherwise} \end{cases} \tag{6.53}$$

where dz is the deadzone.

Thus, when the user is pointing their wrist within a certain threshold of the desired angle, it will be considered to be perfectly accurate. From here, the control command to the valves is calculated as:

$$u_1 = \text{con}\left(50 + k_{au} A_e, 0, 100\right) \tag{6.54}$$

where u_1 is the angle control input, k_{au} is the sensitivity constant, and A_e is the angle error.

We used $k_{au} = 30$ for all experiments. This equation generates a control input between 0 and 100 with 0 resulting in full actuation in one direction and 100 resulting in full actuation in the opposite direction.

We also wanted to investigate the ability of the haptic feedback to help the user regulate their velocity. When the user is following the desired path with a high level of accuracy, we would want them to speed up to reach the end of the path faster. On the other hand, if the user has a large error, we would want them to slow down to better focus on returning to the path. To this end, we implemented the following feedback control law for velocity:

$$V_d = \max\left(18 - \left|A_e\right| k_{vu}, 3\right), \tag{6.55}$$

where V_d is the desired velocity and k_{vu} is the sensitivity constant.

For all experiments performed, $k_{vd} = 12$. Through this equation, the desired velocity ranges between 18 and 3 and is linearly related to the angle error. When the user is pointing in the desired direction, the desired velocity will be high, telling the user to speed up to return to the path or speed along it toward the goal. However, when the user is not pointing in the correct direction, the desired velocity will be low, telling the user to slow down. We calculated the velocity error in a similar manner to Eq. (6.53), with a corresponding deadzone of 3, and calculated the command to the controller by:

$$u_2 = \text{con}\left(50 + k_{vu} V_e, 0, 100\right), \tag{6.56}$$

where u_2 is the velocity control input, V_e is the error between the desired velocity and the actual velocity, and k_{vu} is a sensitivity constant.

This feedback on the lateral motion of the hand was still used when the velocity was being controlled by the angle of the forearm. As the hand was directed to be held sideways, the changes in the angle of the arm would be in the same plane as the lateral motion of the hand. Thus, the lateral haptic cues on the hand would still indicate to the user the correct direction to move their arm, even if the forces were not directly affecting the joint used for control.

Valve Control

An Arduino® Mega control board receives commands from MATLAB® and directly controls 30 Hz pulse width modulation (PWM) signals of the four digital valves connected to the rPAM actuators. Several different control schemes were evaluated, including our previous method of both valves operating antagonistically, as in [92]. The final control scheme we settled on is as follows:

For $u_i > 50$:

$$\begin{aligned} c_1 &= 0, \\ c_2 &= 2\left(u - 50 + c_O\right) \times 255/100, \end{aligned} \tag{6.57}$$

for $u_i < 50$:

$$\begin{aligned} c_1 &= 2\left(50 - u + c_O\right) \times 255/100, \\ c_2 &= 0, \end{aligned} \tag{6.58}$$

and if $u_i = 50$, then:

$$\begin{aligned} c_1 &= 0, \\ c_2 &= 0, \end{aligned} \tag{6.59}$$

where c_1 and c_2 are 8-bit duty cycles sent directly to the two valves and c_O is a command offset. c_O is included because at low duty cycles ($c < 30$) the commands will be too fast to register and the valve will not actuate.

Thus, c_O was set to 15 for all experiments. This saturation behavior also occurs where $c > 70$, but precision is not necessary when the error is high. These equations were used for pairs of actuators on each actuated wrist axis.

6.5.3 Experimental Results

We performed a range of path-following experiments with users. In order to collect data under a wide range of circumstances, we studied all combinations of the type of path, the type of feedback used, and the variability of velocity, for a total of ten experiments (velocity feedback was not used when the velocity of the agent was fixed). When a fixed velocity was used, we set it at 5 units per cycle, or about one twenty-fourth of the horizontal length of the environment. The two paths used were a straight horizontal line and a sine wave with amplitude 300 and a period of 1000. The sine wave trajectory is shown in Fig. 6.39, with two periods taking up the entire length of the environment. Initial positions were on the desired position at $x = 250$

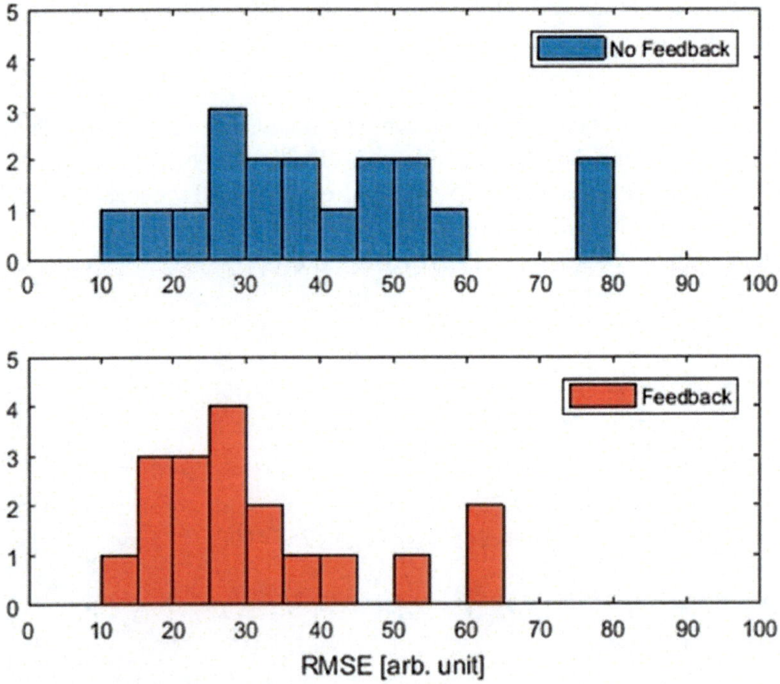

Fig. 6.39 A histogram of user RMS error for users following a sinusoidal path with and without haptic feedback. The agent velocity was fixed and the deadzone $dz = 4.5°$ (0.078 rad)

but with an initial angle $\pi/2$ (i.e., moving in the vertical direction away from the path) forcing the users to correct their trajectory to match the path. This was particularly relevant on the horizontal line-following experiments, where if the initial angle was along the direction of the path, the user wouldn't have to do anything in order to follow it successfully.

We gave the users instructions about the testing scenarios and about how to control the virtual agent and then asked new users to practice control of the agent without feedback. After this, users were given ten experiments in random order, with a brief pause in between each where the properties of the next experiment were listed, allowing the users to prepare themselves.

We performed this set of ten different experiments with nine different volunteers from our research group, with some users performing the random set multiple times. We did an initial run of 18 experiments. In this first set of experiments, we used the lateral motion of the wrist to control the velocity when velocity was variable (see Eq. 6.48) and a deadzone $dz = 4.5°$ (0.078 rad). We measured the root mean square (RMS) error for the user for each of these trials, as well as their average horizontal velocity. The RMS errors (RMSE) were calculated using the distance to the nearest point calculated during the control, while the average velocities were calculated using the final x-coordinate and the final time recorded for a given trial. The means

Table 6.6 The errors for all the path following experiments performed with deadzone $dz = 9°$ (0.156 rad)

	Linear path		Sine path	
	Const. velocity	Var. velocity	Const. velocity	Var. velocity
No feedback	17.3 ± 14.0	43.5 ± 39.5	**40.1 ± 18.2**	88.2 ± 59.0
Angle feedback	15.2 ± 8.5	24.4 ± 16.1	**31.4 ± 14.9**	76.6 ± 52.4
Combined feedback	–	32.9 ± 25.3	–	101.5 ± 58.6

Note: Parameters investigated were the quantity of haptic feedback (no feedback, feedback on the agent angle, and feedback on both the agent angle and velocity), the velocity of the agent (constant (Const.) versus variable (Var.)), and the path type (linear versus sinusoidal). When variable velocity was used, it was controlled using the radial/ulnar deviation of the wrist. Errors are presented in the form mean ± standard deviation and in arbitrary units

and standard deviations for the RMS errors for each experiential type are shown in Table 6.6.

The first observation we can make from this data is that the addition of feedback provides a noticeable improvement for the sine wave experiments, with and without a variable velocity. In particular, there is around a 30% improvement in the RMSE of the fixed-velocity sine wave following, as well as a smaller improvement during variable velocity experiments. Performing a T-test, we can see that this improvement is statistically significant for the fixed-velocity trial ($p < 0.05$).

A histogram comparison of the fixed-velocity trials can be seen in Fig. 6.39. The data without haptic feedback was fairly uniform, with some users struggling to control the agent. With the addition of haptic feedback, the users performed better, though the wrist device was not effective enough to overcome the RMSE = 10° threshold.

One example of a trajectory pair, comparing a single user's performance with and without feedback in a sine-following fixed-velocity scenario, can be seen in Fig. 6.40, where we show both the full trajectories and the error. We can see that the feedback pulls the user away from their initial divergent trajectory much faster. The changing nature of the sine path fosters error on the user, error that the haptic device is able to help correct. In addition to providing correcting forces, some users also reported that the perceived vibration (resulting from the PWM pressure control of the valves) kept them focused on the task when the agent was starting to diverge.

In contrast, an example comparison of a single user's performance following a straight path can be seen in Fig. 6.41. This figure shows both the error (which, as the desired path is straight, represents the trajectory) and the error with respect to the desired angle (see Eq. 6.51). In this case, we can see that the feedback does not help. The feedback seems to drive the user into over-correcting. In addition, because the linear trajectory is easier to follow, much more of the motion occurs within the deadzone. Thus, feedback plays a much smaller role in affecting performance, which is supported by the aggregate data for the straight-line-following experiment.

We saw that the haptic feedback provided a slight improvement for the variable-velocity sine-following experiments, but the variance in the data was too large to show any statistical significance. This applied to other experiments as well and was

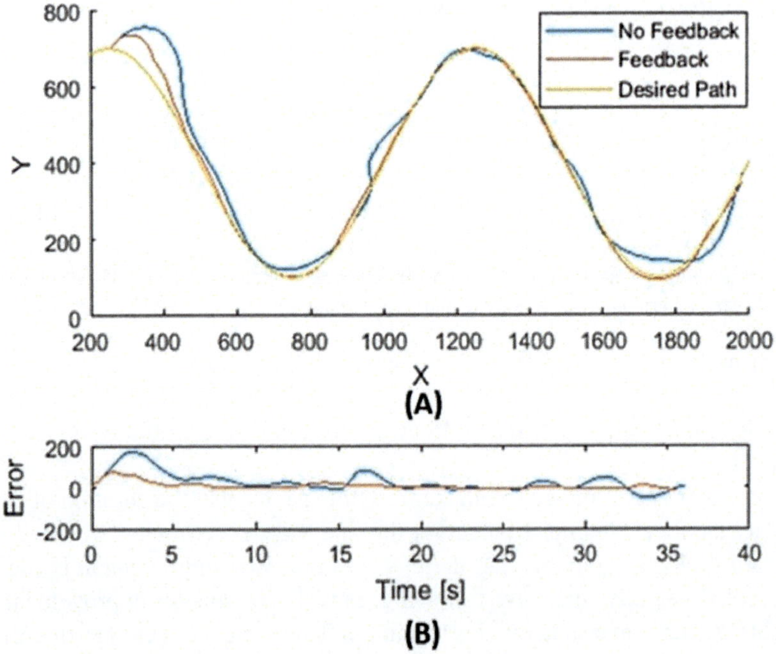

Fig. 6.40 An example of a user's performance with and without haptic feedback when following a sine wave path with a deadzone $dz = 4.5°$ (0.078 rad). (**a**) The trajectory, (**b**) error

not affected by the haptic velocity control. Users reported that using both DoF on the wrist was difficult, such as increasing the velocity without turning and vice versa. This is despite initial calibration experiments that showed that the two degrees of freedom were not coupled. Within the cognitive burden of the path-following task, users found it hard to perform pure motions that are decoupled in the 2 DoF we measure through our soft haptic wrist device.

To circumvent this problem, we investigated an alternate method of controlling the linear velocity of the agent. As discussed previously, we performed experiments using the angle of the forearm to directly control velocity (see Eq. 6.50). In addition, we also shrunk to $dz = 3°$ (0.052 rad), with the goal of improving feedback performance on the straight-line trials. We performed 11 additional experiments under these conditions, with all other aspects of the experimental processes remaining the same. The results of these experiments can be seen in Table 6.7.

From this table, we can see that the changes did improve the effectiveness of the haptic feedback, but not in a statistically significant manner. For the fixed-velocity horizontal line following, the decrease in the size of the deadzone did not seem to improve the effects of the haptic feedback. Even when the rPAMs in the haptic glove are applying a very small amount of force, the vibration resulting from the PWM pressure control may be causing the users to overcorrect. An example plot of the angle error can be seen in Fig. 6.42. While the addition of haptic feedback speeds up the user's ability to approach the desired path, it seems to increase the

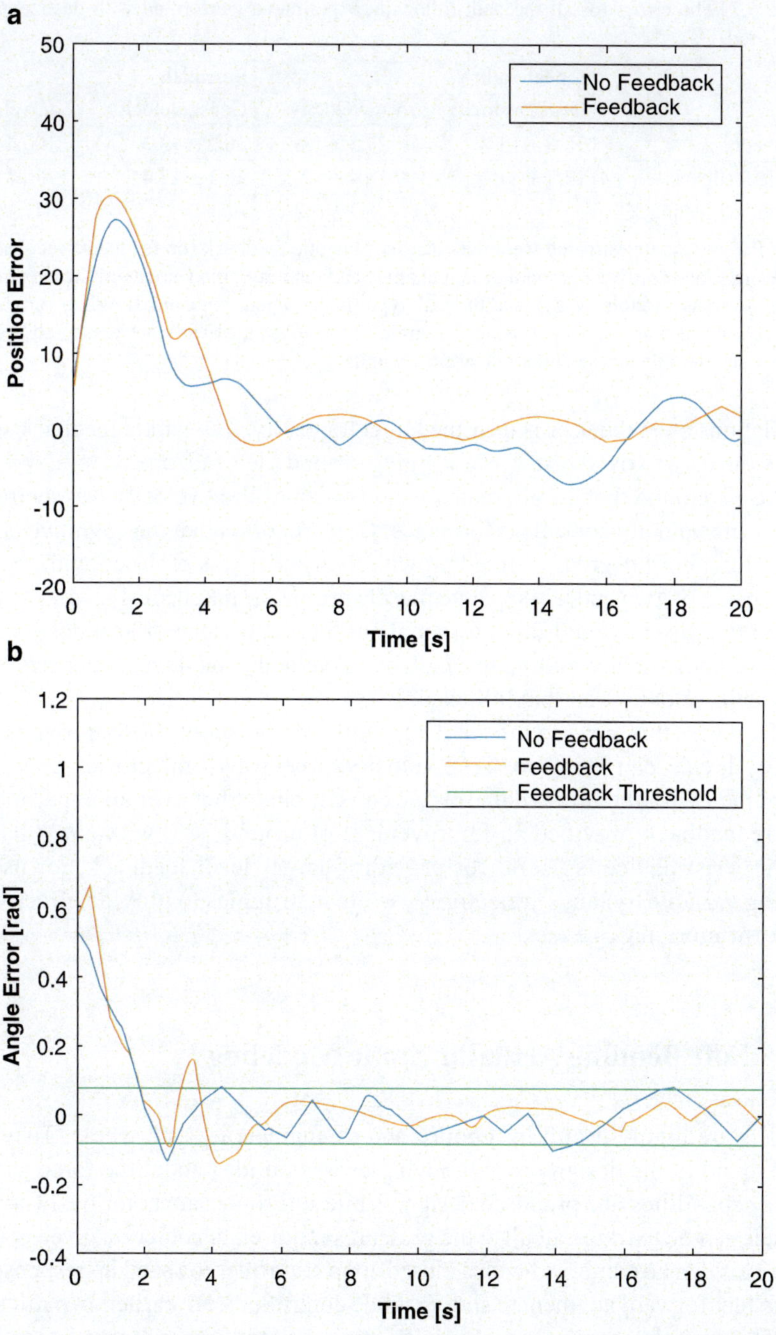

Fig. 6.41 An example of a user's performance with and without haptic feedback when following a linear path with a deadzone $dz = 4.5°$ (0.078 rad). (**a**) The error with respect to the desired position for each path. (**b**) The error with respect to the desired angle chosen by the haptic algorithm

Table 6.7 The errors for all the path following experiments performed with deadzone $dz = 3°$ (0.052 rad)

	Linear path		Sine path	
	Const. velocity	Var. velocity	Const. velocity	Var. velocity
No feedback	18.6 ± 13.7	24.5 ± 13.2	**36.2 ± 13.3**	100.4 ± 121.0
Angle feedback	12.5 ± 4.0	18.4 ± 12.1	**26.1 ± 12.4**	50.2 ± 20.2
Combined feedback	–	16.7 ± 6.2	–	47.1 ± 13.5

Note: Parameters investigated were the quantity of haptic feedback (no feedback, feedback on the agent angle, and feedback on both the agent angle and velocity), the velocity of the agent (constant (Const.) versus variable (Var.)), and the path type (linear versus sinusoidal). When variable velocity was used, it was controlled using the angle of the forearm. Errors are presented in the form mean ± standard deviation and are in arbitrary units

oscillations once the user is near the target trajectory. The haptic feedback seems to cause the user to overcorrect, zig-zagging around the deadzone.

We also found that the alternate method of controlling velocity was ineffective at increasing velocity control performance. Users reported that the cognitive burden of controlling both degrees of freedom with their wrist was replaced with the burden of focusing on two entirely separate muscle groups simultaneously: wrist and elbow. In addition, users would often forget about the importance of arm position during fixed-velocity trials, resulting in a high initial velocity and then greater errors during a variable-velocity trial that followed.

However, this new round of data continues to show the improvements the haptic device can bring to fixed-velocity, sine wave trajectories at $p < 0.05$. Combined with the initial data set, we can conclude that over all experiments the haptic feedback provided an improvement of around 30% with a $p < 0.01$. Our haptic wrist device is useful for providing larger-level feedback, for users following variable dynamic trajectories, while it struggles to provide effective feedback for more precise tasks.

6.6 Soft Bending Actuator Static Modeling

Models are highly useful in robotics and in engineering as a whole. They can be used to aid in the design process, giving us a good idea about the force and workspace capabilities of a proposed design. While soft robots are notoriously imprecise, models can be used to calculate the pressures required to achieve a given configuration. This can be combined with a closed-loop controller to speed up response times, as the feedforward augmented sliding mode controller from earlier. In particular, the model is useful because manually calculating the feedforward function for a multi-DoF manipulator with an arbitrary range of external forces would be prohibitively time-consuming.

The contributions of this section are as follows:

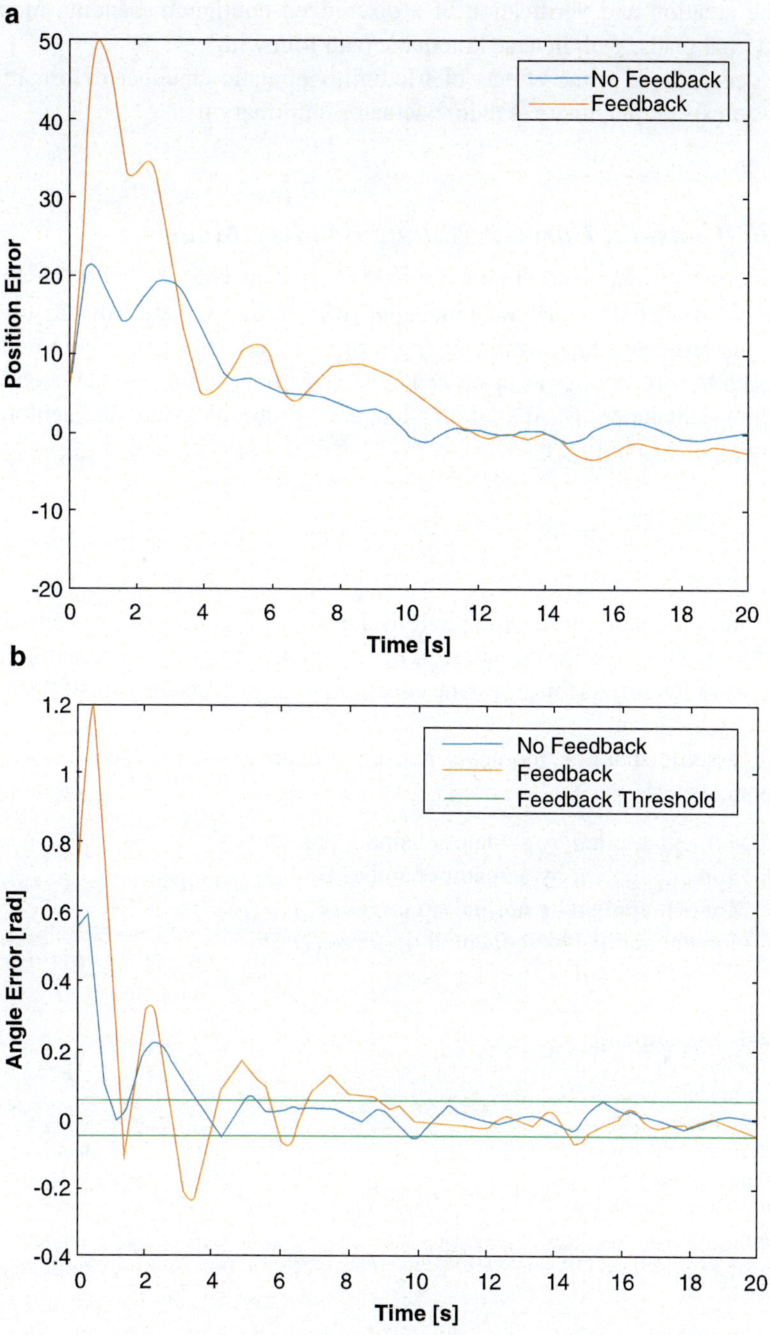

Fig. 6.42 An example of a user's performance with and without haptic feedback when following a linear path with a deadzone $dz = 3°$ (0.052 rad). (**a**) The error with respect to the desired position for each path. (**b**) The error with respect to the desired angle chosen by the haptic algorithm

- The creation and verification of a discretized continuum bending model with external loads, with inverse kinematic path following
- A deeper look at the effects of friction, pneumatic chamber deformation, and hysteresis on pneumatic bending actuator deformation

6.6.1 Constant Curvature Static Bending Model

We first formulated an analytical model of soft bending actuators that assumed constant curvature bending within a single pressurized segment, a commonly used approach in soft robotics as discussed by [93]. Similar to the soft-actuated skeleton modeling, this model was based on a balance of torques around the neutral axis of bending, point O in Fig. 6.43.

$$\frac{M}{EI} = \kappa = \frac{\theta}{L_o} \tag{6.60}$$

where M is the sum moment applied to the beam, E is the elastic modulus of the actuator material, κ is the curvature, θ is the bending angle of the actuator, and L_o is the actuator length. As this is a static equation, M represents the sum of the moments applied to the beam.

We assume that the moments can be broken down into the following four components:

- M_a Moment applied by actuator chamber A (F_a)
- M_b Moment applied by actuator chamber B (F_b)
- M_n Moment applied by normal tip forces (F_n)
- M_t Moment applied by tangential tip forces (F_t)

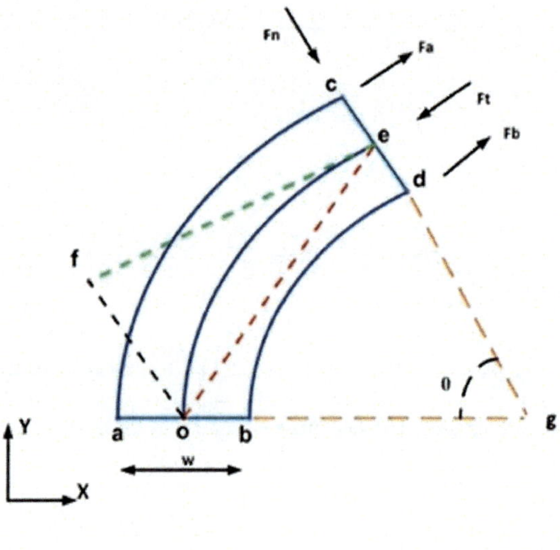

Fig. 6.43 Kinematic model of the 1-DoF bending module. The model focuses on torque balance around the base of the bending actuator (point O) with constant curvature bending

To determine the force applied by each actuator chamber, we assume that the force of each actuator is equal to the force of air at its end face. Thus:

$$F_a = Ap_a \tag{6.61}$$

where A is the cross-sectional area of the actuator chamber and p is the pressure inputted into the actuator chamber.

This equation applied to both pressure chambers on either side of the bending actuator. These forces are each situated $w/2$ away from the bending axis if the actuator, resulting in the following equation:

$$M_a = Ap_a \frac{w}{2} \tag{6.62}$$

To simplify our analysis and maintain our constant curvature assumption, we will examine the sum of moments around point O. We assumed that normal forces were propagated simply along the length of the actuator, regardless of the angle of bending. Thus, the moment caused by normal tip forces can be approximated as:

$$M_n = L_o F_n \tag{6.63}$$

where L_o is the actuator length and F_n is the normal tip force.

More complicated is the moment caused by tangential (comprehensive or tensile) forces F_t, which is as follows:

$$M_t = f_t \left| fo \right| \tag{6.64}$$

where $|fo|$ is the distance between points f and o, the moment arm of the tangential force F_t.

$$\left| fo \right| = \mathrm{sqrt}\left(f_{x^2} + f_{y^2} \right) \tag{6.65}$$

where f_x and f_y are the x- and y-coordinates of point f.

The equations for these are as follows:

$$f_x = \frac{\tan\left(\frac{\pi}{2} - \theta\right) e_x - e_y}{\tan\left(\frac{\pi}{2} - \theta\right) - \tan\left(\pi - \theta\right)} \tag{6.66}$$

$$f_y = \tan\left(\pi - \theta\right) f_x \tag{6.67}$$

where e_x and e_y are the x- and y-coordinates of the point e.

The equations for these are as follows:

$$e_x = \frac{L_o}{\theta}\left(1 - \cos(\theta)\right) \tag{6.68}$$

$$e_y = \frac{L_o}{\theta}\sin(\theta) \tag{6.69}$$

Combining all the moments together, we have that

$$M = F_n L_o + F_t\left|of\right| + Ap_a\frac{w}{2} - Ap_b\frac{w}{2} \tag{6.70}$$

We can use this equation to solve for various parameters, including the normal forces applied by a certain actuator under a certain pressure under static conditions.

6.6.2 Constant Curvature Verification

We performed experiments to verify the mode's ability to predict the actuator force output. We mounted a single chamber bending actuator, provided to us by Boston Engineering, tangent to a force plate. The single-chamber actuator has a line of Kevlar threading along the lower edge, causing it to bend into the force place when pressurized. The setup can be seen in Fig. 6.44.

The results of this experiment can be seen in Fig. 6.45, where we compared the force output of the actuator with the force output predicted for an actuator of that dimensions fixed at a 0-degree bending configuration.

We can see significant agreement between the model and the experimental force output, verifying the usability of this approach. The experimental force output began to outpace the model-predicted force at higher pressures, which we thought

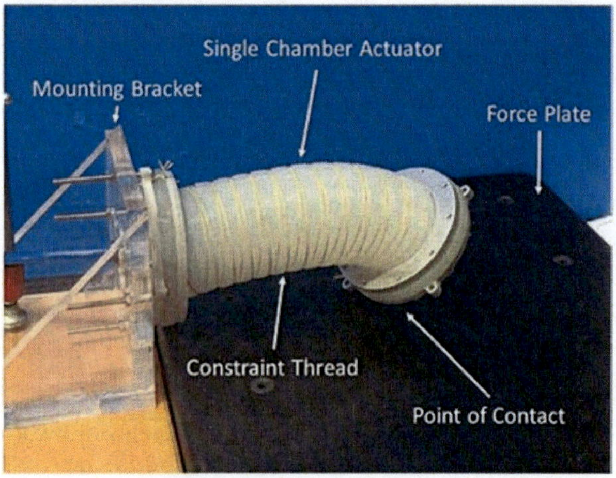

Fig. 6.44 The experimental setup for the soft bending actuator force experiment

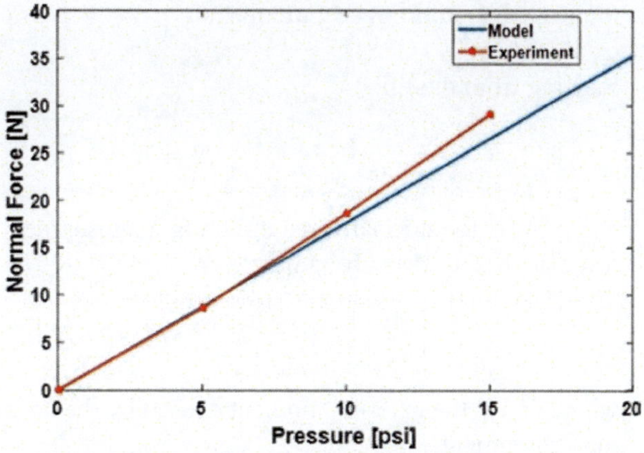

Fig. 6.45 The soft bending actuator force experiment data

may have been caused by the non-fixed nature of the actuator end. From Fig. 6.44, we can see that tip plate of the actuator has already begun to orient itself toward the force plate and at higher pressures they become tangent to each other. However, an alternate explanation is found later in this section.

6.6.3 Discretized Soft Actuator Bending Model

Our existing modeling can predict actuator behavior when subjected to uniform moments, such as those applied by pressures. However, it breaks down under external forces, which result in non-constant moments. To deal with this problem, we divide each segment into a number of subsegments, each with a constant curvature different from those around it. This can be used to approximate a complex continuous using simple equations.

Algorithm 6.1: The forward kinematic model of the bending actuator, which calculates the tip position using the bending angle of each subsegment $\theta_{i,j}$

```
Mn = 0;
ThetaSum = 0;
for i = Segments do
    for j = SubSegments do
        r = L(i)/theta(i,j);
        if r<inf then
            xn = -r*(1-cos(theta(i,j)));
            yn = r*sin(theta(i,j));
        else
            xn = 0;
            yn = L(i);
        end
    [xo yo] = [xo yo] + Rotate([xn,yn], theta(i,j));
    end
end
```

6.6.3.1 Normal Force Calculation

Starting from the tip:

$$N_i = N_{i+1} + F_i \tag{6.71}$$

where N_i is the normal force effecting a subsegment and F_i is the external normal force applied to the subsegment.

$$M_{i,\mathrm{n}} = M_{i+1,\mathrm{n}} + N_i L_i \tag{6.72}$$

where $M_{i,\mathrm{n}}$ is the external moment caused by the normal force and L_i is the length of the subsegment.

$$M_i = P_{\mathrm{a},i} A_i w - P_{\mathrm{w},i} A_i w + M_{i,\mathrm{n}} \tag{6.73}$$

$$\theta_{i,\mathrm{n}} = M_i L_i / (E_i I_i) \tag{6.74}$$

We then use these angles to calculate the state of the manipulator:

Algorithm 6.2: A calculation of the bending angle of each subsegment when under an applied pressure and an external normal force

```
Mn = 0;
Fnormal = 0;
for i = Segments, from the tip do
    Fnormal = Fnormal + Fexternal(i);
    for j = SubSegments do
        Mn = Mn+Fnormal*Length(i);
        Mtotal = Pa(i)*A(i)*w(i)-Pb(i)*A(i)*w(i)+Mn;
        theta(i,j) = Mtotal*L(i)/(E(i)*I(i));
    end
end
```

6.6.3.2 Tangential Force Calculation

Tangential forces are more complicated, as their effect is determined by the state of the manipulator.

We make the assumption that, as each subsegment is small, the moment caused by tangential force is not affected significantly by changes in subsegment bending.

Thus we:

1. Calculate the subsegment angle without tangential forces (using Eq. 6.74)
2. Calculate the segment base position with respect to the tip frame using this angle and all previous
3. Recalculate the subsegment angle incorporating tangential forces

$$r_i = L_i / \theta_{i,\mathrm{n}} \tag{6.75}$$

$$xc = -r_i \left(1 - \cos\left(\theta_{i,\mathrm{n}}\right)\right) \tag{6.76}$$

$$yc = r_i \sin\left(\theta_{i,n}\right) \tag{6.77}$$

$$x_{t,i} = x_{t,i+1}\cos\left(tSum\right) - y_c \sin\left(tSum\right) \tag{6.78}$$

$$M_{i,t} = M_i - F_t x_{t,i} \tag{6.79}$$

We use this to recalculate the subsegment bending angle and position of the next point. A pseudocode version of the algorithm can be found in 3. This method could be generalized to take into account multiple tangential forces, but would require keeping track of a new moment arm for every additional tangential force.

6.6.3.3 Global Forces

This model only takes into account forces defined in the manipulator tip frame. But what about forces fixed in the global frame, such as gravity? Direct conversion to the tip frame is impossible, since the forces, themselves, change the state of the manipulator and thus change the rotation matrix between global and tip. Thus, we need to find the rotation between global and tip coordinate frames that causes the final tip angle to be equal to that rotation. A visualization of this can be seen in Fig. 6.46. Figure 6.46a shows the conversion between the global local forces, while Fig. 6.46b shows the final simulated state of the manipulator, with the local force matching the final tip orientation.

Algorithm 6.3: A calculation of the subsegment angles of the discretized bending model when subjected to tangential forces

```
Mn = 0;
Fnormal = 0;
for i = Segments, from the tip do
    Fnormal = Fnormal + Fexternal(i);
    for j = SubSegments do
        // Normal-Force angle calculation
        Mn = Mn+Fnormal*Length(i);
        Mtotal = Pa(i)*A(i)*w(i)-Pb(i)*A(i)*w(i)+Mn;
        theta(i,j) = = Mtotal*L(i)/(E(i)*I(i));
        // reversed kinematics from the perspective of the tip to
           calculate tangential force moment arm.
        r = L(i)/theta(i,j);
        xn = -r*(1-cos(theta(i,j))); yn = r*sin(theta(i,j));
        [xo yo] = [xo yo] + Rotate([xn,yn], theta(i,j));
        // incorporate tangential moment into angle  calculation
        Mtotal = Mtotal-Ft*xo;
        theta(i,j) = = Mtotal*L(i)/(E(i)*I(i));
        // update kinematics, from the perspective of the tip
        r = L(i)/theta(i,j);
        xn = -r*(1-cos(theta(i,j)));
        yn = r*sin(theta(i,j));
        [xo yo] = [xo yo] + Rotate([xn,yn], theta(i,j));
        tSum = tSum +  theta;
    end
end
```

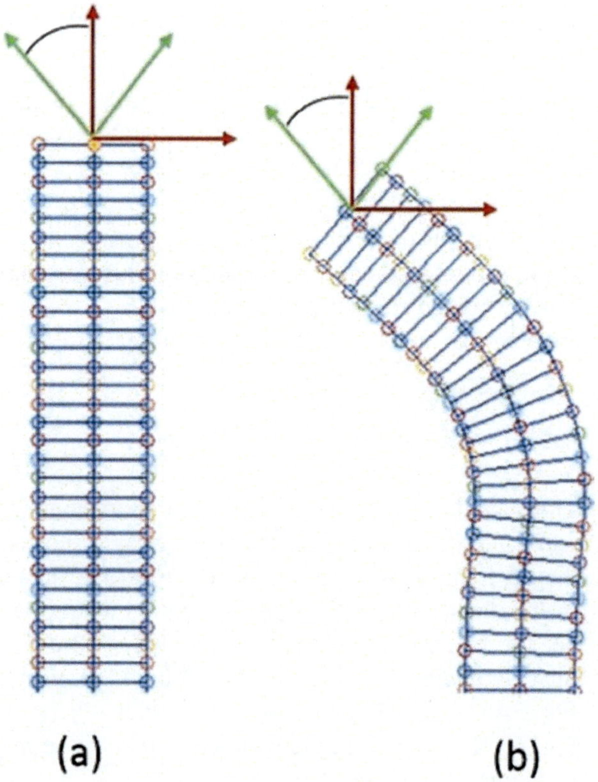

Fig. 6.46 A visualization of the global force algorithm. (**a**) Any forces in the global coordinate frame (red) are converted into the local tip coordinate frame (green). (**b**) The forward kinematic simulation is run using normal/tangential forces, and we check to make sure that the local frame lines up with the final tip angle

We developed an algorithm, Algorithm 6.4, to perform this search, which iterates until the tip orientation is equal to the initial global-local rotation:

Algorithm 6.4: Iterative calculation of the actuator state when subjected to external forces tied to the global reference frame

```
ForceAngle = 0;
LocalForces = Rotate(GlobalForces,ForceAngle);
TipAngle = LocalForceKinematics(LocalForces, Pressures);
Error = TipAngle-ForceAngle;
while | Error |>0.01 do
    // adjust global/local conversion angle
    ForceAngle = ForceAngle+Error*Alpha;
    // calculate local forces
    LocalForces = Rotate(GlobalForces,ForceAngle);
    // Perform simulation
    TipAngle = LocalForceKinematics(LocalForces, Pressures);
    // Does tip angle match local force angle?
    Error = TipAngle-ForceAngle;
end
```

6.6.3.4 Inverse Kinematics

In performing inverse kinematics, we needed to calculate the pressures required for the tip of a given manipulator to reach a desired position and orientation. We minimized the error between the desired tip position and orientation and the actual tip position and orientation returned by the forward kinematic function. We used MATLAB's constrained minimization toolbox with the active-set algorithm to perform this minimization, taking into account the constraints of the pressure that the actuators can withstand. In addition, to simplify the minimization, we reduced the pressure inputs to the minimum necessary to traverse the entire workspace. For a 1-DoF planar bending actuator with two inputs, we only considered a single pressure input ranging from $-p_{\max}$ to $p_{\max}$. When applying these pressures to a physical system, negative pressures were treated as positive pressures for the other pressure chamber.

One difficulty with iterative methods of calculating inverse kinematics is that the method of calculating the forward kinematics of a manipulator under global forces is, itself, iterative. In particular, many cycles are spent trying to find the exact end-effector position for configurations that will not be used. In order to speed up calculations, we can assume that the final tip state will be the desired state. Thus, we can use the final tip angle to rotate global forces into the local tip frame. This reduces the computational complexity of the IK algorithm, but if the desired tip angle cannot be reached then the IK should be recalculated without the fixed tip forces.

6.6.4 Discretized Model Verification

The behavior of the discretized bending model was verified in a planar setting.

6.6.4.1 Model Force Verification

The first step in model verification is to verify a single segment under external load. This was first done by mounting a single bending segment with a fixed base and applying a normal load to its tip. An example of this can be seen in Fig. 6.47 with the configuration of the bending angle.

We first performed experiments using 0.98 N of applied load. Using the data from these experiments, we calculated Young's modulus to be 75 kPa, less than 10% different than the manufacturer specs for the material used in the bending actuator (around 69 kPa). This difference could be the result of discrepancies in the actuator dimensions, the non-uniformity of the physical bending actuator, or the variation in the properties of the bending actuator, itself. The experimental results for this experiment, along with the calibrated model results, can be seen in Fig. 6.48.

To verify the model, we performed the same experiment using 1.47 N, the results of which can be seen in Fig. 6.49. Again, the shape of the experimental data is

Fig. 6.47 One-segment planar model experimental setup

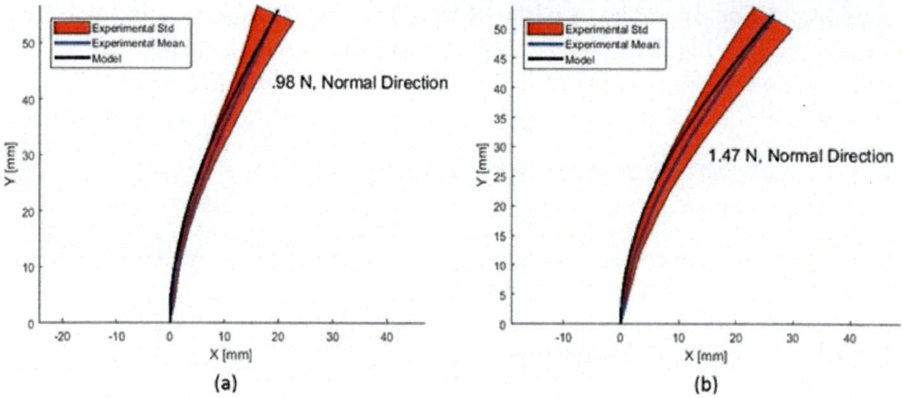

Fig. 6.48 Model verification using a normal force applied to the tip. (**a**) 0.98 N, (**b**) 1.47 N

slightly different than the model. The physical actuators bent more immediately and maintained a shallower bend throughout its length. The model, on the other hand, did not bend as much initially, but caught up to the experimental.

6.6.4.2 Pressure Verification

We performed similar experiments to verify the ability of the model to respond to applied pressures. This experiment was set up similar to that in Fig. 6.47. However, in order to test a wider range of actuator states, we used a motion tracking system to

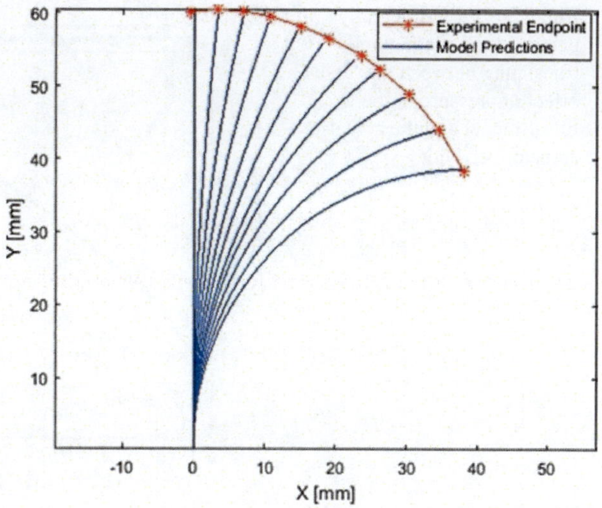

Fig. 6.49 Example of pressure scale calibration. Pressures from 2 to 11 psi

gather data, which only recorded the endpoint of the actuator under pressure. We created a scaling factor to get the model and the experimental results to match. An example of the results of this can be seen in Fig. 6.49.

The correction factors we used were adjusted manually at each pressure tested and can be found in Fig. 6.50. Unfortunately, the necessary correction factors varied widely depending on the pressure applied, meaning that the pressure effected the actuator more the higher it was. This is likely the result of the effect of the pressure on the cross-sectional area. We will discuss this further in Sect. 6.6.5, but in the meantime, we fit a second-order exponential function to these correction factors. This was incorporated into the model to accurately predict the behavior of our actuator using the following equation:

$$P_i = P_{i,\mathrm{r}} S\left(P_{i,\mathrm{r}}\right) \tag{6.80}$$

where P_i is the pressure used in Eq. (6.73); $P_{i,\mathrm{r}}$ is the real pressure, given to the actuator; and $S()$ is the exponential scale factor equation, the data for which is shown in Fig. 6.50. This equation was used independently for pressure chamber *a* and *b*.

6.6.4.3 IK Verification

We performed experiments verifying the inverse kinematic algorithm. In order to create a more complicated workspace, we attached two actuators in series. An example state of this can be seen in Fig. 6.51. To minimize the valving requirements for these experiments, we limited the inputs to a single pressure chamber in each actuator. The interface between the two actuator sections was treated as a subsection with a Young's modulus much higher than the bending segments. In addition, pressure was not applied to this middle segment.

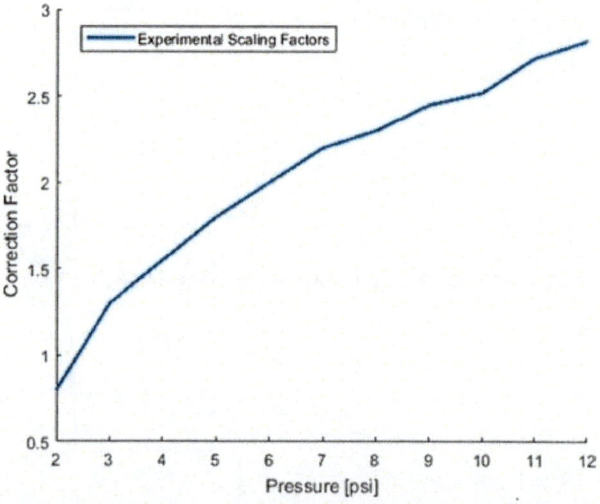

Fig. 6.50 Experimentally determined scale factors translating between the effect of pressure in the linear model and the actual actuator behavior

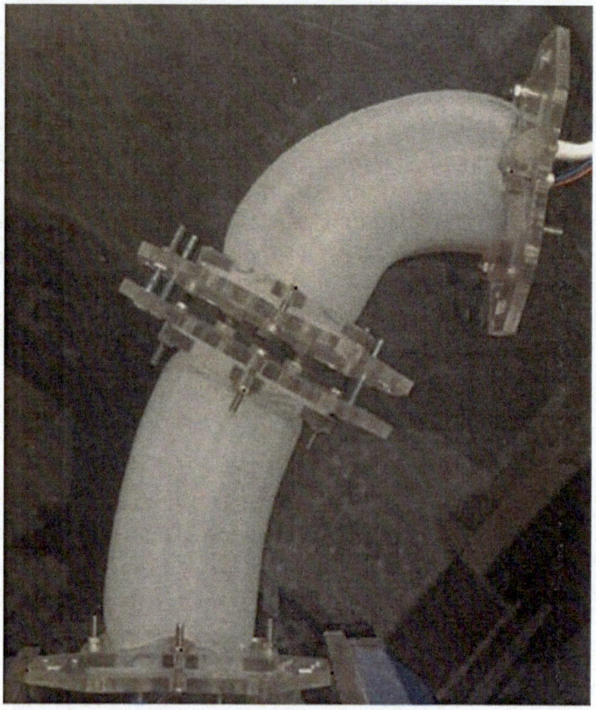

Fig. 6.51 The two DoF inverse kinematic setup. Each actuator is 60 mm long with a 22 mm interface section in between

Our first verification of the IK algorithm was with a series of random points within the workspace. For each point, we used the IK algorithm to calculate the pressures for the actuator to reach that point. Then, we used electronic pressure regulators to apply those pressures to the actuators and used the motion tracking system to track the middle of the interface section as well as the end effector. Results of this experiment can be seen in Fig. 6.52. From this figure, we can see that the IK

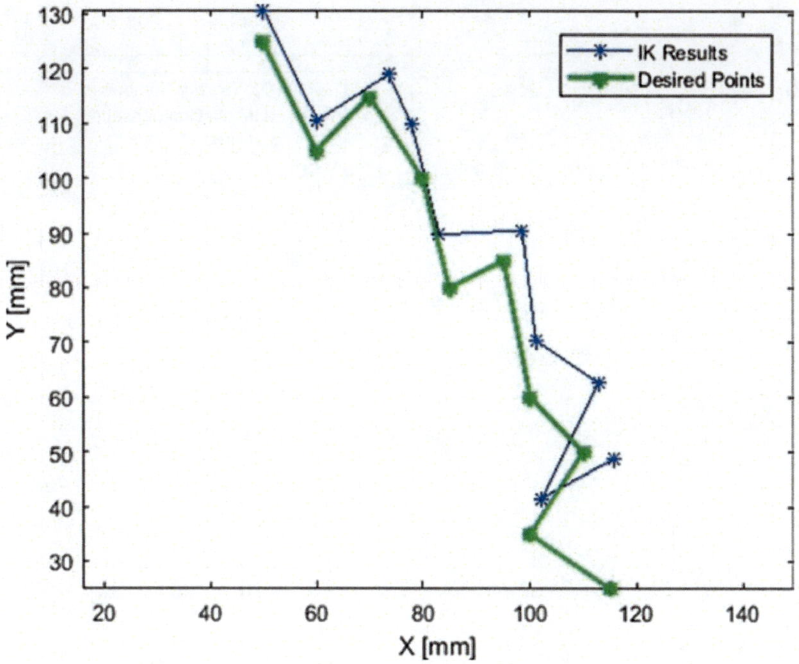

Fig. 6.52 An initial verification of the inverse kinematic model with no external forces and an arbitrary assortment of desired points within the workspace

is fairly accurate, with errors within 10%. The exception is at the very end of the workspace. Here, the actuator begins to buckle under the applied pressure, not quite reaching the desired point. For further tests, we will shrink the size of the workspace when choosing points for the IK.

The next step was to test the IK under load. This would both further verify the IK algorithm and verify the ability of the model to predict actuator behavior under external forces. We first performed this on a similar series of arbitrary points. This series of points was more compact, to ensure that the actuator could still reach them all while under load. For load, a 30 g weight was attached to the end effector and hung off of a pulley. The pulley was situated so that this would cause a constant 0.3 N force vertically with respect to the actuator and was incorporated into the algorithm as such. The IK was performed with and without factoring the weight into (though the weight was included both times), and the results of these experiments can be seen in Fig. 6.53. Maximum pressure was set to be 10 psi to ensure the actuator would not buckle as it had during the previous experiments. From Fig. 6.53a, we can see that both paths show an approximation of the desired points, though the path without gravity compensation is shrunk. Gravity compensation provides a consistently good approximation of the desired point, except for point 4. As travel between point 3 and point 4 requires moving backward and thus letting off pressure, the motion is not actively assisted by the actuator. Both with and without gravity compensation, the end effector did not move backward. The errors can be seen in

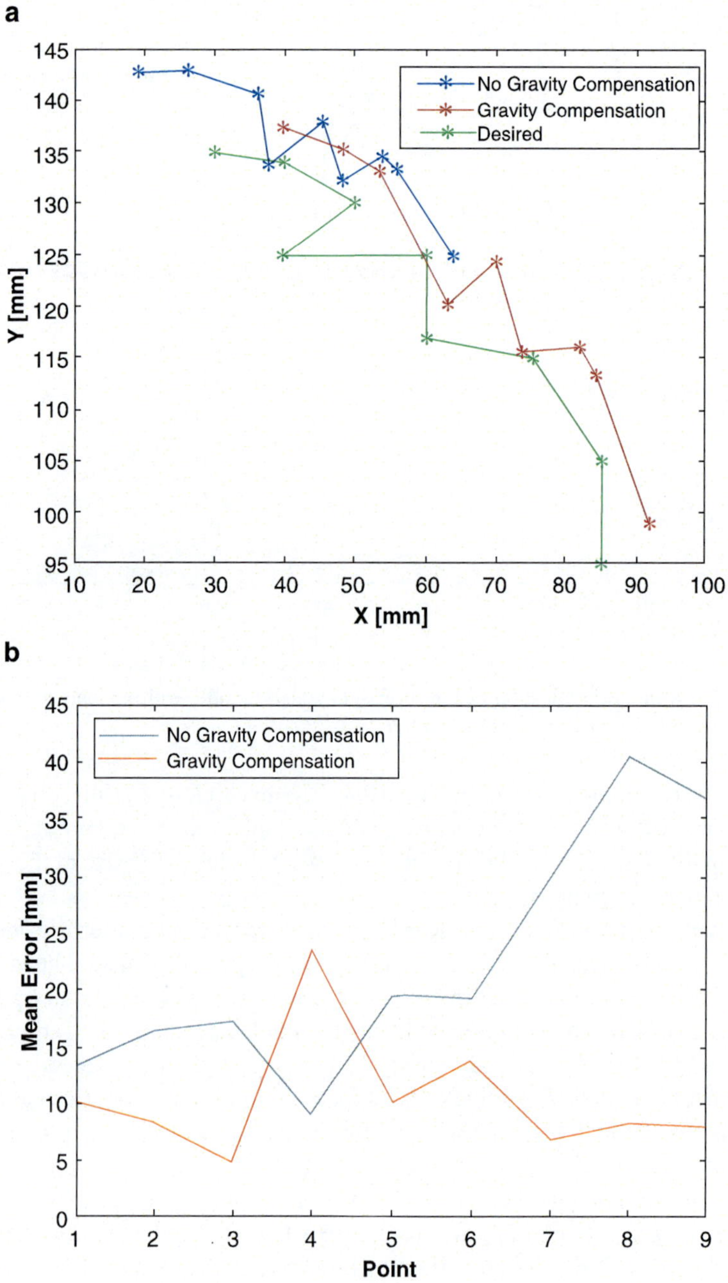

Fig. 6.53 A verification of the inverse kinematic model with a 0.294 N vertical external force and an arbitrary set of desired points within the workspace. Experiments both with and without the force factored into the model were performed. (**a**) Shows the tip trajectories, while (**b**) shows the tip errors

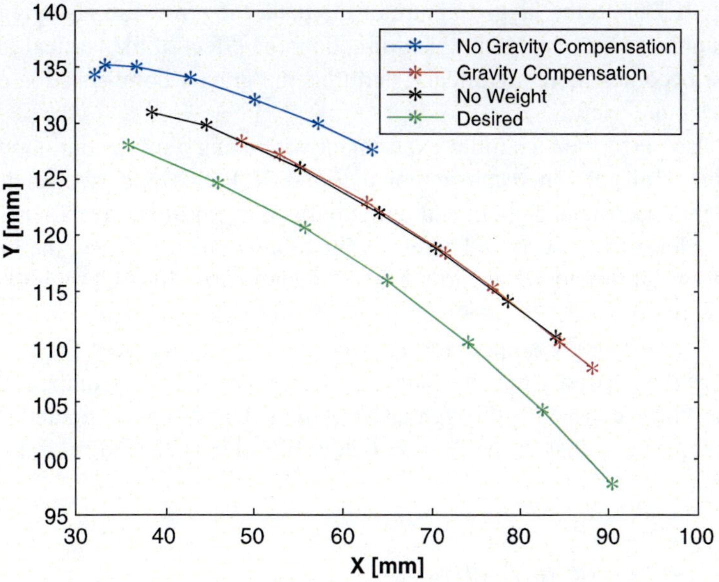

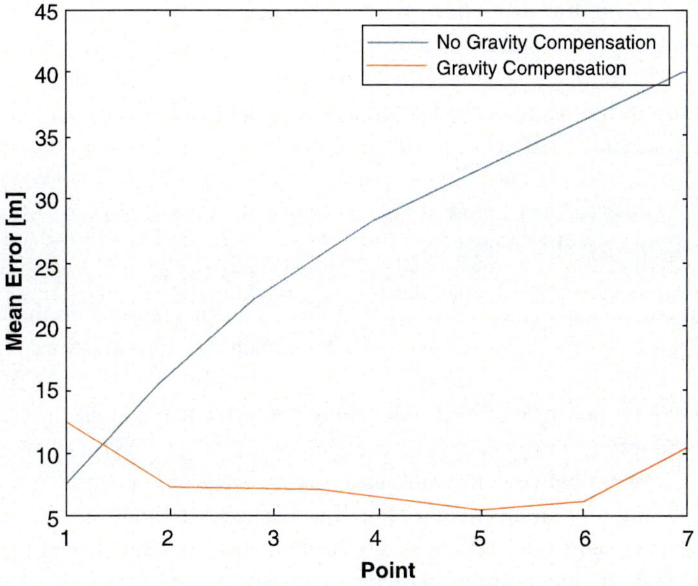

Fig. 6.54 A verification of the effects of gravity compensation in the soft actuator model on the 2-DoF manipulator. The red line shows the manipulator behavior when the model ignores the 0.294 N vertical force, while the blue line shows manipulator behavior with the force taken into account. The black line shows the actuator behavior without the external force. (**a**) Shows the tip trajectories, while (**b**) shows the tip errors

Fig. 6.53b. Due to the jagged nature of the path, the errors are also jagged. As the desired point moves farther in the x-direction, the effect of the vertical force on the actuator becomes more significant, and thus the gravity compensation experiment becomes more useful.

We also performed a similar experiment with using a more continuous series of point, the results of which can be seen in Fig. 6.54. In this case, we also included the no-weight experiment. Again, without gravity compensation, the path is shrunk and situated closer to the y-axis. There was broad agreement between the experiments with the weight factored in and the experiments with no weight, highlighting the accuracy of our model in calculating the effect of external forces.

The "No Gravity Compensation" error increases throughout the seven desired points, as the desired x-position increase, while the "Gravity Compensation" error remains fairly consistent throughout. The total length of the actuator is around 144 mm, meaning that the error stays below 10% during the entire trial.

6.6.5 *Scale Factor Analysis*

We observed a nonlinear relationship between actuator pressure and bending angle. For the control performed, we determined this relationship experimentally and fit an exponential function to it, which we incorporated into the forward kinematics bending actuator. In this section, we looked at two separate causes of the nonlinear relationship: changing cross section and friction.

6.6.5.1 Cross Section Analysis

While the model only takes into account the effect of pressure on the ends of the actuator, the air pushed against all surfaces of each pressure chamber. This distributed load could change the size of the actuator chamber, and thus the force output, much like a modeling balloon, only elongates after it expands radially. We discounted this effect because previous experiments (Fig. 6.45) had shown a near-linear relationship between pressure and force. However, the actuator used for that experiment had a circular cross section and was constructed out of relatively stiff silicone. The actuators used in the discretized bending model verification used much softer silicone and had non-circular cross sections.

We analyzed the changing cross section using resin (Smooth-Cast 305), which has a 7-min cure time. After mixing, we poured the still-liquid resin into the pressure chamber of the bending actuator. We sealed the pressure chamber and pressurized it to the desired pressure. The resin would cure while conforming to the shape of the pressure chamber while pressurized. A series of resin slices from this experiment can be found in Fig. 6.55, along with the accompanying area. We can see that the chamber area increases significantly between 0 and 9 psi. Clearly, changing cross-sectional area plays a non-trivial part in actuator behavior. This increase in

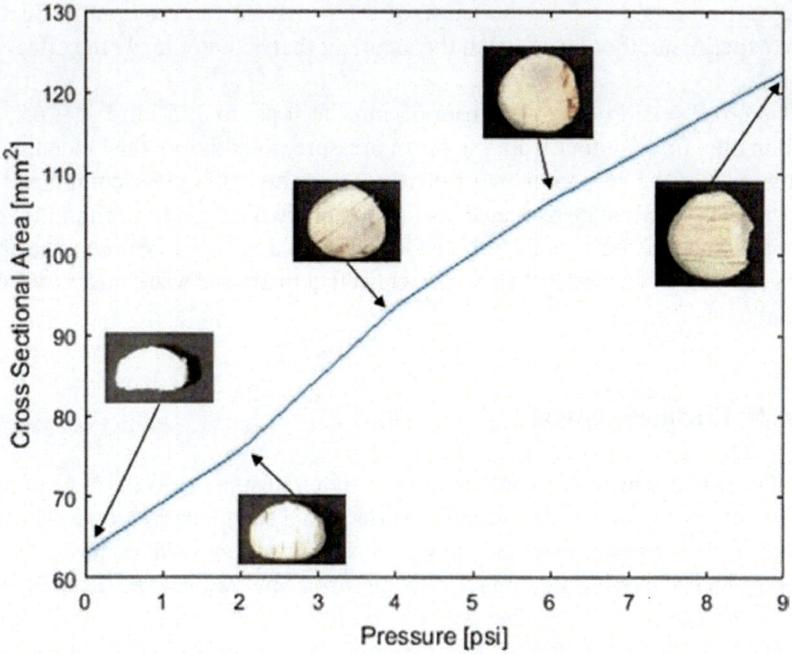

Fig. 6.55 Actuator cross-sectional areas were found by filling the pressure chamber with pressurized resin and extracting the resulting cores after the resin had cured

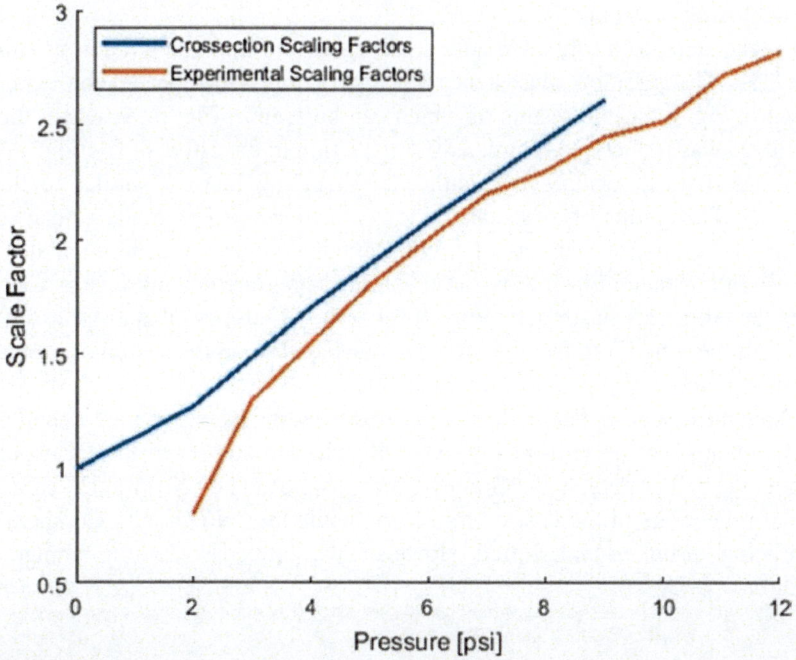

Fig. 6.56 Comparison between the scaling factors determined using the cross-sectional cores and those determined by experiments on actuator behavior

cross section could also be the cause of the increased force output of Sect. 6.6.2 force experiment, though as stated the nature of that actuator made the effects much smaller.

The cross section increases from 65 mm^2 at 0 psi to 125 mm^2 at 9 psi, nearly doubling the force output from the given pressure. In addition, this widening of the cavity reduces the area moment of inertia I_i from Eq. (6.74). We combined both of these effects to calculate expected scaling factors, which can be seen in Fig. 6.56 in comparison with the actual scaling factors from Fig. 6.50. We can see that the changes in cross-sectional area do not entirely explain the scale factors we experimentally determined.

6.6.5.2 Friction Analysis

The planar bending actuator is not strong enough to support itself in a cantilevered configuration. Thus, it must touch the surface in a few places along its length. We had taken steps to minimize the friction caused by the points of contact. The actuator end plates, the points of contact with the surface below, were made out of acrylic, as was the surface, itself. The minimal friction from acrylic-on-acrylic was still non-trivial, as demonstrated by the following experiment. We had the actuator cycle between 0-4-8-4-0 psi and looked at the difference between the actuator position at 4 psi depending on if it was rising or falling, the results of which can be seen in Fig. 6.57. Here, there were 14 mm between the end-effector position depending on if it was rising or falling.

To take steps to fix this, we added ball casters to the manipulator, as shown in Fig. 6.58. The addition of casters reduced friction. We repeated the experiment shown in Fig. 6.57, the results of which can be seen in Fig. 6.59, where the 4 psi difference was reduced to 8 mm, a 40% reduction in the effect of friction.

We used these casters to calculate new scaling factors, similar to those in Fig. 6.50. The results of these calculations are shown in Fig. 6.60, we can see that the casters remove the effects of friction enough to get the experimentally determined scaling factors to follow the theoretical pressure scaling factors calculated using the cross sections much more closely. The differences that crop up are likely the result the remaining friction, noise in our pressure inputs, or minor problems in the measurement of the cross-sectional area.

Using these scaling factors, we performed the same IK experiment as before. The results of this can be seen in Fig. 6.61. Results are similar to the IK experiment without the ball casters, with "Gravity Compensation" and "No Weight" paths being similar reflections of the desired trajectory, while the "No Gravity Compensation" path being shrunken and shifted. However, we noticed something strange in the variation.

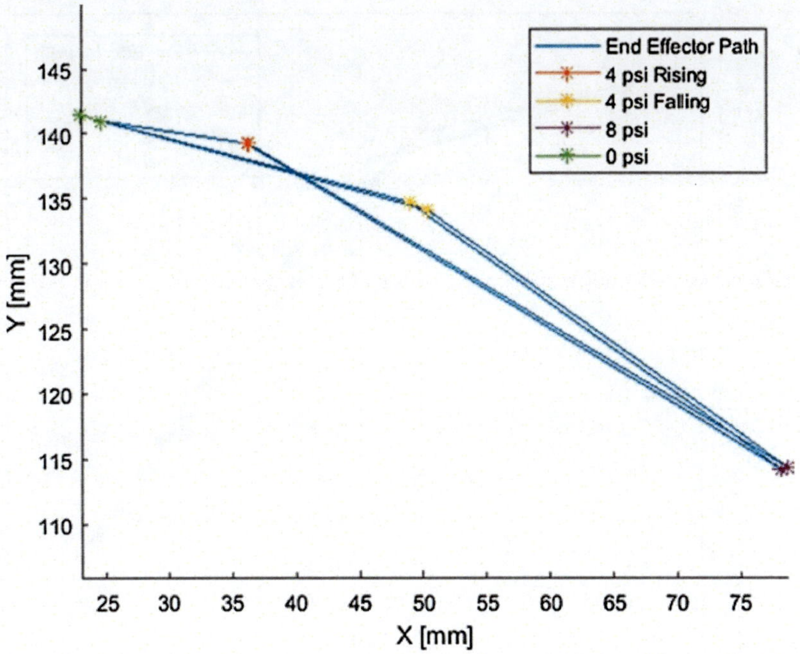

Fig. 6.57 A demonstration of the effects of friction on the two-segment bending actuator. There is on average a 14 mm difference between the end-effector position depending on whether it was rising from 0 psi or falling from 8 psi

Fig. 6.58 We added ball casters to the bottom of the bending actuator to further reduce the effects of friction

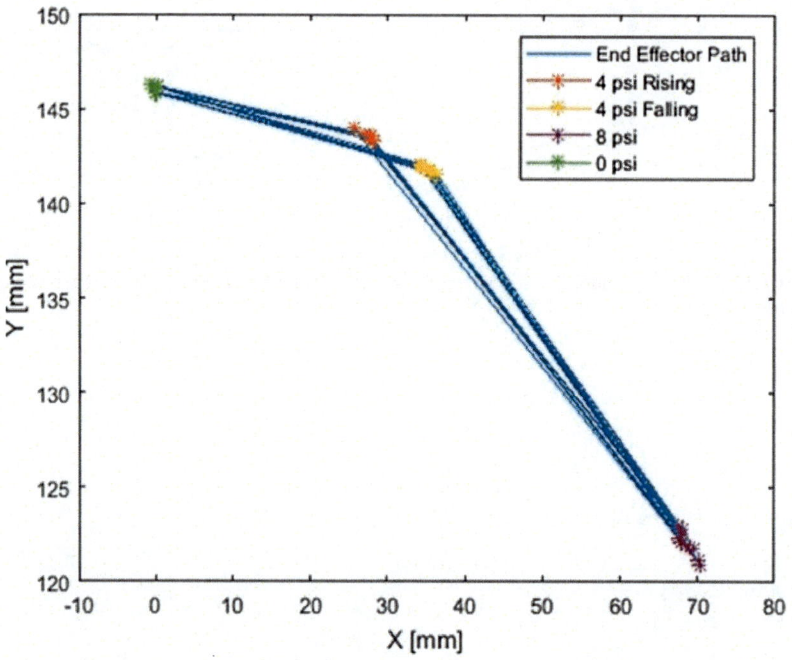

Fig. 6.59 A demonstration of the effects of friction on the two-segment bending actuator with friction-reducing ball casters. There is on average an 8 mm difference between the end-effector position depending on whether it was rising from 0 psi or falling from 8 psi, a 40% reduction from the original configuration

6.6.5.3 Internal Variation

We looked at the variation in the data within this experiment, shown in Fig. 6.62 with standard deviation ellipses. One thing that we noticed was that the variation for the first point was larger than for the others. We observed that this was caused by changes in actuator behavior within an experiment. Each experiment had the actuator do four cycles of traversing the 7 IK-calculated pressure combinations with a reset to (0,0) psi in between each. This set of four cycles was repeated at different time to get the full data set. We observed that the first cycle of each set seemed to have a different behavior than the other three cycles. Separating the points, we found that there was indeed a systemic difference, as shown in Fig. 6.63.

The largest difference can be seen for point 1, where the mean "No Weight" end-effector position changed by about 15 mm depending on if it was during the first cycle of the remaining cycles. The groupings were highly clustered, with small standard deviation ellipses, meaning that this behavior was highly consistent. A similar behavior was seen with the gravity compensation data, though the difference was only around 10 mm and obfuscated by larger standard deviations.

Another interesting aspect of this data was the way that the standard deviations decreased from left to right. This is most pronounced for the "No Weight" experiments, but can be seen a little in the "No Gravity Compensation" experiments. We

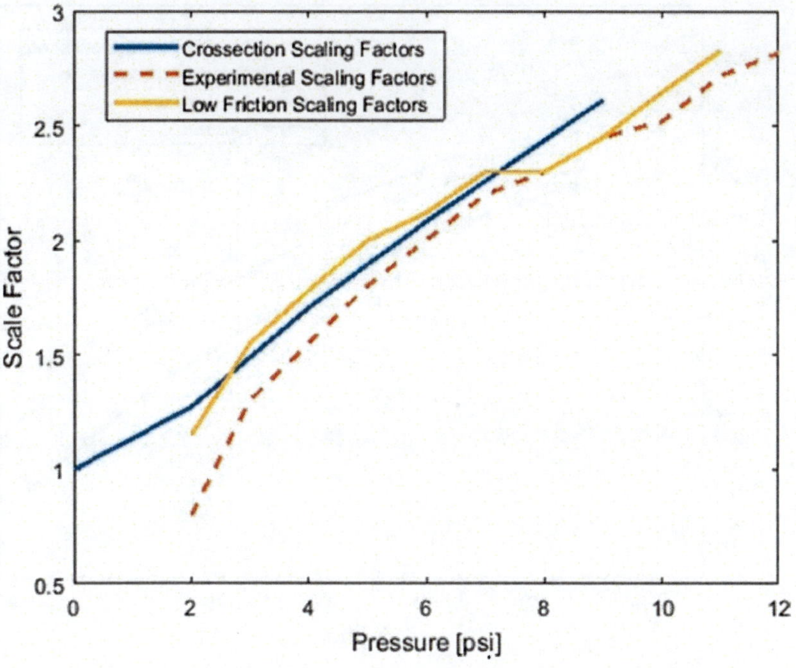

Fig. 6.60 A comparison of the scaling factors that take into account actuator pressure chamber deformation. With the use of low-friction ball casters, there is a general agreement between the scaling factors determined using physical cross sections of the pressure chamber and experimentally determined using actuator behavior

believe that this and the difference in the first cycles have the same cause: the Mullins effect, where rubber materials exhibit different stress-strain curves depending on how much they have been previously loaded. While the pause between cycles was intended to provide a reset for the system, it does not seem to have been enough. Instead, the manipulator is still warming up during the first cycle, resulting in a different behavior than all the others. This is most significant during the first point, but as the cycle continues, the manipulator gets warmed up, and its behavior begins to match the other cycles more closely. This is not seen as much for the experiments with external weight, with and without the gravity compensation in the model, which may have been caused by noise in the direction of the force.

6.6.5.4 Constant Offset

All of these low-friction experiments, regardless of amount of the Mullins effect, overshot the desired trajectory by around 5–7 mm. This was the same behavior seen in the normal-friction environment. One cause of this was the extensibility of the segments. The model assumed that the segments were perfectly inextensible, and while the segments had a plastic sheet running through and anchored to the end plates, there was still some extension that could happen under load. This could be

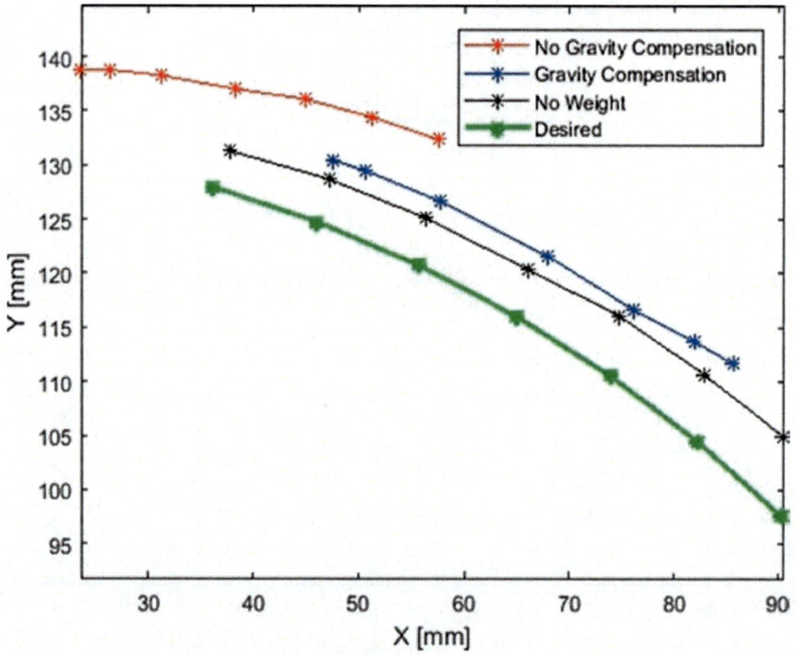

Fig. 6.61 A verification of the effects of gravity compensation in the soft actuator model on the 2-DoF manipulator with ball casters to reduce friction. The red line shows the manipulator behavior when the model ignores the 0.294 N vertical force, while the blue line shows manipulator behavior with the force taken into account. The black line shows the actuator behavior without the external force

taken into account by the model, with an extra DoF added for actuator length, which could be treated as a linear spring with all the chambers pushing against it, with minimums to make sure the actuator doesn't compress too much under external load.

Another cause of this was difficulties in tracking the exact endpoint of the bending segment. The nature of the bending segment meant that the IR tracking beacon could not easily be fixed at the end of the bending segment (where the silicone met acrylic plates), and thus there was a small difference between length of the segment and the length according to the tracking system.

6.6.5.5 Frictionless Analysis

We performed a final experiment with the two-segment planar manipulator with the goal of completely removing the effects of friction. To do this, we hung the actuator upside down, as shown in Fig. 6.64a. In this position, the manipulator would not rest against any surface while actuating. We repeated the friction verification experiment from Fig. 6.59 and found that there was no difference between the 4 psi rising or falling end-effector position. Using this setup, we also performed the IK verification experiment using the smooth set of desired points. Though the hanging

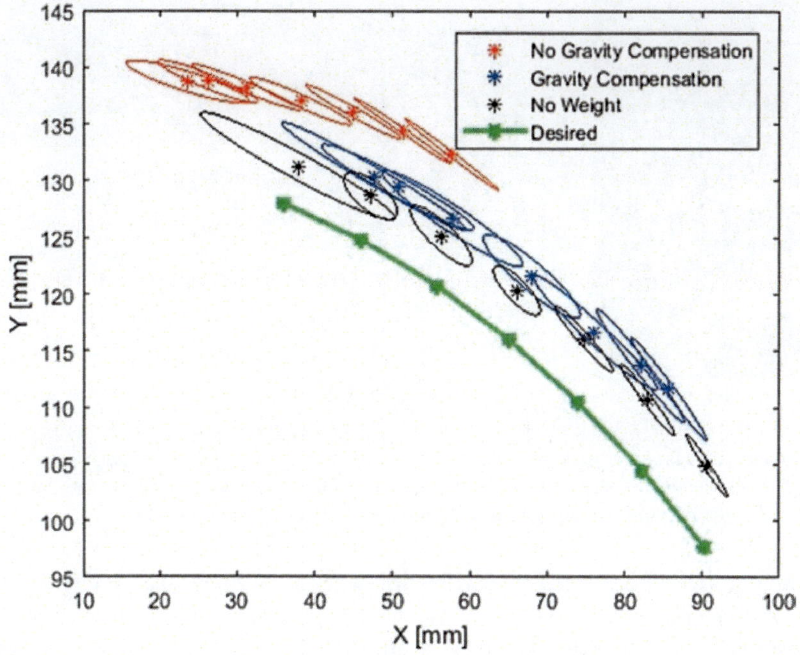

Fig. 6.62 A verification of the effects of gravity compensation in the soft actuator model on the 2-DoF manipulator with ball casters to reduce friction, with standard deviation ellipses included for all the experiments

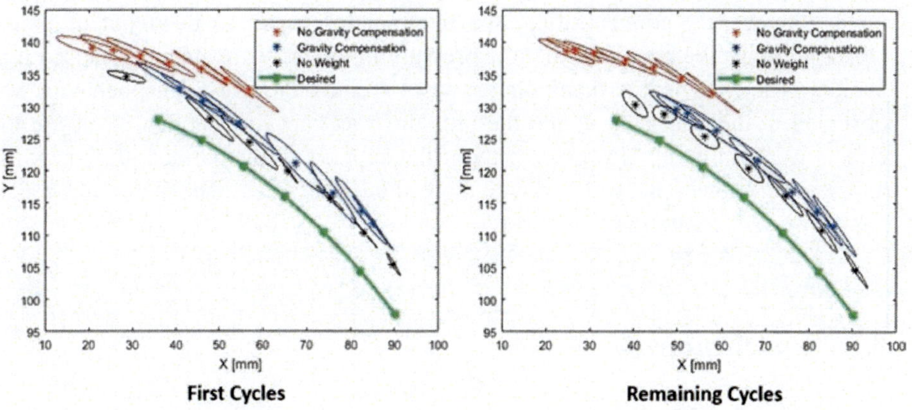

Fig. 6.63 A subset comparison of the data of the previous experiment, looking at the changes in manipulator behavior between cycles in each four-cycle trial

configuration eliminates friction, the force of gravity is now in the plane of motion and directly effects the behavior of the manipulator. To simplify our analysis and because the actuator model was only capable of taking into account a single force at the tip, we modeled gravity as the force from a 30 g mass at the tip of the manipulator, where 30 g was around a quarter of the entire 110 g mass of the 2-DoF

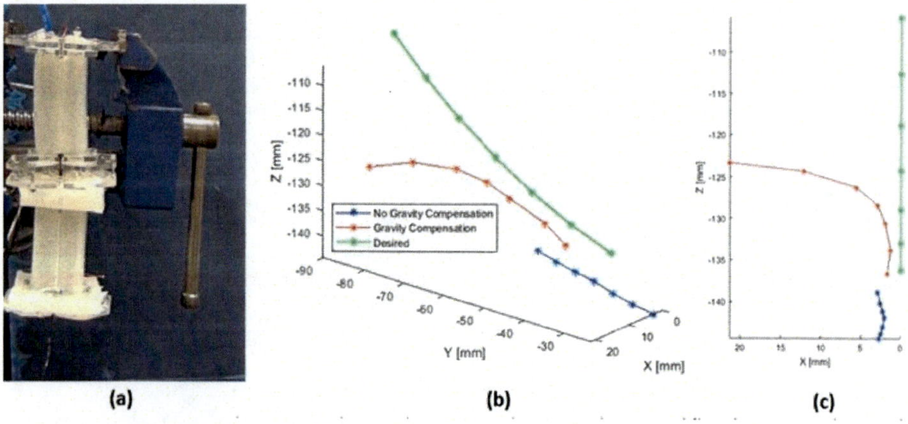

Fig. 6.64 (**a**) Hanging experimental setup. (**b**) Hanging verification experimental data. (**c**) Hanging verification experimental data, second view, where we can see how the actuator bends out of plane as the pressure passes a threshold and the actuator begins to buckle

manipulator. We performed IK verification with and without the gravity compensation. As we could not turn off gravity, we were unable to replicate a "No Weight" experiment. The three-dimensional results of these experiments can be seen in Fig. 6.64b, c.

From the figure, we can see similar results to the planar experiments, with the 30 g approximation doing a surprisingly good job of modeling the effect of gravity on the planar manipulator. However, one difference was that in the hanging experiment with gravity compensation, the manipulator began to bend out of plane. Observing this, we could see that the pressure in the pressure chamber caused it to buckle, making the entire manipulator twist to relieve the stress. This behavior was observed in the planar case, and the pressure was kept low enough to ensure the effects were minimal. However, without the support of the surface, this buckling happened at a lower pressure and with more significant effect. The manipulator as-is has no way of countering this effect.

6.7 Conclusions

This chapter walks through a progression of advances in soft robotics, focusing on using simple models and controllers to aid in the design and controllability of soft pneumatic actuators. We started by discussing the modeling of a soft linear actuator we call the reverse pneumatic artificial muscle (rPAM), which we then used to drive a revolute joint using an iterative sliding mode controller. This controller worked well on its own, but was aided by the use of a feedforward term which predicted the control input required to reach the desired angle. From there, we looked at a different level of control: that of treating the dynamic behavior of the soft actuator as a simple second-order system. We used a model reference adaptive controller to force a series of soft bending actuators with slightly different material properties to behave like the same model and then used that model to develop an optimal control scheme.

Circling back, we expanded upon the revolute joint to control a universal joint using linear rPAMs. The complexity of three dimensions made it difficult to develop a feedforward function, so we used an adapted version of the iterative sliding mode controller to control the end effector of the 2-DoF universal joint and a pair of them as a 4-DoF manipulator. We then used a similar structure as this to provide haptic feedback; only instead of a 2-DoF universal joint, the rPAM actuators were mounted around the user's wrist. Using our rapid response valve system, we were capable of providing haptic feedback to improve user's ability to use their wrist to control a virtual agent performing a line-following task.

From there, we returned to soft bending actuators and used simple constant-curvature beam bending to model the behavior of these infinite passive-DoF structures. We divided each segment into a series of subsegments, each with a constant curvature but a different externally applied moment, which allowed us to use this approach to accurately model bending segment behavior under both normal and tangential external loads. We used this model to perform inverse kinematics in a planar setting, demonstrating its usability in both unloaded and loaded situations. We also investigated the effect of friction and hysteresis on actuator behavior.

This model and the rapid response control methods represent a meaningful step toward the widespread use of robots in everyday human environments. They will be able to move quickly and precisely in complex environments, much as humans do. They will collaborate on tasks without risk, bringing humans and robots closer like never before.

The biggest challenge for the soft actuator discretized bending model moving forward is in dealing with the actuator idiosyncrasies. We incorporated these into the two-segment planar bending experiments, including friction and changes in pressure chamber cross-sectional area. This was a time-consuming process that yielded actuator-specific information. In order to truly generalize this modeling approach, we must be able to analytically take these into account without resorting to physical experiments and function fitting. This would allow the discretized bending model to be applied to all bending actuators, instead of simply as a framework that takes significant effort to adapt to a given system.

References

1. G.K. Klute, J.M. Czerniecki, B. Hannaford, McKibben artificial muscles: Pneumatic actuators with biomechanical intelligence, in *1999 IEEE/ASME International Conference on Advanced Intelligent Mechatronics Proceedings*, (IEEE, 1999), pp. 221–226
2. R.H. Gaylord, Fluid actuated motor system and stroking device. US Patent 2,844,126, 22 July 1958
3. B. Tondu, Modelling of the McKibben artificial muscle: A review. J. Intell. Mater. Syst. Struct. **23**(3), 225–253 (2012)

4. H. Schulte Jr., The characteristics of the McKibben artificial muscle, in *The Application of External Power in Prosthetics and Orthotics*, (National Academy of Sciences – National Research Council, Washington DC, 1961)
5. W. Liu, C. Rahn, Fiber-reinforced membrane models of McKibben actuators. J. Appl. Mech. **70**(6), 853–859 (2003)
6. F. Ilievski, A.D. Mazzeo, R.F. Shepherd, X. Chen, G.M. Whitesides, Soft robotics for chemists. Angew. Chem. **123**(8), 1930–1935 (2011)
7. B. Mosadegh, P. Polygerinos, C. Keplinger, S. Wennstedt, R.F. Shepherd, U. Gupta, J. Shim, K. Bertoldi, C. Walsh, G.M. Whitesides, Pneumatic networks for soft robotics that actuate rapidly. Adv. Funct. Mater. **24**, 04 (2014)
8. J. Bishop-Moser, G. Krishnan, S. Kota, Force and hydraulic displacement amplification of fiber reinforced soft actuators, in *ASME 2013 International Design Engineering Technical Conferences and Computers and Information in Engineering Conference*, (American Society of Mechanical Engineers, 2013), p. V06AT07A031
9. G. Krishnan, J. Bishop-Moser, C. Kim, S. Kota, Evaluating mobility behavior of fluid filled fiber-reinforced elastomeric enclosures, in *ASME 2012 International Design Engineering Technical Conferences and Computers and Information in Engineering Conference*, (American Society of Mechanical Engineers, 2012), pp. 1089–1099
10. J. Bishop-Moser, S. Kota, Towards snake-like soft robots: Design of fluidic fiber-reinforced elastomeric helical manipulators, in *2013 IEEE/RSJ International Conference on Intelligent Robots and Systems (IROS)*, Nov 2013, pp. 5021–5026
11. C.D. Onal, X. Chen, G.M. Whitesides, D. Rus, Soft mobile robots with on-board chemical pressure generation, in *International Symposium on Robotics Research (ISRR)*, 2011
12. C.D. Onal, D. Rus, A modular approach to soft robots, in *2012 4th IEEE RAS & EMBS International Conference on Biomedical Robotics and Biomechatronics (BioRob)*, (IEEE, 2012), pp. 1038–1045
13. C.D. Onal, D. Rus, Autonomous undulatory serpentine locomotion utilizing body dynamics of a fluidic soft robot. Bioinspir. Biomim. **8**(2), 026003 (2013)
14. M. Luo, W. Tao, F. Chen, T. K. Khuu, S. Ozel, C.D. Onal, Design improvements and dynamic characterization on fluidic elastomer actuators for a soft robotic snake, in *Proceedings of the IEEE International Conference on Technologies for Practical Robot Applications*, 2014
15. M. Luo, M. Agheli, C.D. Onal, Theoretical modeling and experimental analysis of a pressure-operated soft robotic snake. Soft Robot. **1**(2), 136–146 (2014)
16. A.D. Marchese, K. Komorowski, C.D. Onal, D. Rus, Design and control of a soft and continuously deformable 2D robotic manipulation system, in *Proceedings of the IEEE International Conference on Robotics and Automation*, 2014
17. R.F. Shepherd, F. Ilievski, W. Choi, S.A. Morin, A.A. Stokes, A.D. Mazzeo, X. Chen, M. Wang, G.M. Whitesides, Multigait soft robot. Proc. Natl. Acad. Sci. **108**(51), 20400–20403 (2011)
18. M.T. Tolley, R.F. Shepherd, B. Mosadegh, K.C. Galloway, M. Wehner, M. Karpelson, R.J. Wood, G.M. Whitesides, A resilient, untethered soft robot. Soft Robot. **1**(3), 213–223 (2014)
19. R.V. Martinez, A.C. Glavan, C. Keplinger, A.I. Oyetibo, G.M. Whitesides, Soft actuators and robots that are resistant to mechanical damage. Adv. Funct. Mater. **24**(20), 3003–3010 (2014)
20. M. Cianchetti, T. Ranzani, G. Gerboni, I. De Falco, C. Laschi, A. Menciassi, Stiff-flop surgical manipulator: Mechanical design and experimental characterization of the single module, in *2013 IEEE/RSJ International Conference on Intelligent Robots and Systems (IROS)*, (IEEE, 2013), pp. 3576–3581
21. J. Bishop-Moser, G. Krishnan, S. Kota, Force and moment generation of fiber-reinforced pneumatic soft actuators, in *2013 IEEE/RSJ International Conference on Intelligent Robots and Systems (IROS)*, (IEEE, 2013), pp. 4460–4465
22. G. Krishnan, Kinematics of a new class of smart actuators for soft robots based on generalized pneumatic artificial muscles, in *2014 IEEE/RSJ International Conference on Intelligent Robots and Systems (IROS 2014)*, (IEEE, 2014), pp. 587–592

23. N. Tsagarakis, D.G. Caldwell, Improved modelling and assessment of pneumatic muscle actuators, in *Proceedings ICRA'00: IEEE International Conference on Robotics and Automation*, vol. 4, (IEEE, 2000), pp. 3641–3646
24. C.S. Kothera, M. Jangid, J. Sirohi, N.M. Wereley, Experimental characterization and static modeling of McKibben actuators. J. Mech. Des. **131**(9), 091010 (2009)
25. B. Kim, S.B. Lee, J. Lee, S. Cho, H. Park, S. Yeom, S.H. Park, A comparison among Neo-Hookean model, Mooney-Rivlin model, and Ogden model for chloroprene rubber. Int. J. Precis. Eng. Manuf. **13**(5), 759–764 (2012)
26. J.C. Case, E.L. White, R.K. Kramer, Soft material characterization for robotic applications. Soft Robot. **2**(2), 80–87 (2015)
27. A.D. Marchese, C.D. Onal, D. Rus, Autonomous soft robotic fish capable of escape maneuvers using fluidic elastomer actuators. Soft Robot. **1**(1), 75–87 (2014)
28. H. Zhao, R. Huang, R.F. Shepherd, Curvature control of soft orthotics via low cost solid-state optics, in *2016 IEEE International Conference on Robotics and Automation (ICRA)*, (IEEE, 2016), pp. 4008–4013
29. E.H. Skorina, M. Luo, S. Ozel, F. Chen, W. Tao, C.D. Onal, Feedforward augmented sliding mode motion control of antagonistic soft pneumatic actuators, in *2015 IEEE International Conference on Robotics and Automation (ICRA)*, (IEEE, 2015), pp. 2544–2549
30. P. Polygerinos, Z. Wang, J.T. Overvelde, K.C. Galloway, R.J. Wood, K. Bertoldi, C.J. Walsh, Modeling of soft fiber-reinforced bending actuators. IEEE Trans. Robot. **31**(3), 778–789 (2015)
31. M. Luo, E.H. Skorina, W. Tao, F. Chen, S. Ozel, Y. Sun, C.D. Onal, Towards modular soft robotics: Proprioceptive curvature sensing and sliding-mode control of soft bidirectional bending modules. Soft Robot. **4**(2), 1–9 (2017)
32. C.-C. Hang, P. Parks, Comparative studies of model reference adaptive control systems. IEEE Trans. Autom. Control **18**(5), 419–428 (1973)
33. J. Schroder, K. Kawamura, T. Gockel, R. Dillmann, Improved control of a humanoid arm driven by pneumatic actuators, in *Proceedings of Humanoids*, 2003
34. Y.-L. Park, J. Santos, K. Galloway, E.C. Goldfield, R.J. Wood, A soft wearable robotic device for active knee motions using flat pneumatic artificial muscles, in *2014 IEEE International Conference on Robotics and Automation (ICRA)*, (IEEE, 2014), pp. 4805–4810
35. F. Renda, M. Cianchetti, M. Giorelli, A. Arienti, C. Laschi, A 3D steady-state model of a tendon-driven continuum soft manipulator inspired by the octopus arm. Bioinspir. Biomim. **7**(2), 025006 (2012)
36. T. Nakamura, Y. Akamatsu, Y. Kusaka, Development of soft manipulator with variable rheological joints and pneumatic sensor for collision with environment. J. Robot. Mechatron. **20**(4), 634 (2008)
37. K. Boku, T. Nakamura, Development of 3-DoF soft manipulator with ER fluid clutches. J. Intell. Mater. Syst. Struct. **21**(15), 1563–1567 (2010)
38. B. Tondu, S. Ippolito, J. Guiochet, A. Daidie, A seven-degrees-of-freedom robot-arm driven by pneumatic artificial muscles for humanoid robots. Int. J. Robot. Res. **24**(4), 257–274 (2005)
39. G. Tonietti, A. Bicchi, Adaptive simultaneous position and stiffness control for a soft robot arm, in *2002 IEEE/RSJ International Conference on Intelligent Robots and Systems*, vol. 2, (IEEE, 2002), pp. 1992–1997
40. S. Eskiizmirliler, N. Forestier, B. Tondu, C. Darlot, A model of the cerebellar pathways applied to the control of a single-joint robot arm actuated by McKibben artificial muscles. Biol. Cybern. **86**(5), 379–394 (2002)
41. P. van der Smagt, F. Groen, K. Schulten, Analysis and control of a rubbertuator arm. Biol. Cybern. **75**(5), 433–440 (1996)
42. T. Amemiya, H. Gomi, *Distinct Pseudo-Attraction Force Sensation by a Thumb-Sized Vibrator that Oscillates Asymmetrically* (Springer Berlin Heidelberg, Berlin, Heidelberg, 2014), pp. 88–95

43. A.V. Dowling, D. Fisher, T. Andriacchi, Gait modification via verbal instruction and an active feedback system to reduce peak knee adduction moment. J. Biomech. Eng. **132**(7), 071007 (2010)
44. L.A. Johnson, C.M. Higgins, A navigation aid for the blind using tactile-visual sensory substitution, in *2006 International Conference of the IEEE Engineering in Medicine and Biology Society*, Aug 2006, pp. 6289–6292
45. M. Chciuk, A. Milecki, P. Bachman, *Comparison of a Traditional Control and a Force Feedback Control of the Robot Arm During Teleoperation* (Springer International Publishing, Cham, 2017), pp. 277–289
46. C. Riecke, J. Artigas, R. Balachandran, R. Bayer, A. Beyer, B. Brunner, H. Buchner, T. Gumpert, R. Gruber, F. Hacker, K. Landzettel, G. Plank, S. Schatzle, H.-J. Sedlmayr, N. Seitz, B.-M. Steinmetz, M. Stelzer, J. Vogel, B. Weber, B. Willberg, A. Albu-Schaffer, Kontur-2 mission: The DLR force feedback joystick for space telemanipulation from the ISS, in *The Proceedings of the International Symposium on Artificial Intelligence, Robotics and Automation in Space (i-SAIRAS 2016)*, 2016
47. C.R. Wagner, N. Stylopoulos, R.D. Howe, The role of force feedback in surgery: Analysis of blunt dissection, in *Proceedings 10th Symposium on Haptic Interfaces for Virtual Environment and Teleoperator Systems. HAPTICS 2002*, 2002, pp. 68–74
48. J.C. Metzger, O. Lambercy, R. Gassert, High-fidelity rendering of virtual objects with the ReHapticKnob-novel avenues in robot-assisted rehabilitation of hand function, in *2012 IEEE Haptics Symposium (HAPTICS)*, March 2012, pp. 51–56
49. D. Margineanu, E.-C. Lovasz, C.M. Gruescu, V. Ciupe, S. Tatar, *5 DoF Haptic Exoskeleton for Space Telerobotics – Shoulder Module* (Springer International Publishing, Cham, 2018), pp. 111–120
50. M.A. Zhou, P. Ben-Tzvi, RML glove: An exoskeleton glove mechanism with haptics feedback. IEEE Trans. Mechatron. **20**(2), 641–652 (2015)
51. J. Blake, H.B. Gurocak, Haptic glove with MR brakes for virtual reality. IEEE/ASME Trans. Mechatron. **14**(5), 606–615 (2009)
52. M. Bouzit, G. Burdea, G. Popescu, R. Boian, The Rutgers Master II-new design force-feedback glove. IEEE/ASME Trans. Mechatron. **7**(2), 256–263 (2002)
53. A. Erwin, M.K. O'Malley, D. Ress, F. Sergi, Kinesthetic feedback during 2DoF wrist movements via a novel MR-compatible robot. IEEE Trans. Neural Syst. Rehabil. Eng. **25**(9), 10127–10134 (2016). [Online]. Available: IEEE Transactions on Neural Systems and Rehabilitation Engineering
54. S.B. Schorr, Z.F. Quek, R.Y. Romano, I. Nisky, W.R. Provancher, A.M. Okamura, Sensory substitution via cutaneous skin stretch feedback, in *2013 IEEE International Conference on Robotics and Automation*, May 2013, pp. 2341–2346
55. C.-P. Chou, B. Hannaford, Measurement and modeling of McKibben pneumatic artificial muscles. IEEE Trans. Rob. Autom. **12**(1), 90–102 (1996). [Online]. Available: http://ieeexplore.ieee.org/abstract/document/481753/
56. S. Jadhav, V. Kannanda, B. Kang, M. Tolly, J.P. Schulze, Soft robotic glove for kinesthetic haptic feedback in virtual reality environments, in *The Engineering Reality of Virtual Reality 2017*, (Society for Imaging Science and Technology, 2015), pp. 19–24. [Online]. Available: http://www.ingentaconnect.com/contentone/ist/ei/2017/00002017/00000003/art00004
57. P.E. Patterson, J. Katz, Design and evaluation of a sensory feedback system that provides grasping pressure in a myoelectric hand. J. Rehabil. Res. Dev. **29**(1), 1–8 (1992)
58. I.M. Koo, K. Jung, J.C. Koo, J.D. Nam, Y.K. Lee, H.R. Choi, Development of soft-actuator-based wearable tactile display. IEEE Trans. Robot. **24**(3), 549–558 (2008)
59. F. Ilievski, A.D. Mazzeo, R.F. Shepherd, X. Chen, G.M. Whitesides, Soft robotics for chemists. Angew. Chem. Int. Ed. 50(8), 1890–1895 (2011). [Online]. Available: https://doi.org/10.1002/anie.201006464
60. P. Polygerinos, Z. Wang, K.C. Galloway, R.J. Wood, C.J. Walsh, Soft robotic glove for combined assistance and at home rehabilitation. Rob. Auton. Syst. **73**, 135–143 (2015)

Wearable Robotics. [Online]. Available: http://www.sciencedirect.com/science/article/pii/S0921889014001729
61. D. Sasaki, T. Noritsugu, M. Takaiwa, Development of active support splint driven by pneumatic soft actuator (ASSIST), in *Proceedings of the 2005 IEEE International Conference on Robotics and Automation (ICRA)*, 2005
62. H. Al-Fahaam, S. Davis, S. Nefti-Meziani, Wrist rehabilitation exoskeleton robot based on pneumatic soft actuators, in *2016 International Conference for Students on Applied Engineering (ICSAE)*, (IEEE, 2016)
63. H. Al-Fahaam, S. Davis, S. Nefti-Meziani, The design and mathematical modelling of novel extensor bending pneumatic artificial muscles (EBPAMs) for soft exoskeletons. Rob. Auton. Syst. **99**, 63–74 (2018). [Online]. Available: http://www.sciencedirect.com/science/article/pii/S0921889017302592
64. A.K. Mishra, A. Mondini, E. Del Dottore, A. Sadeghi, F. Tramacere, B. Mazzolai, Modular continuum manipulator: Analysis and characterization of its basic module. Biomimetics **3**(1), 3 (2018). [Online]. Available: http://www.mdpi.com/2313-7673/3/1/3
65. F. Renda, M. Cianchetti, M. Giorelli, A. Arienti, C. Laschi, A 3D steady-state model of a tendon-driven continuum soft manipulator inspired by the octopus arm. Bioinspir. Biomim. **27**(2), 334–345 (2012)
66. M. Mahvash, P.E. Dupont, Stiffness control of surgical continuum manipulators. IEEE Trans. Robot. **27**(2), 334–345 (2011)
67. B.A. Jones, I.D. Walker, Kinematics for multisection continuum robots. IEEE Trans. Robot. **22**(1), 43–55 (2006)
68. A.A. Alqumsan, S. Khoo, M. Norton, Robust control of continuum robots using Cosserat rod theory. Mech. Mach. Theory **131**, 48–61 (2019). [Online]. Available: http://www.sciencedirect.com/science/article/pii/S0094114X18311777
69. W. McMahan, V. Chitrakaran, M. Csencsits, D. Dawson, I.D. Walker, B.A. Jones, M. Pritts, D. Dienno, M. Grissom, C.D. Rahn, Field trials and testing of the OctArm continuum manipulator, in *Proceedings 2006 IEEE International Conference on Robotics and Automation, 2006. ICRA 2006*, May 2006, pp. 2336–2341
70. D. Trivedi, A. Lotfi, C.D. Rahn, Geometrically exact models for soft robotic manipulators. IEEE Trans. Robot. **24**(4), 773–780 (Aug 2008)
71. D. Braganza, D.M. Dawson, I.D. Walker, N. Nath, A neural network controller for continuum robots. IEEE Trans. Robot. **23**(6), 1270–1277 (2007)
72. R. Kang, D.T. Branson, T. Zheng, E. Guglielmino, D.G. Caldwell, Design, modeling and control of a pneumatically actuated manipulator inspired by biological continuum structures. Bioinspir. Biomim. **8**(3), 1270–1277 (2013)
73. J. Frás, J. Czarnowski, M. Maciás, J. Główka, Static modeling of multi-section soft continuum manipulator for stiff-flop project, in *Recent Advances in Automation, Robotics and Measuring Techniques*, ed. by R. Szewczyk, C. Zieliński, M. Kaliczyńska, (Springer International Publishing, Cham, 2014), pp. 365–375
74. T. Mahl, A. Hildebrandt, O. Sawodny, A variable curvature continuum kinematics for kinematic control of the bionic handling assistant. IEEE Trans. Robot. **30**(4), 935–949 (2014)
75. C.-P. Chou, B. Hannaford, Measurement and modeling of McKibben pneumatic artificial muscles. IEEE Trans. Robot. Autom. **12**(1), 90–102 (1996)
76. C. Vergari, P. Pourcelot, L. Holden, B. Ravary-Plumiöen, G. Gerard, P. Laugier, D. Mitton, N. Crevier-Denoix, True stress and Poisson's ratio of tendons during loading. J. Biomech. **44**, 719–724 (2011)
77. R. Ogden, Large deformation isotropic elasticity-on the correlation of theory and experiment for incompressible rubberlike solids. Proc. R. Soc. Lond. A Math. Phys. Sci. **326**(1567), 565–584 (1972)
78. L. Crocker, B. Duncan, J. Urquhart, R. Hughes, A. Olusanya, The application of rubber material models to analyse flexible adhesive joints. Proc. Adhes. **99** (2007)

79. P. Martins, R. Natal Jorge, A. Ferreira, A comparative study of several material models for prediction of hyperelastic properties: Application to silicone-rubber and soft tissues. Strain **42**(3), 135–147 (2006)
80. S. Youssefian, N. Rahbar, E. Torres-Jara, Contact behavior of soft spherical tactile sensors. IEEE Sens. J. **14**(5), 1435–1442 (2013)
81. American Society for Testing and Materials, *Standard Test Methods for Vulcanized Rubber and Thermoplastic Elastomers-Tension. ASTM International Standards, D412-06* (ASTM International, West Conshohocken, 2006)
82. American Society for Testing and Materials, *Standard Test Methods for Rubber Properties in Compression. ASTM Standard, D575-91* (ASTM International, West Conshohocken, 2012)
83. D. Brown, Tracker video analysis and modeling tool @ONLINE (2015). [Online]. Available: http://www.cabrillo.edu/dbrown/tracker/
84. M. Luo, E. Skorina, W. Tao, F. Chen, C.D. Onal, Optimized design of a rigid kinematic module for antagonistic soft actuation, in *IEEE Conference on Technologies for Practical Robot Applications*, 2015
85. V. Utkin, Variable structure systems with sliding modes. IEEE Trans. Autom. Control **22**(2), 212–222 (1977)
86. E. Skorina, W. Tao, F. Chen, M. Luo, C.D. Onal, Motion control of a soft-actuated modular manipulator, in *IEEE International Conference on Robotics and Automation (ICRA)*, 2016
87. S. Ozel, N. Keskin, D. Khea, C. Onal, A precise embedded curvature sensor module for soft-bodied robots. Sensors Actuators A Phys. **236**, 349–356 (2015)
88. E. Skorina, M. Luo, W. Oo, W. Tao, F. Chen, S. Youssefian, N. Rahbar, C. Onal, Reverse pneumatic artificial muscles (rPAMs): Modeling, integration, and control. *PLoS One* **13**(10), e0204637 (2018)
89. M. Luo, E. Skorina, W. Tao, F. Chen, C.D. Onal, Optimized design of a rigid kinematic module for antagonistic soft actuation, in *2015 IEEE International Conference on Technologies for Practical Robot Applications (TePRA)*, (IEEE, 2015), pp. 1–6
90. W. Tao, E.H. Skorina, F. Chen, J. McInnis, M. Luo, C.D. Onal, Bioinspired design and fabrication principles of reliable fluidic soft actuation modules, in *2015 IEEE International Conference on Robotics and Biomimetics (ROBIO)*, (IEEE, 2015), pp. 2169–2174
91. J.L. Morse, M.-C. Jung, G.R. Bashford, M.S. Hallbeck, Maximal dynamic grip force and wrist torque: The effects of gender, exertion direction, angular velocity, and wrist angle. Appl. Ergon. **37**(6), 737–742 (2006). [Online]. Available: http://www.sciencedirect.com/science/article/pii/S0003687005001675
92. M. Luo, E.H. Skorina, W. Tao, F. Chen, S. Ozel, Y. Sun, C.D. Onal, Toward modular soft robotics: Proprioceptive curvature sensing and sliding-mode control of soft bidirectional bending modules. Soft Robot. **4**(2), 117–125 (2017). [Online]. Available: http://online.liebertpub.com/doi/pdf/10.1089/soro.2016.0041
93. D. Rus, M.T. Tolley, Design, fabrication and control of soft robots. Nature **521**, 467475 (2015)

Erik Howard Skorina is a Postdoctoral Associate in the Soft Robotics Lab at Worcester Polytechnic Institute. His research in robotics focuses on improving the understanding and performance of soft robots operating in real-world environments. Approaches include improving the speed and precision of soft robotic actuators, better understanding the deformation of soft actuators under load, and improving soft actuator stiffness without compromising their compliance. He received his PhD from the Robotics Department of Worcester Polytechnic Institute in 2018 and dual Bachelor's Degrees in Mechanical Engineering and Mathematics from Union College in 2012.

Prof. Cagdas D. Onal is the Dean's Associate Professor and Arsenault Faculty Fellow in Robotics Engineering and Mechanical Engineering at Worcester Polytechnic Institute. He received his PhD in Mechanical Engineering from Carnegie Mellon University in 2009. Before joining WPI, he was a Postdoctoral Associate in the Computer Science and Artificial Intelligence Laboratory at MIT. His research interests include soft robotics, printable robotics, alternative actuation/sensing mechanisms, bio-inspiration, control theory, and wearable robotics. He is the director of WPI Soft Robotics Lab and leads the NSF Future of Robots in the Workplace-Research and Development (FORW-RD) NRT Program at WPI. Prof. Onal is the recipient of an NSF CAREER award on origami-inspired soft robotic systems.

Chapter 7
Active Thermal Control of Satellites with Electroactive Materials

Frédéric Vidal, Gwendoline Petroffe, Layla Beouch, Sophie Cantin, Claude Chevrot, Pierre-Henri Aubert, and Jean-Paul Dudon

Abstract In orbit, satellites are subjected to strong temperature variations depending on the environment they encounter (hot or cold phases of the mission). Thus, one of the challenges in thermal control of artificial satellites is to develop new coatings, with variable emissivity, light, inexpensive, and requiring little or no electricity consumption. In this context, several research groups have developed electroemissive devices based on organic or inorganic materials with tunable infrared (IR) optical properties. This chapter is also devoted to a bibliographical study addressing the concept of emissivity on the one hand and describing the state of the art of electroemissive devices (EED) on the other hand. The most advanced EED fully described in the literature is reported as well as the EED developed by CY Cergy Paris Université and Thales Alenia Space. The latter was evaluated in thermal-vacuum conditions close to the space environment for a proof-of-concept-level testing, and the thermal regulation performances were reported.

Keywords Electroactive materials · Electroactive polymers · EAP · Electroemissive devices · EED · Electrochemically active species · EAS · Satellite · Thermal control · Thermal regulation

F. Vidal (✉) · G. Petroffe · L. Beouch · S. Cantin · C. Chevrot · P.-H. Aubert
Laboratoire de Physicochimie des Polymères et des Interfaces, Institut des Matériaux, CY Cergy Paris Université, Neuville-sur-Oise, France
e-mail: frederic.vidal@cyu.fr

J.-P. Dudon
Thales Alenia Space (TAS), Cannes, France

© Springer Nature Switzerland AG 2022

L. Rasmussen (ed.), *Smart Materials*,
https://doi.org/10.1007/978-3-030-70514-5_7

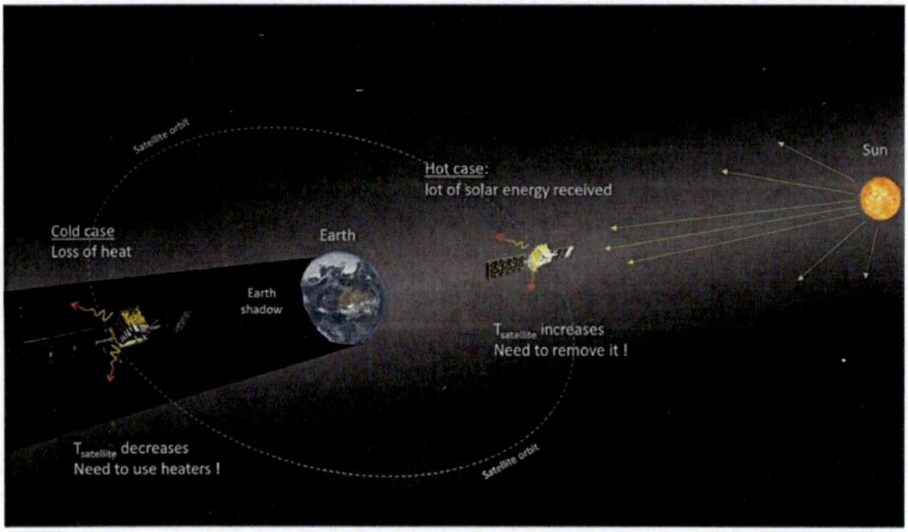

7.1 Introduction

In orbit, a satellite can suffer from drastic temperature changes, depending on sun (referred to as "hot environment") or shadow (referred to as "cold environment") exposures. This can damage the satellite components. In order to ensure the lifetime and optimal operation of the various elements of the satellite, the temperature inside the satellite needs to be controlled. Currently, the temperature regulation is achieved using (i) active means (heaters, fluid loop, cryo-cooler) and (ii) passive means (heat pipes, thermal coatings, paints, optical solar reflector, multilayer insulation). Unlike the passive coatings, which do not require any external stimuli to operate, active coatings operate from electrical power supplies. Specifically, passive coatings have constant properties in the infrared (IR) and in the visible spectrum. Generally, they are chosen to be adapted first to the "hot environment," so they directly influence the heat rejection from the satellite to the space. On the other hand, active means are used when the satellite is facing a "cold environment." This active thermal control, mainly composed of Joule effect heaters, requires electrical power in order to increase the temperature within the satellite. The combination of these passive coatings and active heaters allows generally the elaboration of a thermal control system adapted to the satellite.

Active means like heaters, however, have major drawbacks: a high energy consumption and the associated impact on the on-board weight (through the need of solar panels and batteries in cold phases). Therefore, research of new systems to control the heat flow, into and out of the spacecraft during the extreme hot and cold phases of the mission, are being much further investigated. As a result, coatings with variable emissivity have been increasingly studied such as IR-thermochromic [1–3] or IR-electrochromic [4–6] materials since the 1990s.

The emissivity corresponds to the radiative flux of thermal radiation emitted by a surface element at a given temperature, related to the reference value which is the flux emitted by a black body at this same temperature (see Sect. 7.2). IR-thermochromic materials maintain the same emissivity over a temperature range. There is a temperature, referred to as critical temperature (T_c), for which the emissivity of these materials changes. Thus, IR-thermochromic materials enable to display two levels of emissivity, one for a temperature below T_c and the other for a temperature higher than T_c, with a certain range around this value depending on the IR-thermochromic material considered. For an interesting application to satellite thermal control, the aim is thus to get a low emissivity below T_c and a high emissivity above. The most widely used IR-thermochromic material is based on vanadium dioxide (VO_2).

On the other hand, IR-electrochromic or IR-electroemissive materials, when they operate in the infrared, can modulate their IR emissivity under the application of a bias voltage. Thus, the emissivity of an electroemissive device (EED) can be tuned according to the environment encountered by the satellite and its own thermal dissipation related to the operating of its system (payload and platform). When the satellite is in a hot phase, facing the sun and in "full drive" mode, a high emissivity would lead to the rejection of the heat toward space in order to keep the temperature below +90 °C. On the other hand, when the satellite is in a cold phase, in the shadow of the Earth and with "low electrical operating" mode, a low emissivity would allow the conservation of the heat. By this direct control of the coating emissivity, these materials would significantly reduce the energy consumption dedicated to maintaining the temperature in a good operational range and the on-board weight (solar panel, batteries). Therefore, IR-electroemissive devices (EEDs) are suitable candidates for the thermal control of satellites and are the subject of this chapter.

Only one of the EED components, referred to as the active layer, is able to modulate its emissivity. The active layer may be inorganic (such as tungsten oxide or nickel oxide) or organic (such as electronically conducting polymers, ECP). The emissivity variation of the active layer results from a modification of its IR properties by applying a bias voltage. For this, EEDs are similar to electrochemical cells with two electrodes separated by an electrolyte. The latter is necessary to ensure the ionic conduction within the electrochemical cell. One of the electrodes, on which the active layer is coated, is the working electrode, while the second electrode acts as a counter-electrode (Fig. 7.1).

The chapter is devoted to a bibliographic study addressing the concept of emissivity and the relationship between the reflectivity and the emissivity of a body. A description of the architecture of electroemissive devices and the state of the art of devices are exposed according to the nature of the EED active layer. The latest developed devices reported in the literature are detailed, and a description of the electroemissive devices developed in our laboratory is presented.

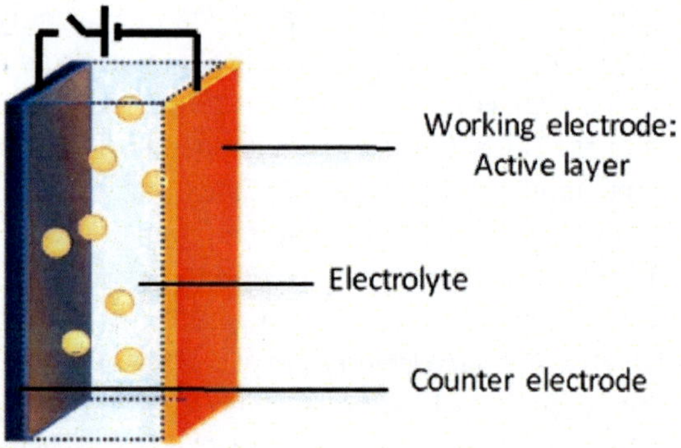

Fig. 7.1 Schematic drawing of the common design of EED: electrochemical cell, including a working electrode, a counter-electrode, and an electrolyte

7.2 Concepts Related to the Emissivity of a Body

7.2.1 Emissivity Concept

The black body is an ideal object, which absorbs the entire electromagnetic energy. At thermal equilibrium, the energy conservation implies that the capacity of the black body to emit radiation is equal to its capacity to absorb the same radiation. The emitted radiation is referred to as "black-body radiation" whose spectrum, described by Planck's law, depends only on its temperature. The luminance L is defined as the elementary radiative flux dϕ per unit of area dS' normally to the propagation direction and per unit of solid angle dΩ. The spectral luminance L_λ, expressed in $Wm^{-2}sr^{-1}\mu m^{-1}$, represents the luminance per wavelength unit. These definitions are shown in Fig. 7.2. The emission spectra of the black body at different temperatures are represented by the spectral luminance in relation to the wavelength (Fig. 7.3).

The spectral luminance spectra exhibit a maximum at a wavelength λ_{max}, which only depends on the temperature according to Wien's law (Eq. 7.1).

$$\lambda_{max} \times T = 2.9 \times 10^{-3}\ \text{K.m} \tag{7.1}$$

The spectral luminance maximum shifts to lower wavelengths when the temperature increases. For temperatures close to ambient temperature, the black body emits in the spectral region of the middle infrared. For example, a black body at 300 K (27 °C) has a maximum spectral luminance for wavelengths ranging from 5 to 11 μm (Fig. 7.3, red curve). To obtain the maximum emission in the visible, very high temperatures must be reached, for example, 3000 K. Nevertheless, even at such temperatures, the emission remains important in the infrared since it is more important than that of a black body at 300 K. From the definition of the spectral luminance, the emittance M is defined as the radiated power per unit of emitting surface into the half-space. As for the luminance, the spectral emittance corresponds to the emittance per wavelength unit. These definitions are illustrated in Fig. 7.4.

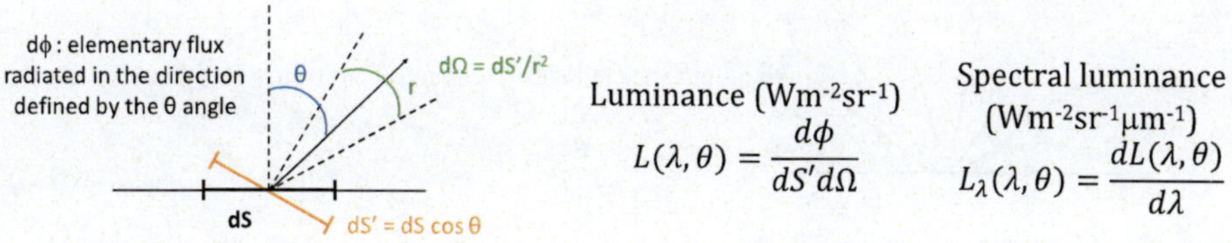

Fig. 7.2 Definitions of the luminance and the spectral luminance

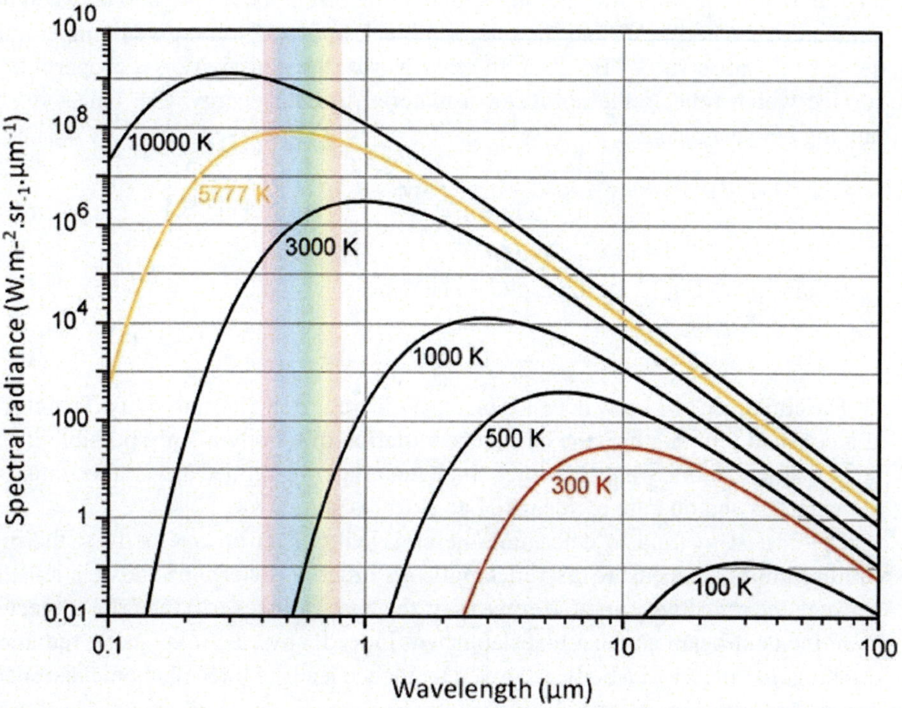

Fig. 7.3 Black body spectral luminance spectra for temperatures ranging from 100 to 10,000 K [7]

According to Planck's law, the emission of the black body is isotropic. In this case, the spectral luminance and the spectral emittance are proportionally linked according to Lambert's law (Eq. 7.2):

$$M_\lambda = \pi L_\lambda \tag{7.2}$$

Moreover, Stefan-Boltzmann's law establishes that the total emittance of the black body, M^0, is expressed by Eq. (7.3):

$$M^0 = \sigma T^4 \tag{7.3}$$

where σ is Stefan-Boltzmann's constant equal to 5.67×10^{-8} W.m^{-2}K^{-4}.

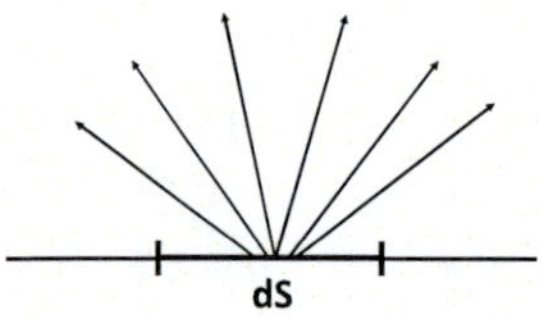

$$M = \frac{d\phi}{dS} \qquad M_\lambda = \frac{dM}{d\lambda}$$

Fig. 7.4 Definitions of the emittance and the spectral emittance

"Real bodies," however, do not behave like the black body. They do not follow Planck's law and do not absorb all the electromagnetic radiation. They are characterized by an emittance lower than the one of the black body. The ratio between the emittance of a body (M) and the one of a black body (M^0) at a given temperature defines the emissivity ε (Eq. 7.4). In other words, the emissivity is a property of a surface which reflects the ability of a material to emit energy. The emissivity is ranging between 0 and 1.

$$\varepsilon = \frac{M}{M^0} \tag{7.4}$$

According to Eq. (7.3), the Eq. (7.4) becomes:

$$M = \varepsilon\sigma T^4 \tag{7.5}$$

The emittance of a body depends on the temperature and its emissivity. Therefore, at a constant temperature, the emittance variation of a body is only possible if the body is able to modify its emissivity. Such materials are defined as variable emissivity materials and operate by means of an external stimulus.

The easiest method to determine the emissivity of an opaque body at thermal equilibrium is to measure its reflectivity. In order to determine the relationship between the reflectivity and the emissivity, the energy fluxes of a material subjected to an incident beam ϕ_i must be established. Indeed, any electromagnetic radiation reaching the surface of a body at a given incidence leads to three phenomena: reflection ϕ_r, transmission ϕ_t, and absorption ϕ_a (Fig. 7.5).

The energy balance is written according to Eq. (7.6):

$$\phi_i = \phi_r + \phi_a + \phi_t \tag{7.6}$$

By introducing the factors of reflection R_λ, absorption α_λ (or monochromatic absorptivity), and transmission T_λ, respectively, defined as the ratios of the reflected, absorbed, and transmitted beam with respect to the incident beam, the Eq. (7.7) can be established [8]:

$$R_\lambda + \alpha_\lambda + T_\lambda = 1 \tag{7.7}$$

For a body in thermodynamic equilibrium with its environment, Kirchhoff's law implies that the energy absorbed by the body is re-emitted in a radiative form, i.e., $\alpha_\lambda = \varepsilon_\lambda$.

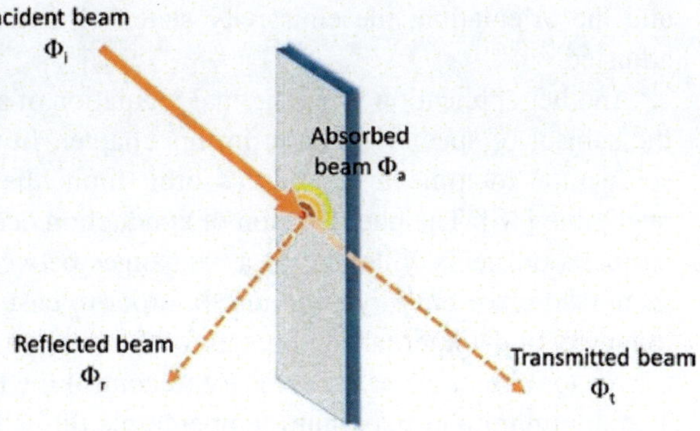

Fig. 7.5 Illustration of an absorbed ϕ_a, reflected ϕ_r, and transmitted ϕ_t beam of an incident ϕi beam on a body

For an opaque body, namely, a body whose received radiation is not transmitted ($T_\lambda = 0$), the emissivity is equal to (Eq. 7.8):

$$\varepsilon_\lambda = \alpha_\lambda = 1 - R_\lambda \tag{7.8}$$

One can see that a method to determine the emissivity of an opaque body at a given temperature is to measure its reflectivity in an extended IR-wavelength range. Therefore, a highly absorbing body is also highly emissive. Conversely, a reflecting body absorbs only a small part of the energy of the incident radiation and has therefore a low emissivity. Thus, for an opaque body, modulating the emissivity amounts to the modulation of the reflectivity.

7.2.2 Applications to the Thermal Regulation

Thermal exchanges are carried out by conduction, convection, and/or radiation. For thermal regulation, the IR emissivity control is related to radiative phenomena (IR radiation). Thermal regulation has an interest in energy saving, which is currently one of society's major concerns. From this point of view, the control of thermal fluxes received by a system is very important [9, 10]. In buildings, transparent or opaque materials with variable emissivity, such as "smart windows," can control the heat fluxes between the indoors and outdoors [11–14]. In order to limit the excessive use of heaters and air conditioners to regulate the inside temperature, variable emissivity coatings may save energy. Indeed, in winter, windows and building's facades exposed to the sun could be in a high IR-absorbing (IR-emissive) state so the IR radiations of the sun would warm up the building. In contrast, the windows and building's facades in the shade could be in a low emissivity state to prevent heat loss. Conversely, in the summer, surfaces exposed to the sun could be in a reflective (low emissivity) mode, limiting the penetration of heat from IR radiations into the building. The shaded surfaces could be in a high emissive state to evacuate the heat flux from the inside of the building to the outside. Thus, depending on the season and the orientation, the emissivity states of different building facades could be adjusted.

Another application is the thermal regulation of artificial satellites, which will be the subject of special attention in this chapter. In the case of a satellite (or other spacecraft) rotating in the Earth's orbit limit, the phenomenon of convection is negligible [15]. The phenomenon of conduction occurring only between the different parts of the satellite, the heat exchanges between the satellite and its environment take place only by radiation. In order to ensure maximum performances and longevity of the internal components of the satellite (such as instruments, electronic boards, or structure), it is necessary to control the temperature within the satellite. The minimum and maximum temperatures differ for the various elements inside and outside the satellite. Usually the temperature inside the satellite must be maintained in the range of −30 °C to +90 °C and in the range of −150 °C to +150 °C for the outside of the satellite.

During the rotations around the Earth, the artificial satellite sides are exposed to various thermal environments and heat loads, leading to two "extreme cases." The latter are denoted as the "hot" and the "cold" environments (Fig. 7.6) [16].

The "hot" case occurs when the satellite panels receive their maximal dissipation from the payload units ("full drive" mode) but also when the satellite is facing the sun (the outside temperature on certain parts can reach ≈150 °C) (Fig. 7.6). The amount of thermal energy is then so high that the on-board electronics may be damaged by an excessive temperature within the satellite. Therefore, the heat dissipated by the internal subsystems and the absorbed solar flux must be evacuated as much as possible so that the temperature within the satellite should not exceed a

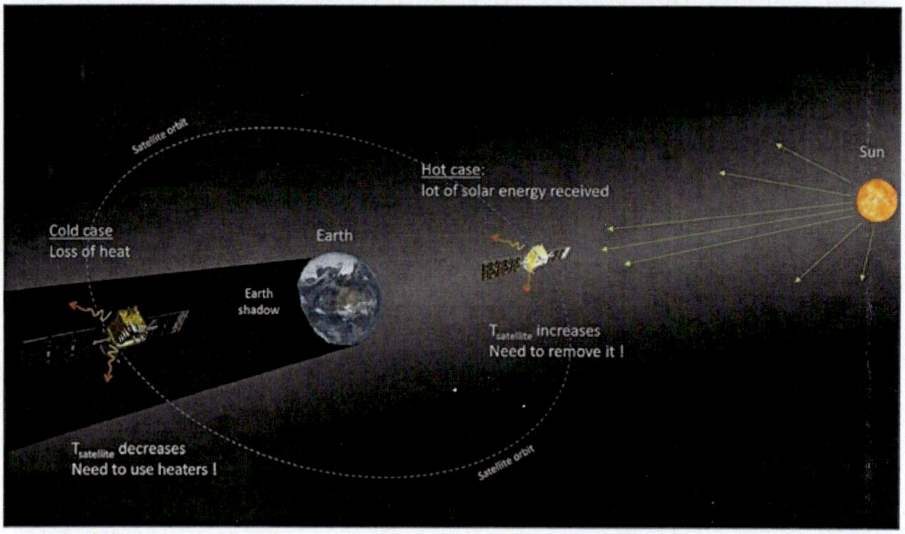

Fig. 7.6 Representation of the environments to which a satellite is exposed during the rotations around the Earth. In front of the sun, the environment is "hot." In the shadow of the Earth, the environment is "cold" [16]

maximum temperature of +90 °C. In this case, some faces of the satellite are insulated, while others are covered with a dedicated external thermal radiative coating in order to ensure the required heat rejection capacity. This radiative coating is highly emissive to reject the heat toward space. Furthermore, its solar absorptivity (α_s), in the visible and near infrared, is very low in order to avoid overheating.

Conversely, the "cold" case is the situation when the satellite is in the shadow of the Earth (Fig. 7.6), during the equinox time for a geostationary satellite and when the heat dissipated by the internal components is minimal. Since the emissivity of the external thermal radiative coating is high as it has been set for the "hot" case, the satellite continues to reject heat leading to a sharp decrease of the internal temperature. An active thermal control, such as heaters or fluid loop, is then used to increase the temperature within the satellite in the range of −30 °C to +90 °C. These active thermal control systems have the disadvantage of requiring a significant electrical power provided by batteries and solar panels. The latter have an on-board weight, which is not negligible, and entail high costs.

To summarize, the regulation of the thermal exchanges, within the satellite and between the satellite and the space, is currently carried out in two ways:

- By a passive thermal control (heat pipes, thermal coatings, paints, optical solar reflectors, multilayer insulation)
- By an active thermal control requiring an energy supply such as heaters, fluid loop, and cryo-cooler

For the passive thermal control, the equilibrium temperature of coatings can be evaluated by assimilating the artificial satellite to a sphere of radius R. By only considering the solar flux, the thermal energy received by the satellite on the area equivalent to a disk is equal to the energy losses over the entire surface of the satellite. The equilibrium temperature of the coatings is then given by Eq. (7.9):

$$\alpha_s \phi_s \pi R^2 = \varepsilon \sigma T_{\mathrm{eq}}^4 \times 4\pi R^2 \tag{7.9}$$

where α_s is the visible solar absorptivity, σ Stefan-Boltzmann's constant, and ϕ_s the solar constant equal to 1360 W/m^2.

Therefore, the equilibrium temperature of the coating can be expressed by Eq. (7.10):

$$T_{\mathrm{eq}} = \left(\frac{\alpha_s \phi_s}{4\varepsilon\sigma} \right)^{1/4} \tag{7.10}$$

The thermal control coatings are listed in Table 7.1 according to their thermo-optical parameters, especially the ratio α_s/ε. The latter determines, at equilibrium, the surface temperature of the material which can be obtained [17].

According to Table 7.1, it can be noted that a "cold" coating ($\alpha_s/\varepsilon < 1$), which exhibits a high IR emissivity, reaches an equilibrium temperature of −115 °C. This coating is therefore adapted to a "hot" environment. On the other hand, an "extremely

Table 7.1 Equilibrium temperatures and thermo-optical properties of the different coatings used on satellites [17]

Coatings	Solar absorptivity (α_s)	IR emissivity (ε)	Equilibrium temperature
"Cold" ($\alpha_s/\varepsilon < 1$) Ex: White paints, optical solar reflectors	0.2	0.8	−115 °C
"Hot" ($\alpha_s/\varepsilon = 1$) Ex: Black paints	0.90	0.90	7 °C
"Extremely hot" ($\alpha_s/\varepsilon > 1$) Ex: Gold	0.25	0.04	170 °C

hot" coating ($\alpha_S/\varepsilon > 1$), exhibiting a low IR emissivity, reaches an equilibrium temperature of 170 °C. Thus, this coating is suitable for a "cold" environment.

Therefore, variable emissivity devices are suitable candidates to actively control the heat flux. It is possible to imagine that the coating could be in a high emissivity state in a "hot" environment. Then the coating could switch to a low emissivity state when the satellite faces away from the sun. It would allow either to reject the heat toward space or to conserve the heat, limiting the drop of the internal temperature of the satellite. The visible solar absorptivity (α_s) should be low to pass the hot extreme case. Obviously, in the cold extreme case, the visible solar absorptivity does not impact the temperature since there is no sun illumination.

Such an active radiative material would thus provide a considerable electrical power saving and an advantage in terms of associated on-board weight as a significant part of the heaters could be removed [18]. Currently, electroemissive and IR-thermochromic materials are increasingly developed for such application.

IR-thermochromic materials are part of the innovative technologies of interest for the thermal control of satellites because their emissivity is variable depending on the temperature. Indeed, these materials have a critical temperature (T_c) for which their emissivity changes. The most known IR-thermochromic materials are based on $NdNiO_3$ and VO_2. For example, films based on VO_2 have a high IR emissivity at high temperatures, a low IR emissivity at low temperatures, and a critical temperature around 68 °C. However, due to the high critical temperature, they seem unsuitable for this type of application. To overcome this problem, some authors inserted dopants into VO_2 films in order to decrease T_c [19]. Although this method is effective, it is often done to the detriment of the efficiency of the IR-thermochromic effect. For thermal regulation applications, manganese perovskites are increasingly being used. In fact, various studies showed that compositions such as $R_xA_{1-x}MnO_3$ – where R is a rare-earth (e.g., La, Pr, Nd, Sm) and A is an alkaline earth (e.g., Ba, Sr, Ca) – have a critical temperature around room temperature [20]. These materials have been the subject of numerous investigations over the past years [21–23].

On the other hand, electroemissive materials have the ability to modify their IR emissivity through redox process. By applying a bias voltage to the material, its emissivity can be tuned. This chapter focuses on this type of technology. Besides, some of these systems are sufficiently advanced to be studied in real conditions. An overview of the state of the art of variable emissivity devices reported in the literature is proposed in the following section.

7.3 Electroemissive Devices: From Principle to State of the Art

7.3.1 Common Design of Electroemissive Devices

Electroemissive devices (EEDs) are systems whose IR-optical properties (emissivity/reflectivity) of the active layer can be modified by applying a bias voltage. These devices operate on the same principle as an electrochemical cell and consist of two electrodes immersed into an electrolyte (Fig. 7.7). The first electrode, acting as a working electrode, provides the optical properties to the EED. This electrode constitutes the active layer, while the second one acts as a counter-electrode. By applying a potential difference (ΔV) between the electrodes, a change of the optical properties of the active layer occurs, especially in the infrared range. These electrodes are separated by an electrolyte, which ensures an ionic conduction in the electrochemical cell. Also, ion transport is necessary to balance the charges created at the electrodes upon commutation.

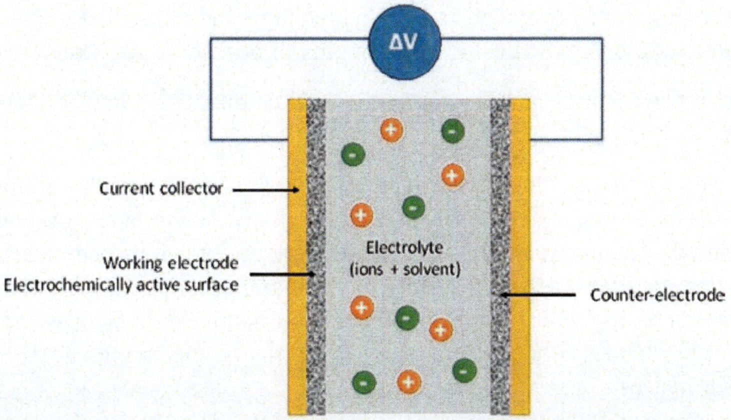

Fig. 7.7 Schematic drawing of an electroemissive device (EED) architecture

7.3.2 *Performance Indicators and Characteristic Parameters*

Before presenting the EEDs referenced in the literature, it is useful to define the various parameters, which will allow the evaluation of the performances of such devices.

Reflectivity Variation

In the case of an opaque active surface (zero transmission, $T_\lambda = 0$), the emissivity and the reflectivity are related according to the equation: $\varepsilon_\lambda = 1 - R_\lambda$. Reflectivity measurements are simpler to implement using a spectrophotometer operating in reflection and equipped with an integrating sphere compared to emissivity measurements that require specific equipment. Therefore, the optical properties of the EEDs are often characterized by their reflectivity. The reflectivity variation ($\%\Delta R$) is defined as the difference between the maximum value ($\%R_{max}$) and the minimum value ($\%R_{min}$) of the IR-optical reflectivity responses obtained from the device during a switch. The greater the difference between these two states, the more the device/material is effective. Generally, it is difficult to compare the reflectivity variations from one device to another since the spectral domains usually differ according to the intended application. For infrared stealth, the authors focus on the IR-bands II (3–5 μm) and III (8–12 μm), while for a thermal regulation application in the space domain, the entire infrared spectrum is often considered (2.5–45 μm).

Memory Effect

The memory effect of a device represents the time during which the device can remain in one optical state (reflecting or absorbing state) without applying a bias voltage.

The memory effect provides information on the energy required to keep the emissivity/reflectivity constant. With a low memory effect, a voltage should be applied to keep constant the optical state of the EED. As a result, the energy needs will be important. Conversely, an EED with a high memory effect will need only a small source of energy to ensure its autonomy of operation over time.

Two other parameters also allow the evaluation of the performances of the devices:

- *The switching time*: it is the time required to switch from one limit optical state to the other. It can vary widely depending on the active layer (organic or inorganic), its geometry, and its size. In the literature, it is reported that the switching time is often measured at 95% of the maximum optical variation [24].
- *Cyclability or lifetime*: it is the number of consecutive state changes that a device can withstand without significant loss of its electrochemical or optical performances.

7.3.3 *Overview of State-of-the-Art Electroemissive Devices*

The works of P. Chandrasekhar et al. [25] and C.G. Granqvist et al. [26] provide a complete presentation of devices operating in transmission in the IR range. However, in the purpose of the thermal regulation of satellites, the overview is focused on devices able to modulate their IR reflectivity (or IR emissivity).

The electroemissive devices operating in reflection can be classified into two families depending on the design of the active layer.

In the first family, the modulation of the active layer reflectivity is obtained from a coating of an electrochemically active species (EAS) on a reflective surface (e.g., gold). The active layer is thus similar to a dual material, i.e., EAS + gold. The thin electroactive layer, deposited on a reflective surface, switches from an IR-transparent to an IR-absorbing state through a redox reaction. When the EAS is IR-transparent, the active layer has a high reflectivity due to the IR reflection of the gold layer. In contrast, when the EAS is IR-absorbing, the active layer displays a low reflectivity since the IR reflection of the gold layer is prevented due to the strong IR absorption of the EAS.

In the second family, the electroactive layer is thick, i.e., whatever the redox state, the IR transmission is null. In this configuration, the intrinsic properties of this layer allow the modulation of the reflectivity through a redox process. As no gold layer is needed anymore to provide the IR modulation, the electrochemical active species is not only the electroactive layer but also the whole active layer. Hence, the electroemissive device (EED) works in full reflection/absorption mode. Therefore, the reflectivity can be tuned depending on the electrical potential of the electrochemically active species. Thus, according to the potential applied, the EAS is either in a high IR-reflectivity state (low IR-emissivity state) or in a low IR-reflectivity state (high IR-emissivity state).

7.3.3.1 Active Layers as Dual Material Including EAS and Reflective Surfaces

The active layers of the EEDs can be obtained from electronically conducting polymers (ECPs) (Fig. 7.8). Their p-orbital overlapping creates a system of delocalized π-electrons, which can result in interesting and useful electronic properties. The electronic conjugation results in the merging of energy levels, leading to a band structure similar to the one of inorganic semiconductors. This band structure explains the singular behavior of ECPs: semiconductors in the neutral state, they

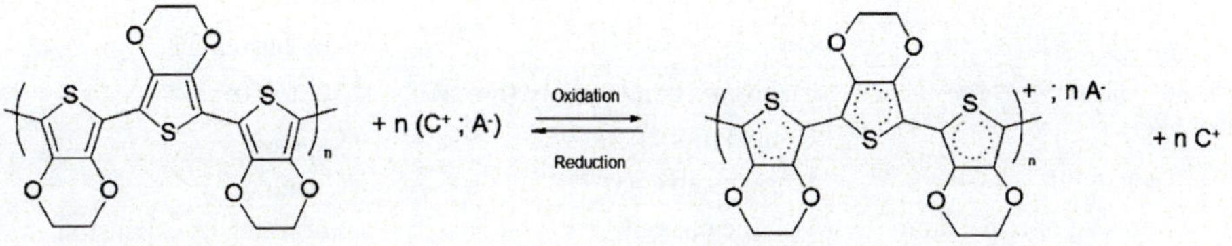

Fig. 7.8 Redox reactions of an ECP: poly(3,4-ethylenedioxythiophene) (PEDOT)

become electronic conductors after doping. The doping of the ECPs corresponds to an oxidation (p-doping) or reduction (n-doping) reaction of repeat units. The modification of the redox state can be carried out both chemically or electrochemically and generally corresponds to a reversible process. It implies that the electrical charges (positive or negative) created on the chain are compensated by the incorporation of ions to ensure the electro-neutrality of the polymer (Fig. 7.8). Therefore, it is essential to have a reserve of ions to balance electrical charges resulting from the redox reaction.

During doping/dedoping, through a redox process, the structure of the electronically conductive polymers is modified, resulting in modifications of their properties, e.g., volume change or optical modulation (Table 7.2).

According to its oxidation degree, a thin ECP layer exhibits a different IR-optical behavior, from a IR-transparent state to an IR-absorbing state. Specifically, in the neutral state, thin ECP layers are semiconducting and exhibit an IR-transparent character. Conversely, in the oxidized state, thin ECP layers exhibit a conducting character and are IR-absorbing.

Chandrasekhar et al. [27] reported in 1995 the first dynamic variation of IR reflectivity of an electronically conducting polymer (ECP). As mentioned above, the device has a multilayer architecture similar to the conventional schema of an electrochemical cell as shown in Fig. 7.9a. The electrolyte combines poly(vinyl alcohol), poly(ethylene glycol), a glycerol derivative, and an inorganic Li salt. The active layer and the counter-electrode have the same composition, namely, a microporous membrane, a gold layer, and an ECP. The latter is a copolymer based on two electronically conducting polymers, the polyaniline (PANI) and the poly(diphenylamine) (PDPA) [28–33]. It is worth mentioning that the polyaniline has a singular behavior. In fact, according to its oxidation degree, the PANI presents three distinct forms. In the neutral and fully oxidized states, the PANI is semiconducting and thus IR-transparent, whereas in a partially oxidized state, the polyaniline exhibits a conducting character and is IR-absorbing. Thus, by applying a voltage of −1.1 V (neutral state) or +0.8 V (fully oxidized state) to the active layer, the copolymer is IR-transparent, and the gold layer reflects the IR incident radiation. So, the device is highly reflective (low emissive). On the other hand, by applying a bias voltage of 0 V (partially oxidized state), the copolymer is IR-absorbing, preventing the incident radiation from being reflected by the gold layer. In this configuration, the device has a low reflectivity (high emissivity) as observed in Fig. 7.9b. For such devices, the authors reported in 2002 reflectivity variations of 0.3–0.4 in the 2.5–45 μm range [29].

Table 7.2 Modifications of the ECP properties depending on its redox state

Neutral ECP		Oxidized ECP
10^{-15} to 10^{-5}	Electronic conductivity (S/cm)	10^{3} to 10^{-1}
Color 1	Optical properties	Color 2
Discharged	Storage capacity	Charged
Shrinking or expansion	Volume change	Shrinking or expansion

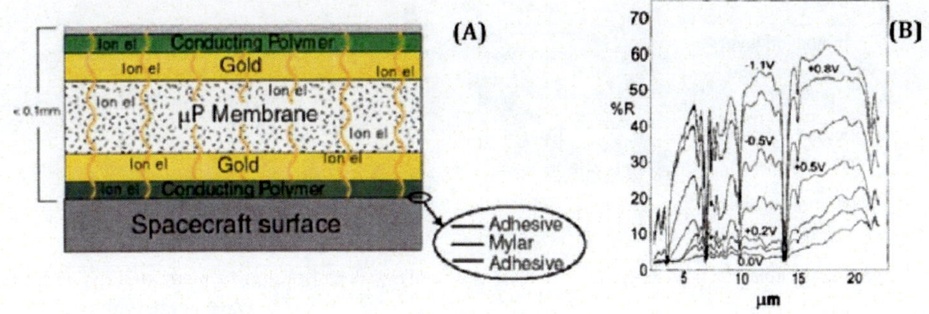

Fig. 7.9 (**a**) Descriptive schema of the architecture of the electroemissive device based on electronically conducting polymer (ECP) proposed by P. Chandrasekhar [33]. (**b**) Reflection spectra on the IR-wavelength range of 1.1–25 μm of the EEDs of P. Chandrasekhar according to the bias voltage applied [29]

Following the work of Chandrasekhar et al., the design of other electroemissive devices based on polyaniline was reported in the literature [34, 35]. For example, Li et al. [35] described an electroemissive device with a similar architecture (Fig. 7.10a) in which the active layer is also composed of a microporous membrane, a gold layer, and an ECP. However, for the second half of the device, the gold layer and the ECP layer are inverted with respect to the device designed by Chandrasekhar [33]. This inversion of the layers was made in order to allow a good diffusion of the electrolyte between the two ECP layers. The electrolyte used is also different since it is a mixture of poly(vinyl alcohol), sulfuric acid, and deionized water in 12/6/82% weight ratio. By applying a bias voltage of −0.8 V, the active layer is in a high reflective state (low emissivity) due to the IR-transparent properties of the PANI in the neutral state. Conversely, by applying a bias voltage of 0 V, the ECP is in an IR-absorbing state leading to a device with low reflective properties. These observations are consistent with the singular behavior of the PANI described previously. However, by applying a bias voltage of +0.8 V, the optical properties of the device observed by the authors are quite different from those reported by Chandrasekhar et al. Indeed, at +0.8 V, the ECP remains in an IR-absorbing state resulting in low reflective properties. The authors suggested that the PANI is still in a partially oxidized state at +0.8 V and that the fully oxidized state of the PANI was not reached. With this system, the authors obtained an average reflectivity variation of 0.24 in the infrared region of 8–12 μm (Fig. 7.10b).

This type of architecture inspired many groups to develop EEDs differing mostly by the nature of the ECP used but also by the nature of the electrolyte incorporated. These ECPs are mainly the poly(diphenylamine) (PDPA) [36], the poly(3,4-ethylenedioxythiophene) (PEDOT) [36, 37], the polyaniline (PANI) [35], and the poly(3,4-propylenedioxythiophenedimethyl) (PProDOT-Me_2) [37–40]. Nevertheless, most of these studies were focused on a small part of the infrared spectrum, close to the near infrared. For instance, Reynolds' group has investigated the IR emittance of the PProDOT-Me_2 as active layer over the near infrared up to 5 μm [40]. They obtained a reflectivity variation around 0.50.

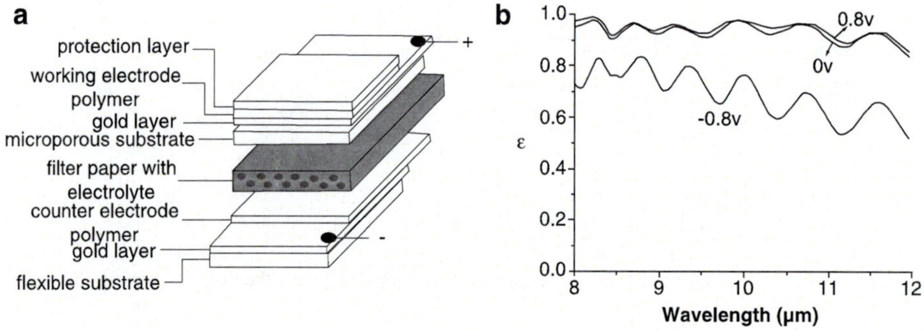

Fig. 7.10 (**a**) Architecture of the EEDs proposed by Li et al. and (**b**) the emissivity (ε) spectra recorded in the wavelength range of 8 to 12 μm as function of the bias voltage applied [35]

$$WO_3 + xe^- + xH^+ \rightleftharpoons H_xWO_3$$

Fig. 7.11 Insertion/disinsertion of protons and electrons from a tungsten oxide (WO_3) film upon electrochemical cycling [48]

EEDs can also be designed with inorganic electrochemically active species. Metal oxides such as tungsten oxide (WO_3) [6, 41–48] and niobium oxides (Nb_xO_y) [49] are used in the same way as the electronically conducting polymers. The modulation of the IR transparency of these species is driven by a redox reaction which leads to a change of the oxidation degree of the metallic cation. Specifically, this change is due to an insertion or a disinsertion of protons and electrons, represented in Fig. 7.11 for tungsten oxide.

In 1999, Woollam et al. [41] presented a multilayer device based on tungsten oxide (WO_3). The electrolyte used is tantalum oxide (Ta_2O_5), and the counter-electrode is nickel oxide (NiO) (Fig. 7.12a). The modulation of the emissivity relies on the insertion/disinsertion of the protons of the WO_3 film, which is, respectively, linked to its disinsertion/insertion of the NiO film. By applying a positive voltage to the top electrode, the protons move out of the WO_3 film into the NiO film, which leads to a high reflective state of the device. Indeed, in this state, the WO_3 and the NiO films are IR-transparent, and the gold electrode is responsible for the high reflectivity. On the contrary, by applying a negative voltage to the top electrode (conductive grid), the protons move out of the NiO film and into the WO_3 film, leading to a less reflective device. By optimizing the thicknesses of the NiO and WO_3 layers as well as the portion of the device covered by the top conductive grid, the authors obtained a reflectivity variation of 0.54 over an infrared wavelength range of 2 to 13.8 μm (Fig. 7.12b).

A. Rougier et al. [46–48] developed electroemissive devices with a similar structure for IR-stealth application (Fig. 7.13a). They incorporated, into their device, thin optimized layers of WO_3 with a reflectivity variation of 0.73 in band II and 0.50 in band III. The assembled device exhibits a reflectivity modulation of 0.27 in band II and 0.18 in band III (Fig. 7.13b).

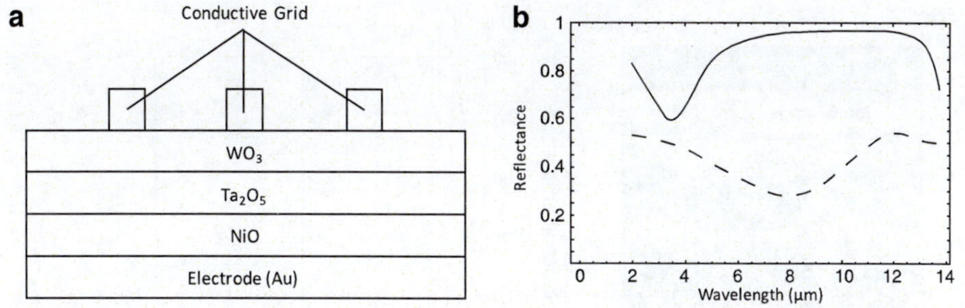

Fig. 7.12 (**a**) Schematic drawing of the EED based on WO_3 proposed by Woollam et al. and (**b**) the associated reflectance on the IR-wavelength range of 2–13.8 μm depending on the insertion/disinsertion of protons and electrons [41]

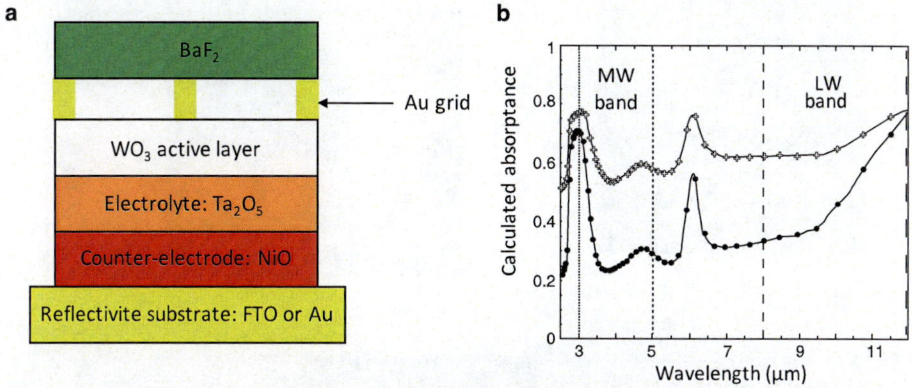

Fig. 7.13 (**a**) Architecture of the EED proposed by Rougier et al. and (**b**) the calculated absorptance in the IR-wavelength range of 2.5–12 μm of the corresponding EED [48]

For thermal regulation application, Demiryont et al. [45, 50–52] developed very promising EEDs, referred to as Eclipse Variable Emissivity Device (EclipseVED™). The architecture of these EEDs is well-known (Fig. 7.14a); nevertheless, the components of their system are not specified. The authors reported reflectivity variations of 0.70 over an infrared range of 4 to 26 μm [50] (Fig. 7.14b). Despite the high optical performances, their device has a switching time estimated at 30 minutes.

To demonstrate the efficiency of their systems, they have placed four EclipseVED™ devices on a hotplate at 55 °C, two samples being in a high IR-emissivity state and the two others in a low IR-emissivity state. Visually, the four samples appear nearly identical (Fig. 7.14c). Using an infrared thermal camera, the two samples with a low emissivity active layer are much cooler than the temperature of the heating plate (Fig. 7.14d).

Due to their high optical performances, these devices were the subject of a study program in 2007 (Midshipman Space Technology Applications Research) within the MidSTAR-1 satellite. The full functionality of the testing module including two EclipseVED™ devices (Fig. 7.14e) was demonstrated in space, and the modulation of the system emissivity was checked by heat dissipation tests.

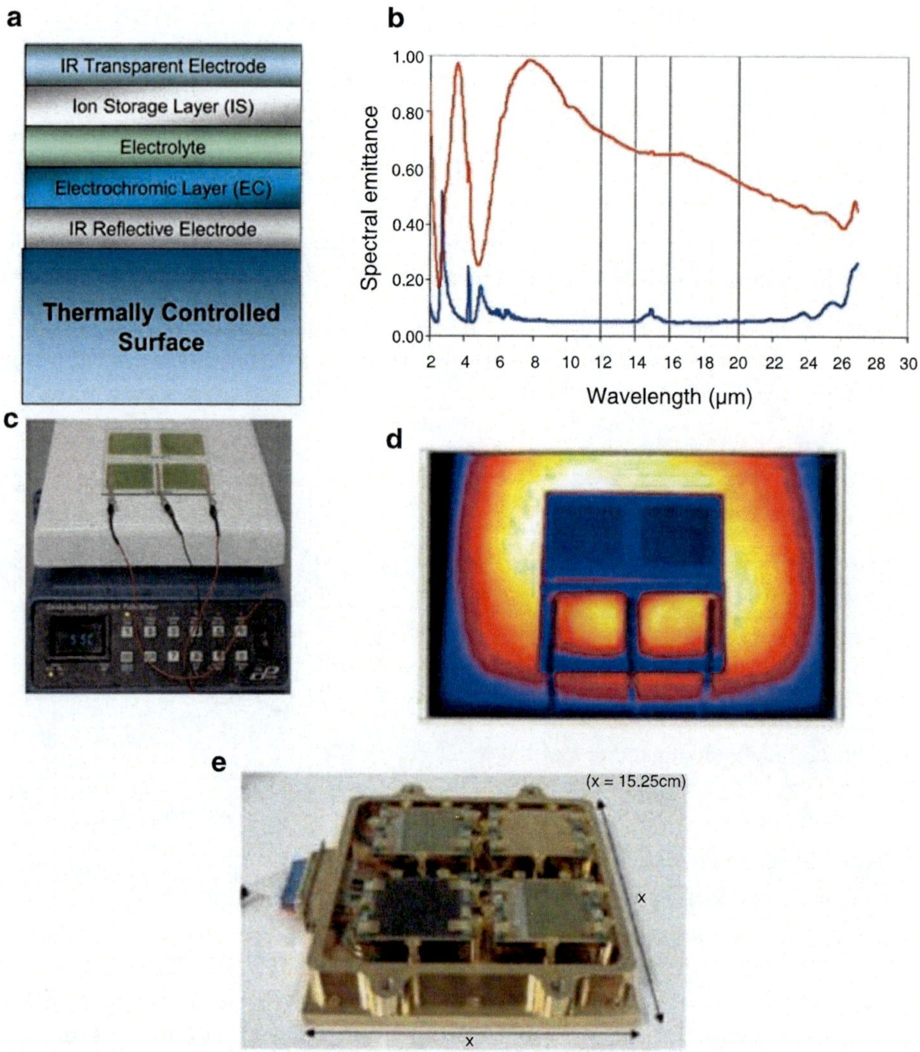

Fig. 7.14 (**a**) Illustration of the EclipseVED™ architecture. (**b**) Spectral emittance reported by H. Demiryont et al. before the testing on the MidSTAR-1 satellite. (**c**) Image of four EEDs on a hot plate. (**d**) Image of thermal infrared camera showing the four samples on the hot plate at 55 °C, the two dark samples are in a low emissivity state, and the two yellow/red samples are in a high emissivity state. (**e**) Testing module including two EclipseVED™ devices, used for the MidSTAR program [50]

7.3.3.2 Opaque Materials with Variable Emissivity in the IR

In the previous section, the active layers of the electroemissive devices (EEDs) were similar to a dual material consisting of an electrochemical active species (EAS) and a reflective surface. The emissivity modulation was obtained by a change of the IR transparency of the EAS. In this part, the active layer is only composed of the electrochemical active species. The common design of the EED is similar to the

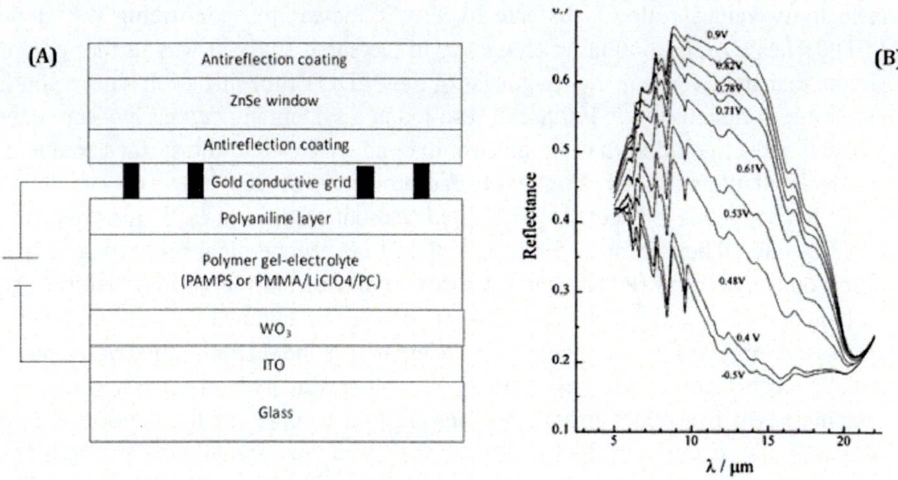

Fig. 7.15 (**a**) Architecture of the EED proposed by Topart and Hourquebie [53] and (**b**) the associated reflectance depending on the potential applied to the gold grid

previous devices presented, except they do not require an additional reflective surface such as a gold layer. Indeed, in the case of IR-opaque layers, i.e., thick layers, the active layer switches directly from a high reflectivity state to a low reflectivity state. Some electronically conducting polymers (ECP) have the ability to modulate their reflectivity in the infrared. Specifically, in the oxidized state, IR-opaque ECP layers are conductive and exhibit IR-optical properties close to metallic substrates, namely, a high IR reflectivity. Conversely, in the neutral state, IR-opaque ECP layers are in a semiconducting state and exhibit a low IR reflectivity.

Topart and Hourquebie [53] showed that the change in the oxidation degree of the polyaniline (PANI) induces a change of its optical properties in the infrared. In the light of that, they elaborated an EED based on PANI in 1999. The latter, operating either in the visible (in transmission mode) or in the infrared (in a reflection mode), is based on a multilayer architecture (Fig. 7.15a). When the device operates in reflection and by applying +0.9 V, the polyaniline switches from a low reflectivity state to a high reflectivity state (Fig. 7.15b). In the same way, by applying −0.5 V, the polyaniline switches from a high reflectivity state to a low reflectivity state.

This interesting property was also applied to other ECPs such as PEDOT [54, 55] or polypyrrole [56]. For example, Heeger et al. [54] demonstrated that the neutral PEDOT, in its semiconducting state, exhibits a low reflectivity level. On the contrary, in its conductive state (corresponding to the oxidized state), PEDOT exhibits a high reflectivity level. In 2015, Louet et al. [55] showed the dependence of the PEDOT IR reflectivity not only with the electronic conductivity but also with the ECP layer roughness. The reflectivity of the PEDOT films was about 0.20 with an electronic conductivity varying between 10^{-1} and 10^{-3} S/cm. In contrast, in the oxidized state, the PEDOT films having an electronic conductivity of the order of 10^2 S/cm had a reflectivity of approximately 0.65. In addition, it was demonstrated that when the electronic conductivity of the films reaches at least 85 S/cm, the reflectivity values follow satisfactorily the Drude model, confirming that doped PEDOT has a pseudo-metallic character. In the same study, it was further demonstrated that an increase in the roughness of the PEDOT film surface, having a similar electronic conductivity of 100 S/cm, resulted in a systematic decrease of the reflectivity. It was concluded that the electronic conductivity and the surface roughness are the key factors influencing the infrared properties of ECP.

In 2013 and 2017, Jia et al. [57, 58] carried out studies on ECP films deposited on ITO glass. They compared the optical and electrochemical properties of films based on polyaniline (PANI) and on a copolymer of aniline and o-anisidine. The authors demonstrated that the use of the copolymer based on aniline and o-anisidine allows the improvement of the performances such as the response time of the device and the electroactivity with regard to a polyaniline film. The reflectivity variation is also increased from 0.29 for the polyaniline film to 0.41 for the copolymer film, over an infrared range of 8 to 14 μm. In the same way, the authors compared the optical properties of films based on the poly(carbazoyltriphenylamine) (poly(CBZ-TPA)) and on poly(carbazoyltriphenylaminethiophene) (poly(CBZ-TPA-Th)). The poly(CBZ-TPA-Th) and the poly(CBZ-TPA) exhibit similar optical properties, with a reflectivity variation of 0.41 and 0.37, respectively. It therefore seems that in this case the copolymerization with a thiophene unit did not significantly improve the optical properties.

Since 2008, our laboratory has developed PEDOT-based EEDs for two applications, infrared stealth [59–65] and thermal regulation of satellites [18, 66]. The EEDs are designed by an original approach. By mimicking a trilayer structure (working electrode/electrolyte/counter-electrode), a "one-block" electroemissive device was developed. A more detailed description of the EEDs designed in our laboratory during recent years is presented in §7.4.

In 2017, Brooke et al. [67] presented two types of flexible electroemissive devices based on PEDOT. The PEDOT is obtained by a vapor-phase polymerization on a poly(ethylene) substrate, leading to PEDOT:Tosylate (PEDOT:Tos). The latter acts as both a working electrode and a counter-electrode. In the case of the first device, referred to as lateral, the two electrodes of PEDOT:Tos are separated by a gel-like electrolyte (Fig. 7.16a). The gel is based on an ionic liquid, the 1-ethyl-3-methylimidazolium ethyl sulfate (EMI-ES), with 7.5 wt% of hydroxyl ethyl cellulose. For the second device, referred to as vertical, the two PEDOT electrodes are separated by a propylene membrane soaked in the gel electrolyte (Fig. 7.16b). Only the optical properties of the lateral device were studied. The authors have reported a reflectivity variation of about 0.15 over an infrared range of 7.5 to 13 μm. However, they compared the two devices by studying their thermal modulation. A thermal modulation of 5.6 °C (−8.9 °C to −3.3 °C) was reported for the vertical device by applying a potential difference of ±1.5 V. By applying the same potential difference, a higher temperature modulation was observed, 9.7 °C (−0.9 °C to −10.6 °C), for the lateral device. Nevertheless, the authors have reported that the lateral device switches, between the two redox states, in a few minutes, whereas the vertical one switches in just a few seconds. Therefore, the authors studied in more detail the vertical device and reported a memory effect up to 10 h. They also studied the

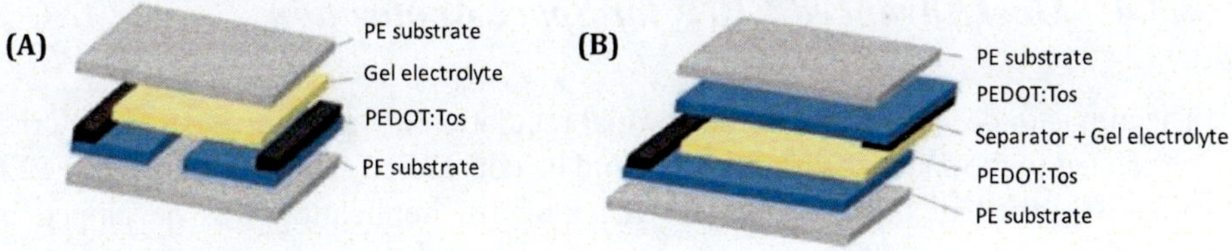

Fig. 7.16 (**a**) Lateral electroemissive device architecture and (**b**) vertical electroemissive device architecture [67]

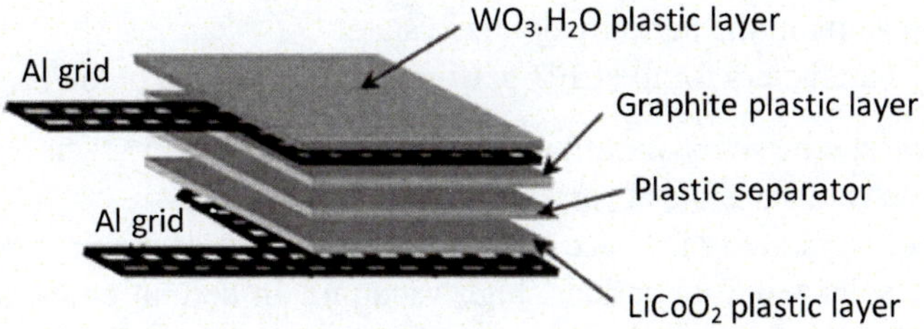

Fig. 7.17 EED designed by Bessière et al. using the plastic lithium-ion technology [73]

cycling stability of this device and observed significant degradation after only 1000 cycles.

Tungsten oxides are also used to design EEDs without an additional metal layer [16, 68–70]. In this case, the tungsten oxide layer is thicker [15, 71]. The first device using only WO_3 as active layer was reported in 1995 by S.F. Cogan [68]. Later, Bessière et al. [72, 73] studied flexible devices based on hydrated WO_3 powder ($WO_3.H_2O$). To obtain flexible devices, the authors used the plastic lithium-ion (PLiON) technology developed initially for Li-ion batteries. The active components, e.g., $WO_3.H_2O$, were synthesized in a powder form and then embedded in a plastic matrix, following the Bellcore-Telcordia PLiON plastic battery fabrication process [74, 75] as shown in Fig. 7.17.

In this configuration, the active electrochemical species is the $WO_3.H_2O$, while $LiCoO_2$ is used as an ion storage layer. Aluminum grids are placed on each side of the device as current collectors. For a better electronic conduction from the aluminum grid to the active layer, a graphite plastic layer was added. This additional layer allows a better insertion/extraction of the lithium within the system. With this multilayer architecture, the authors were able to obtain a modulation of the reflectivity of 0.32 in the IR-wavelength range of 2.5 to 25 μm.

7.3.4 Most Advanced EEDs for Space Application

The aim is to design EEDs for the thermal regulation of satellites; thus, some criteria related to the space environment should be considered. The most advanced EED fully described in the literature for the intended application was developed by Chandrasekhar et al. [33].

The active layer of their device is a dual material, including a thin electroactive layer – an electronically conducting polymer (ECP) – and a reflective surface, a gold layer. The optical properties of this device are obtained by a modulation of the IR transparency of the ECP. The latter, which is a copolymer of polyaniline and poly(diphenylamine), is either IR-transparent in the neutral and fully oxidized states and IR-absorbing in the partially oxidized state.

The key parameters for the EED performances are listed below:

- The highest emissivity variation.
- The lowest solar visible absorptivity.
- The full functionality in space conditions (temperatures ranging from −30 °C to +90 °C, solar wind, radiations, high vacuum). In certain cases, it requires the encapsulation of the device to prevent an electrolyte leakage due to the high vacuum and to allow the device to resist space radiations.
- The anti-electrostatic discharge.

Chandrasekhar et al. [33], in collaboration with the National Aeronautics and Space Administration (NASA), considered all these parameters in their latest generation of devices (Fig. 7.18). As previously described, the device is based on a multilayered architecture (ECP//Au/membrane/electrolyte/Au//ECP).

All the layers constituting the device were developed with the aim of a spatial application.

- *Layer 1* is the top layer. Nearly IR-transparent, this layer is composed of a doped-Ge-Si layer and indium tin oxide (ITO) deposited on a poly(ethylene) (PE) film. The ITO layer was optimized in order to eliminate the electrostatic discharge. The doped-Ge-Si layer is used to reduce the solar absorption, limiting overheating when the satellite is facing the sun. Indeed, generally ECPs are quite dark,

Fig. 7.18 Latest generation of the multilayered EED architecture reported by Chandresekhar et al. for thermal regulation of satellite [33]

thus displaying a high solar absorption. This protection layer is heat-bonded directly to the ECP surface and has therefore multiple purposes: (i) eliminate the electrostatic discharge, (ii) reduce the solar absorptivity, and (iii) prevent the permeation of the electrolyte onto the ECP surface, which otherwise would influence the IR properties of the layer.

- *Layer 2* is the active layer. The electronically conducting polymer (ECP) is electrosynthesized from aniline and diphenyl amine monomers in a 95/5% molar proportion. The copolymer obtained is electrodeposited on the Au/membrane (layer 3) with a thickness ranging from 0.25 to 2 μm.
- *Layer 3* is the multi-task layer. First, an Au layer is deposited on a microporous membrane with a thickness of 500 nm obtained via e-beam thermal evaporative deposition. The porosity of the Au/membrane allows the ions provided by the layer 4 to circulate via the layer 3 to the ECP (layer 2). The Au/membrane acts as support for the electropolymerization of the ECP, due to its conductive character. But this layer also provides the IR-reflectivity properties to the device when the ECP is IR-transparent.
- *Layer 4* is the ion reservoir membrane. It is a microporous membrane comprising poly(sulfone) and poly(carbonate). The electrolyte, loaded into the membrane, is a formulation of two ionic liquids (1-(2-hydroxyethyl)-3-methylimidazolium bis (trifluoromethanesulfonylimide) and 1-butyl-3-methylimidazolium tetrafluoroborate) and a lithium salt (lithium tetrafluoroborate). Generally, the electrolyte is an organic solvent or gel containing salt. However, ionic liquids have appeared as an interesting alternative to conventional electrolytes [76, 77]. Their wide temperature range and their non-volatile property make them very well suitable for space application.
- *Layer 5* is a gold layer obtained in the same conditions as the layer 3. In the same way, this layer acts as support material for the electropolymerization of the monomers, layer 6. Nevertheless, this gold layer does not provide optical properties to the device.
- *Layer 6* is the counter-electrode. It is identical to the layer 2, except this layer is five times thicker than the active layer. The counter-electrode does not provide optical properties. It only allows a complementary redox reaction between the active layer and the counter-electrode during the application of a potential difference.

The authors measured reflectivity variations of 0.43–0.49 for the complete device in the 2.5–45 μm wavelength range. The reflectivity ranges between 0.61 and 0.77 when the ECP is IR-transparent and 0.23 and 0.36 when the ECP is IR-absorbing. The top layer allows the reduction of the solar absorptivity from 0.94 to 0.46. The entire device was also subjected to environmental testing, vacuum ultraviolet tests, electrostatic discharge tests, and solar wind tests. The authors claim that their device has passed all the tests. Therefore, the device complies with all the key parameters required by the intended application.

7.4 Overview of State-of-the-Art EEDs Designed at CY Cergy Paris Université

7.4.1 EEDs Designed at CY Cergy Paris Université

In the case of the devices elaborated in our group, the latter operates without any reflective metallic sublayer. The electroactive layer is thick and constitutes the whole active layer of the device. From a doping/dedoping process, the active layer switches from a high IR-reflective state (oxidized state) to a low IR-reflective state (neutral state). Therefore, the potential applied allows the modulation of the reflectivity of the device. These EEDs have the advantage of being easy to implement compared to the fabrication of other EEDs reported in the literature, and their fabrication process is relatively inexpensive. The architecture of these EEDs, based on polymer network [78] and two PEDOT electrodes interpenetrated on both sides of the polymer matrix, leads in two steps only to a one-block architecture. The first electroemissive devices designed by Verge et al. were based on a poly(ethylene oxide) (PEO) network and two PEDOT electrodes (Fig. 7.19) [59–61].

The role of each layer is described as follows:

- *Layer 1* is the active layer. The PEDOT is locally interpenetrated underneath both faces of the PEO membrane. The PEDOT is obtained by a chemical polymerization of EDOT. This layer is always IR-opaque and is either IR-absorbing or IR-reflective depending on the applied voltage.
- *Layer 2* is the host matrix. The PEO network plays the role of an ion reservoir membrane. This matrix is swollen with an ionic liquid, typically the 1-ethyl-3-methylimidazolium bis(trifluoromethanesulfonylimide) (EMITFSI).
- *Layer 3* is the counter-electrode. The PEDOT is obtained using the same experimental conditions as layer 1. The presence of the PEDOT at the counter-electrode ensures the redox reaction of the active layer without over-oxidation or over-reduction.

This type of architecture has the advantage of preventing the delamination's risks, which is not the case with conventional EED architectures. Indeed, the latter are multilayered, and some adhesion issues between the different layers could entail delamination. From an optical point of view, these devices have interesting properties in terms of reflectivity modulation with variations of 0.30 and 0.37 in band II and band III, respectively, and 0.40 in the 10–25 μm infrared wavelength range [59] (Fig. 7.20).

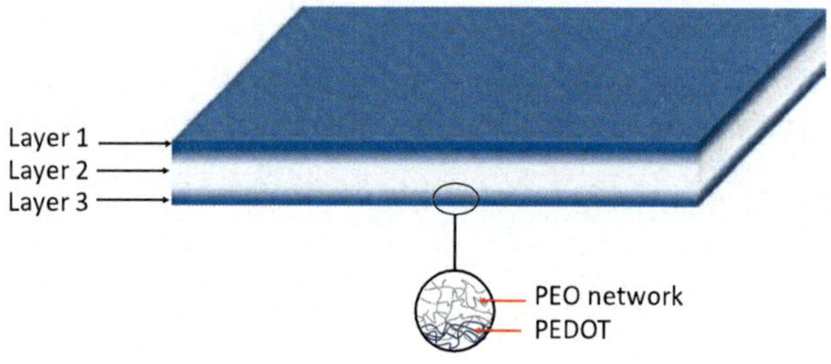

Fig. 7.19 EEDs based on a polymer network and two ECP electrodes, designed by P. Verge et al. [60]

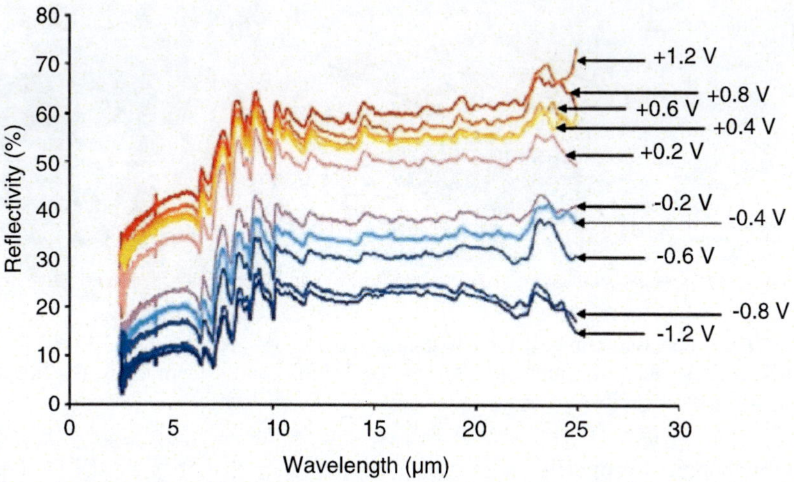

Fig. 7.20 Reflectivity spectra, on the IR-wavelength range of 2.5 to 25 μm, of the EED designed by Verge et al. for various potentials applied to the active layer [60]

These EEDs based on a PEO/PEDOT material swollen with EMITFSI were the subject of a simulated spatial test campaign carried out in 2012 by Teissier et al. [18] at Thales Alenia Space. By applying a potential difference of ±1.5 V to the EEDs, temperature variations of 4 °C were then measured in a high thermal vacuum chamber at a pressure of 10^{-6} mbar. It demonstrated that these EEDs can be used for the thermal regulation of satellites. Nevertheless, several drawbacks were highlighted. Firstly, the solar absorptivity of the active layer is high (≈0.9), which would lead to an overheating issue when the satellite would face the sun. Secondly, mechanical weaknesses were detected. Indeed, during the doping/dedoping process, some EEDs cracked due to an actuator behavior [81]. Furthermore, during the thermal aging, hydroxyl radicals resulting from the PEO degradation are formed [79], causing the degradation of the PEDOT hence to a decrease of the optical performances over time. Thus, the improvement of the thermomechanical properties and, at the same time, of aging was the object of further studies.

To overcome the PEO matrix issues and for the development of EEDs mechanically more robust and more stable over time, a new approach to design EEDs was investigated by Fabre-Francke et al. [80]. First, an elastomer substitutes the PEO matrix. The nitrile butadiene rubber (NBR) has numerous advantages, such as its flexibility and its strength. Then, the PEDOT synthesis was carried out by an electrochemical process. The electropolymerization was performed on one side of a NBR film, and the full device was prepared by the association of two NBR/PEDOT films. Low reflectivity variations of 0.18 and 0.17 were obtained in [3–5 μm] region and [8–12 μm] region, respectively. Despite interesting mechanical properties of the devices, in the presence of an ionic liquid, NBR ionic conductivity is too low (~10^{-6} S/cm), which greatly limits the switching time of the device.

Therefore, Goujon et al. [63] and Festin et al. [81] designed interpenetrating polymer network (IPN)-based matrixes for the development of EEDs associating

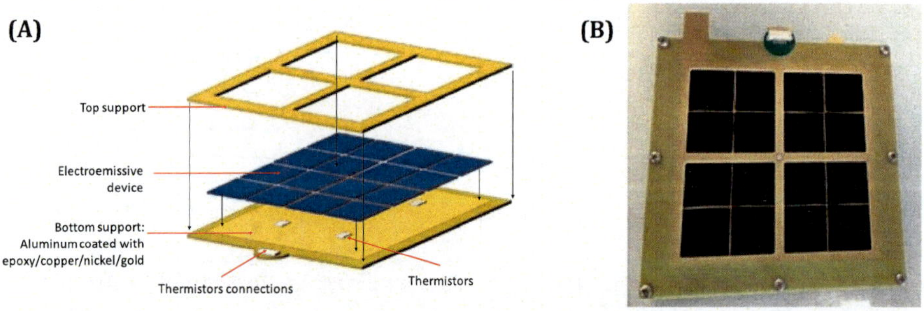

Fig. 7.21 (**a**) Radiators composed of a bottom support, a top support, and 16 EEDs – NBR/PEO/PEDOT IPNs swollen to saturation with neat EMITFSI – and 4 thermistors. (**b**) Image of the assembled radiator [82]

good mechanical properties with high ionic conductivity once the EED is swollen with its electrolyte. IPNs based on NBR as first component and PEO as second component were synthesized and used as ion reservoir membrane for the EEDs. The PEDOT has been prepared from oxidative polymerization of the corresponding heterocyclic monomer within the NBR/PEO IPN matrix. A pseudo-trilayer configuration is obtained with PEO/NBR IPN sandwiched between two interpenetrated PEDOT electrodes at each side.

7.4.2 Control of Thermal Performances in Simulated Environmental Conditions

A test campaign was carried out at TAS [82] in 2019. Radiators composed of 16 EEDs based on NBR/PEO/PEDOT IPNs were designed (Fig. 7.21) with a total surface of 81 cm^2. The EEDs were swollen to saturation with EMITFSI before testing. To simulate conditions close to the space environment, electroemissive radiators were placed into a vacuum chamber under 10^{-6} mbar at −60 °C. The cold (−25 °C), intermediate (20 °C), and hot (+60 °C) environment equilibrium temperatures were set through heating rubbers placed behind the electroemissive radiators. By controlling the heat flux injected though electrical heaters for a given emissivity, it was possible to control the temperature of the radiators. The experiment in the thermal vacuum chamber was performed during 12 days after reaching a vacuum of 10^{-6} mbar. Bias voltages of +1.2 V and −1.2 V were applied to the 16 EEDs of the radiators, and the temperature variation, resulting from an emissivity change, was thus monitored and measured. The results were compared with optical solar reflector (OSR) which exhibits a passive behavior with a constant infrared emissivity. OSRs are commonly used for radiators on spacecraft. They consists in a top layer made out of quartz, over a reflecting metal layer. Therefore, for a given heating power injected, OSR temperature, once stabilized, remains constant.

Before starting the experiment, the active layers are set in an intermediate oxidation state, i.e., a bias voltage of 0 V is applied. Then, the heating power is set in order to obtain a temperature around 20 °C. Once this temperature is reached, the heat flux injected is kept constant during the electrical switching. The starting temperature of the radiator is around 20 °C. By applying a bias voltage of +1.2 V, the latter is in a low emissivity state (Fig. 7.22a), and the temperature of the radiator increases to reach about 25 °C. By applying a bias voltage of −1.2 V, this one reaches a high emissivity state, and the temperature decreases to 13 °C, since this state leads to the rejection of the heat toward vacuum chamber. Thus, the electroemissive behavior of the radiators causes a temperature variation of 12 °C. As expected, the OSR radiator experiences a constant temperature of approximately 12 °C. The hot and cold cases were also simulated with temperature steps of 60 °C and −25 °C, respectively. The temperatures were obtained by increasing or decreasing the heating power injected via electrical heaters. The environment temperature is kept at −60 °C. The temperature evolutions for both cases are presented in Fig. 7.22b, c. For both temperature steps, an increase in temperature is observed when a bias voltage of +1.2 V is applied to the active layer of the EED radiators. A

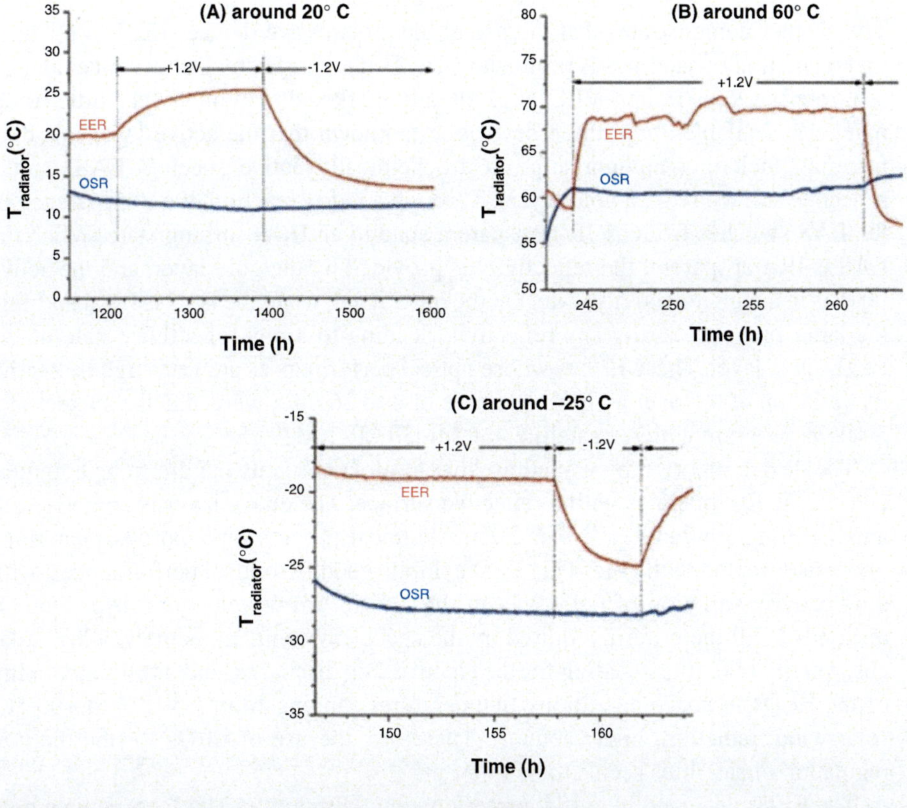

Fig. 7.22 Evolution of the OSR (blue curve) and EED radiator (red curve) temperatures for a simulated temperature of (**a**) 20 °C, (**b**) 60 °C, and (**c**) –25 °C. The bias voltage applied is indicated: +1.2 V (low emissivity); −1.2 V (high emissivity) [82]

decrease in temperature is then recorded when a bias voltage of −1.2 V is applied. These behaviors are similar to the one observed for the intermediate case. Thus, the EED radiators have a similar behavior regardless of the temperature step applied between −25 °C and +60 °C. This means that the system is able to operate at low temperatures, e.g., −25 °C, which is an interesting result, and that the system can function under vacuum over the wide temperature range required for the intended application.

Furthermore, it was determined during the campaign test that the electrical consumption required with the use of such a system is low. An energy saving of about 25% was evaluated with regard to the electrical consumption required when an OSR radiator is used. Besides, a memory effect of 47 hours was observed under vacuum at a temperature step of 0 °C. This memory effect has not reached its limit and could therefore be significantly longer. So, the energy saving could be even more important using the memory effect of the device.

7.5 Conclusion

The chapter demonstrates that the use of electroemissive devices (EEDs) for thermal regulation of satellites is promising. The EED reported by Demiryont et al. [44, 45], referred to as EclipseVED™, is already marketed. Although very little information is available about these devices, it is known that the active layer is a dual material, including an inorganic electrochemically active species (EAS) and a reflective surface (e.g., a gold layer). Through a redox reaction, the transparency of the EAS switches from an IR-transparent state to an IR-absorbing state. When the EAS is IR-transparent, the reflectivity is provided by the gold layer, and the active layer is in a high reflectivity state. On the other hand, when the EAS is IR-absorbing, the latter provides hardly any reflectivity leading to a low reflectivity state of the EED active layer. These EEDs, whose optical performances are very high (reflectivity variation of 0.7 over an infrared range of 4 to 26 μm), were also the subject of a research program within a satellite, the MidSTAR-1.

The device reported by Chandrasekhar et al. [33] seems mature to be commercialized. It also operates with a reflective surface. The active layer is comprised of an ECP film, synthesized by an electrochemical process on a porous membrane previously coated with gold. This EED exhibits good IR-optical performances, with a reflectivity variation of 0.43–0.49 in the 2.5–45 μm wavelength range. Thus, it responds to all the criteria required by the space environment, namely, a low solar absorptivity (≈0.46), a coating for the electrostatic discharge, and a full functioning of the EEDs in space conditions (temperatures ranging from −30 °C to +90 °C, solar wind, radiation, high vacuum). Therefore, the use of EEDs for the thermal regulation of satellites seems to be relevant.

The EEDs designed in our laboratory operate differently. The EAS is the whole active layer, and no reflective surface is required to tune the IR reflectivity of the active layer. In fact, the electroactive layer is thick, and its intrinsic properties enable a direct modulation of the IR reflectivity. Elaborated from an interpenetrated PEDOT active layer within a host matrix, these EEDs have a very simple operating principle. Through a doping/dedoping process (redox reactions), the active layer changes from a high IR-reflectivity state to a low IR-reflectivity state. So, the reflectivity can be tuned by changing the potential applied to the device. These systems have the advantage of being easy to implement, due to the originality of the architecture. The system is "mono-block" avoiding any delamination issues between the layers. Finally, the devices were evaluated under conditions close to the space environment – vacuum chamber under 10^{-6} mbar at −60 °C for 12 days [82]. Temperature variations ranging from 6 to 12 °C were determined for the various radiator equilibrium temperatures ranging from −25 °C to +60 °C. These performances are interesting because they can result in reduced electrical power needs for the thermal regulation in cold cases and associated reduced on-board weight with respect to the use of the current technology, i.e., constant emissivity coating and heaters.

References

1. C.G. Granqvist, Transparent conductors as solar energy materials: A panoramic review. Sol. Energy Mater. Sol. Cells **91**, 1529–1598 (2007)
2. K. Lewis, A. Pitt, T. Wyatt-Davies, J. Milward, Thin film thermochromic materials for non-linear optical devices. Mater. Res. Soc. **374**, 105–116 (1995)
3. C. Technology, Thin film design for advanced thermochromic. Chin. Phys. **16**, 1704–1709 (2007)
4. T.D. Swanson, G.C. Birur, NASA thermal control technologies for robotic spacecraft. Appl. Therm. Eng. **23**, 1055–1065 (2003)
5. D.M. Douglas, T. Swanson, R. Osiander, J. Champion, A.G. Darrin, W. Biter, P. Chandrasekhar, Development of the variable emittance thermal suite for the space technology. AIP Conf. Proc **608** (2002)
6. E.B. Franke, C.L. Trimble, J.S. Hale, M. Schubert, J.A. Woollam, Infrared switching electrochromic devices based on tungsten oxide. J. Appl. Phys. **88**, 5777–5784 (2000)
7. T. Kogure, K.-C. Leung, *The Astrophysics of Emission-Line Stars* (Springer, New York, 2007)
8. B.P. Jelle, Solar radiation glazing factors for window panes, glass structures and electrochromic windows in buildings – Measurement and calculation (2013)
9. C.G. Granqvist, Solar energy materials. Adv. Mater. **15**, 1789–1803 (2003)
10. C.G. Granqvist, Oxide electrochromics: Why, how, and whither. Sol. Energy Mater. Sol. Cells **92**, 203–208 (2008)
11. R.D. Rauh, Electrochromic windows: An overview. Electrochim. Acta **44**, 3165–3176 (1999)
12. D.R. Rosseinsky, R.J. Mortimer, Electrochromic systems and the prospects for devices. Adv. Mater. **13**, 783–793 (2001)
13. J.C. Gustafsson-Carlberg, O. Inganäs, M.R. Andersson, C. Booth, A. Azens, C.G. Granqvist, Tuning the bandgap for polymeric smart windows and displays. Electrochim. Acta **40**, 2233–2235 (1995)
14. R.J. Mortimer, Organic electrochromic materials. Electrochim. Acta **44**, 2971–2981 (1999)
15. A.-L. Larsson, All-thin-film electrochromic devices for optical and thermal modulation. Acta Univ. Ups., 1–62 (2004)
16. L. Beluze, Thesis: Matériaux et dispositifs électroactifs dans l'infrarouge, Université Pierre et Marie Curie (2004)

17. J. Beigbeder, Etudes des propriétés physiques de nanocomposites à matrice polysiloxane: Application au développement d'un revêtement de contrôle thermique froid et antistatique, Université de Toulouse III (2009)
18. A. Teissier, J.P. Dudon, P.H. Aubert, F. Vidal, S. Remaury, J. Crouzet, C. Chevrot, Feasibility of conducting semi-IPN with variable electro-emissivity: A promising way for spacecraft thermal control. Sol. Energy Mater. Sol. Cells **99**, 116–122 (2012)
19. D. Liu, H. Cheng, X. Xing, C. Zhang, W. Zheng, Thermochromic properties of W-doped VO2 thin films deposited by aqueous sol-gel method for adaptive infrared stealth application. Infrared Phys. Technol. **77**, 339–343 (2016)
20. A. Boileau, Elaboration de films minces d'oxydes de nickel et de manganèse et terres rares et caractérisation des propriétés thermo-émissives, Université de Lorraine (2013)
21. M. Soltani, M. Chaker, X.X. Jiang, D. Nikanpour, J. Margot, Thermochromic La1−xSrxMnO3 (x=0.1, 0.175, and 0.3) smart coatings grown by reactive pulsed laser deposition. J. Vac. Sci. Technol. A Vac. Surf. Film. **24**, 1518–1523 (2006)
22. A. Boileau, F. Capon, S. Barrat, P. Laffez, J.F. Pierson, Thermochromic effect at room temperature of Sm 0.5Ca 0.5MnO 3 thin films. J. Appl. Phys **111**, 1–4 (2012)
23. D. Fan, Q. Li, Y. Xuan, P. Dai, Variable emissivity property of magnetron sputtering thermochromic film. Thin Solid Films **570**, 123–128 (2014)
24. A.A. Argun, P.-H. Aubert, B.C. Thompson, I. Schwendeman, C.L. Gaupp, J. Hwang, N.J. Pinto, D.B. Tanner, A.G. MacDiarmid, J.R. Reynolds, Multicolored electrochromism in polymers: Structures and devices. Chem. Mater. **16**, 4401–4412 (2004)
25. P. Chandrasekhar, *Conducting Polymers, Fundamentals and Applications* (Kluwer Academic Publishers, Boston, 1999)
26. C.G. Granqvist, *Handbook of Inorganic Electrochromic Materials* (Elsevier, Amsterdam\New York, 1995)
27. P. Chandrasekhar, T.J. Dooley, Far IR transparency and dynamic infrared signature control with novel conducting polymer systems. Proc. SPIE **2528**, 169–180 (1995)
28. P. Chandrasekhar, G.C. Birur, P. Stevens, S. Rawal, E.A. Pierson, K. Miller, Far infrared electrochromism in unique conducting polymer systems. Synth. Met. **119**, 293–294 (2001)
29. P. Chandrasekhar, B.J. Zay, G.C. Birur, S. Rawal, E.A. Pierson, L. Kauder, T. Swanson, Large, switchable electrochromism in the visible through far-infrared in conducting polymer devices. Adv. Funct. Mater. **12**, 95–103 (2002)
30. P. Chandrasekhar, B.J. Zay, T. McQueeney, A. Scara, D.R. Rosseinsky, G.C. Birur, S. Haapanen, L. Kauder, T. Swanson, D. Douglas, Conducting polymer (CP) infrared electrochromics in spacecraft thermal control and military applications. Synth. Met. **135–136**, 23–24 (2003)
31. P. Chandrasekhar, B.J. Zay, T. McQueeney, G.C. Birur, V. Sitaram, R. Menon, M. Coviello, R.L. Elsenbaumer, Physical, chemical, theoretical aspects of conducting polymer electrochromics in the visible, IR and microwave regions. Synth. Met. **155**, 623–627 (2005)
32. P. Chandrasekhar, B.J. Zay, S. Barbolt, R. Werner, G.C. Birur, A. Paris, High performance variable emittance devices for spacecraft application based on conducting polymers coupled with ionic liquids. AIP Conf. Proc. **1103**, 101–104 (2009)
33. P. Chandrasekhar, B.J. Zay, D. Lawrence, E. Caldwell, R. Sheth, R. Stephan, J. Cornwell, Variable-emittance infrared electrochromic skins combining unique conducting polymers, ionic liquid electrolytes, microporous polymer membranes, and semiconductor/polymer coatings, for spacecraft thermal control. J. Appl. Polym. Sci. **40850**, 1–15 (2014)
34. Y. Tian, X. Zhang, S. Dou, L. Zhang, H. Zhang, H. Lv, L. Wang, J. Zhao, Y. Li, A comprehensive study of electrochromic device with variable infrared emissivity based on polyaniline conducting polymer. Sol. Energy Mater. Sol. Cells **170**, 120–126 (2017)
35. H. Li, K. Xie, Y. Pan, M. Yao, C. Xin, Variable emissivity infrared electrochromic device based on polyaniline conducting polymer. Synth. Met. **159**, 1386–1388 (2009)
36. H. Pagès, P. Topart, D. Lemordant, Wide band electrochromic displays based on thin conducting polymer films. Electrochim. Acta **46**, 2137–2143 (2001)

37. P.-H. Aubert, A.A. Argun, A. Cirpan, D.B. Tanner, J.R. Reynolds, Microporous patterned electrodes for color-matched electrochromic polymer displays. Chem. Mater. **16**, 2386–2393 (2004)
38. A.A. Argun, A. Cirpan, P.-H. Aubert, J.R. Reynolds, Multi-color electrochromic polymers on reflective devices. Polym. Mater. Sci. Eng. **90**, 40 (2004)
39. A.A. Argun, *Patterning of Conjugated Polymers for Electrochromic Devices* (University of Florida, Gainesville, 2004)
40. I. Schwendeman, J. Hwang, D.M. Welsh, D.B. Tanner, J.R. Reynolds, Combined visible and infrared electrochromism using dual polymer devices. Adv. Mater. **13**, 634–637 (2001)
41. J.S. Hale, J.A. Woollam, Prospects for IR emissivity control using electrochromic structures. Thin Solid Films **339**, 174–180 (1999)
42. E.B. Franke, C.L. Trimble, J.S. Hale, M. Schubert, J.A. Woollam, All-solid-state electrochromic reflectance device for emittance modulation in the far-infrared spectral region. Appl. Phys. Lett. **77**, 930–932 (2000)
43. N. Kislov, H. Gorger, R. Ponnappan, All-solid state electrochromic variable emittance coatings for thermal management in space. AIP Conf. Proc **654** (2003)
44. H. Demiryont, K. Shannon III, R. Ponnappan, Electrochromic devices for satellite thermal control. AIP Conf. Proc **813** (2006)
45. H. Demiryont, D. Moorehead, Electrochromic emissivity modulator for spacecraft thermal management. Sol. Energy Mater. Sol. Cells **93**, 2075–2078 (2009)
46. K. Sauvet, A. Rougier, L. Sauques, Electrochromic WO3 thin films active in the IR region. Sol. Energy Mater. Sol. Cells **92**, 209–215 (2008)
47. K. Sauvet, L. Sauques, A. Rougier, IR electrochromic WO3 thin films: From optimization to devices. Sol. Energy Mater. Sol. Cells **93**, 2045–2049 (2009)
48. K. Sauvet, L. Sauques, A. Rougier, Electrochromic properties of WO3 as a single layer and in a full device: From the visible to the infrared. J. Phys. Chem. Solids **71**, 696–699 (2010)
49. M. Macek, B. Orel, Electrochromism of sol-gel derived niobium oxide. Sol. Energy Mater. Sol. Cells **22**, 67–72 (1998)
50. H. Demiryont, K. Shannon III, A. Williams, Emissivity modulating electro-chromic device. Proc. SPIE **6939** (2008)
51. H. Demiryont, K. Shannon III, J. Sheets, Emissivity modulating electrochromic device. Proc. SPIE **7331** (2009)
52. H. Demiryont, Emissivity-modulating electrochromic device for satellite thermal control. SPIE-Newsroom (2015)
53. P. Topart, P. Hourquebie, Infrared switching electroemissive devices based on highly conducting polymers. Thin Solid Films **352**, 243–248 (1999)
54. Y. Chang, K. Lee, R. Kiebooms, A. Aleshin, A.J. Heeger, Reflectance of conducting poly(3,4-ethylenedioxythiophene). Synth. Met. **105**, 203–206 (1999)
55. C. Louet, S. Cantin, J.P. Dudon, P.-H. Aubert, F. Vidal, C. Chevrot, A comprehensive study of infrared reflectivity of poly(3,4-ethylenedioxythiophene) model layers with different morphologies and conductivities. Sol. Energy Mater. Sol. Cells **143**, 141–151 (2015)
56. K. Lee, A.J. Heeger, Optical reflectance studies of conducting polymers on the metal-insulator boundary. Synth. Met. **84**, 715–718 (1997)
57. S. Wu, C. Jia, X. Fu, X. Weng, J. Zhang, L. Deng, A novel electrochromic and broad infrared emissivity modulation film based on the copolymer of aniline and o-anisidine. Electrochim. Acta **88**, 322–329 (2013)
58. X. Weng, S. Wu, Y. Liu, Z. Wan, C. Jia, J. Xie, Novel electrochromic and infrared emissivity modulation films based on poly (carbazoyltriphenylamine) and poly(carbazoyltriphenylamine-thiophene). Org. Electron. **51**, 190–199 (2017)
59. P. Verge, P.-H. Aubert, F. Vidal, L. Sauques, F. Tran-Van, S. Peralta, D. Teyssié, C. Chevrot, New prospects in the conception of IR electro-tunable devices: The use of conducting semi-interpenetrating polymer network architecture. Chem. Mater. **22**, 4539–4547 (2010)
60. P. Verge, PhD, Réseaux Interpénétrés de Polymères électroactifs à signature infrarouge modulable, Université de Cergy-Pontoise (2008)

61. P. Verge, L. Beouch, P.-H. Aubert, F. Vidal, F. Tran-Van, D. Teyssié, C. Chevrot, Symmetrical electrochromic and electroemissive devices from semi-interpenetrating polymer networks. Adv. Sci. Technol. StafaZuerich Switz. **55**, 18–23 (2008)
62. F. Vidal, C. Plesse, P.-H. Aubert, L. Beouch, F. Tran-Van, G. Palaprat, P. Verge, P. Yammine, J. Citerin, A. Kheddar, L. Sauques, C. Chevrot, D. Teyssié, Poly(3,4-ethylenedioxythiophene)-containing semi-interpenetrating polymer networks: A versatile concept for the design of optical or mechanical electroactive devices. Polym. Int. **59**, 313–320 (2010)
63. L. Goujon, Elaboration d'un dispositif électroémissif flexible à base de réseaux interpénétrés de polymères, Université de Cergy-Pontoise (2011)
64. L. Goujon, A. Khaldi, A. Maziz, C. Plesse, G.T.M. Nguyen, P.-H. Aubert, F. Vidal, C. Chevrot, D. Teyssié, Flexible solid polymer electrolytes based on nitrile butadiene rubber/poly(ethylene oxide) interpenetrating polymer networks containing either LiTFSI or EMITFSI. Macromolecules **44**, 9683–9691 (2011)
65. C. Chevrot, D. Teyssié, P. Verge, L. Goujon, F. Tran-Van, F. Vidal, P.-H. Aubert, S. Peralta, L. Sauques, Electroactive semi-interpenetrating polymer networks architecture with tunable IR reflectivity. Proc. SPIE **7976** (2011)
66. C. Louët, Compréhension des propriétés électro-réflectrices dans l'infrarouge de poly(3,4-éthylènedioxythiophène) électropolymérisé: Des couches modèles aux premiers dispositifs, Université de Cergy-Pontoise (2015)
67. R. Brooke, E. Mitraka, S. Sardar, M. Sandberg, A. Sawatdee, M. Berggren, X. Crispin, M.P. Jonsson, Infrared electrochromic conducting polymer devices. J. Mater. Chem. C **5**, 5824–5830 (2017)
68. S.F. Cogan, R.D. Raugh, J.D. Klein, T.D. Plante, Electrochromic materials II. Electrochem. Soc. Proc (1995)
69. M.G. Hutchins, N.S. Butt, A.J. Topping, J. Gallego, P. Milne, D. Jeffrey, I. Brotherston, Infrared reflectance modulation in tungsten oxide based electrochromic devices. Electrochim. Acta **46**, 1983–1988 (2001)
70. L. Beluze, M. Morcrette, B. Viana, J.-C. Badot, N. Baffier, J.M. Tarascon, Infrared electroactive materials and devices. J. Phys. Chem. Solids **67**, 1330–1333 (2006)
71. A.-L. Larsson, G. Niklasson, L. Stenmark, Thin film coatings with variable emittance. Proc. SPIE **3738**, 486–492 (1999)
72. A. Bessière, C. Marcel, M. Morcrette, J.-M. Tarascon, V. Lucas, B. Viana, N. Baffier, Flexible electrochromic reflectance device based on tungsten oxide for infrared emissivity control. J. Appl. Phys. **91**, 1589–1594 (2002)
73. A. Bessière, L. Beluze, M. Morcrette, V. Lucas, B. Viana, J.C. Badot, Control of powder microstructure for improved infrared reflectance modulation of an electrochromic plastic device. Chem. Mater. **15**, 2577–2583 (2003)
74. C. Marcel, J. Tarascon, An all-plastic WO3.H2O/polyaniline electrochromic device. Solid State Ionics **143**, 89–101 (2001)
75. J.M. Tarascon, A.S. Gozdz, C. Schmutz, F. Shokoohi, P.C. Warren, Performance of Bellcore's plastic rechargeable Li-ion batteries. Solid State Ionics **86–88**, 49–54 (1996)
76. M. Galinski, A. Lewandowski, I. Stepniak, Ionic liquids as electrolytes. Electrochim. Acta **51**, 5567–5580 (2006)
77. M. Armand, F. Endres, D.R. MacFarlane, H. Ohno, B. Scrosati, Ionic-liquid materials for the electrochemical challenges of the future. Nat. Mater. **8**, 621–629 (2009)
78. A.D. Jenkins, P. Kratochvil, R.F.T. Stepto, U.W. Suter, International union of pure glossary of basic terms in polymer. Pure Appl. Chem. **68**, 2287–2311 (1996)
79. P. Verge, F. Vidal, P.-H. Aubert, L. Beouch, F. Tran-Van, F. Goubard, D. Teyssié, C. Chevrot, Thermal ageing of poly(ethylene oxide)/poly(3,4-ethylenedioxythiophene) semi-IPNs. Eur. Polym. J. **44**, 3864–3870 (2008)
80. I. Fabre-Francke, P.-H. Aubert, S. Alfonsi, F. Vidal, L. Sauques, C. Chevrot, Electropolymerization of 3,4-ethylenedioxythiophene within an insulating nitrile butadiene

rubber network: Application to electroreflective surfaces and devices. Sol. Energy Mater. Sol. Cells **99**, 109–115 (2012)
81. N. Festin, A. Maziz, C. Plesse, D. Teyssié, C. Chevrot, F. Vidal, Robust solid polymer electrolyte for conducting IPN actuators. Smart Mater. Struct. **22**(10), 104005 (2013)
82. G. Petroffe, L. Beouch, S. Cantin, C. Chevrot, P.-H. Aubert, J.-P. Dudon, F. Vidal, Thermal regulation of satellites using adaptive polymeric materials. Sol. Energy Mater. Sol. Cells **200**, 110035 (2019)

Sophie Cantin is a Full Professor at CY Cergy Paris Université since 2011. After receiving her Ph.D. degree in Solid-State Physics from Paris 6 University (1995), she was appointed as an Assistant Professor at CY Cergy Paris Université in 1996. She joined the Laboratory of Polymer Physical Chemistry and Interfaces (LPPI) in 2000. She is responsible of a 1-year professional degree in "Health, Security, and Environment" since 2013 and is the head of Sciences and Engineering doctoral school at CY Cergy Paris Université since October 2018. Her research currently deals with interfacial phenomena in polymer materials and ultrathin films.

Pierre-Henri Aubert is a Faculty Member of the Chemistry Department at CY Cergy Paris Université. He received his Ph.D. degree in Physicochemistry of Materials in 1999. He has been appointed Associate Professor (2005) and then full Professor (2017) in CY Cergy Paris Université to develop fundamental research programs designing and utilizing electroactive and conducting polymers or nanomaterials for either energy, electrochromics, or biosensing systems. In addition, he is co-responsible for 1st/2nd year of a Biotechnology and Chemistry Engineering cursus at CY Tech and co-supervises the Sensor Energy International Associated Laboratory (SENERGY Lab).

Jean-Paul Dudon received his Ph.D. in Energetics and Mechanics from Provence University in 1997, after a Research Masters in Plasma and Radiations. He then joined Alcatel Space, which became Thales Alenia Space (TAS), as a Research and Technology Engineer in satellite platform development. At TAS, he has specialized in the development of new technologies and methodologies for the thermal control of spacecraft. Among other topics, his research includes development of smart materials such as variable emissivity coatings or self-regulating heaters, based on both organic and inorganic materials. In particular, he has been leading since 2008 R&D activities aiming to increase the maturity of thermochromic and electrochromic devices for space applications.

Frédéric Vidal received his Ph.D. in Polymer Chemistry from Lyon University in 1995. He then joined Eindhoven Technology University as a Postdoctoral Fellow. He moved to CY Cergy Paris Université as an Associate Professor in 1997. Currently, he is Full Professor, a position he has held since 2008. From 2009 to 2018, he was appointed Director of the Laboratory of Physicochemistry of Polymers and Interfaces. He has been appointed Vice-President for Research of CY Cergy Paris Université since May 2018. His research includes development of interpenetrating polymer networks (IPN), conducting polymers, and conducting IPNs based electrochemical devices such as electrochromic devices or artificial muscles.

For more information, please take a look at the web sites of CY Cergy Paris Université (https://www.cyu.fr/) and at the Laboratoire de Physicochimie des Polymères et des Interfaces (https://lppi.cyu.fr/).

Index

A
Accelerate restoration, 53
Active means, 222
Active thermal control, 222, 229
Actuation pressure, 108, 110, 113, 114, 116–118, 120, 122
Actuator configurations, 64, 65
Aerospace, 86, 95
Alkaline earth, 230
Aluminum grids, 241
Amplified muscle restoration, 53
Artificial intelligence (AI), 41, 42
Artificial muscle, 84
Artificial satellites, 228, 229
Automatic sensors, 59
Automation, 9

B
Bending actuator control, 132
Biomimetic design, 10
Biomimetic Venus flytrap, 4
Black-body radiation, 224
Bladder expansion, 104
Blocked force, 106, 108, 110, 114, 115, 117, 118, 120
Bluecell Velcro® strap, 176
Bone density, 56

C
Cable-driven continuum manipulators, 134, 135
CAD model, 143, 144, 171
Cobots, 10, 40, 41
Cold environment, 222, 228, 230
Collaborative robots, 10, 40
Communication skills, 57
Compact Robotic Flow Control (CRFC) Valve, 84–87, 89–92, 94, 95
Compact Synthetic Muscle™-based actuators, 42
Computational science skills, 57
Constant acceleration discretization, 159
Constant curvature static bending model, 190–192
Continuum manipulators, 134
Contractile PAMs, 104–109, 111, 113–122
Corrected actuation pressure, 113
Correction factors, 199
Cosserat rod theory, 134

D
Degree-of-freedom (DoF)
- bending actuator, 156
- 4-DoF manipulator, 170–172
- 6-DoF space, 170
- kinematic model, 2-DoF module, 167

1-Degree-of-freedom (DoF), 132, 136, 144–147, 150
Dielectric elastomeric (DEAs), 8
Dielectric elastomers
- actuators, 63, 64, 74
- advantages, 65
- candidate approaches, 66
- commercially available silicone rubbers, 68
- deflection, 75
- and EAP, 63, 65
- electrostatic actuation, 64
- good-quality mirrored surfaces, 70
- heuristic analysis, 70

© Springer Nature Switzerland AG 2022
L. Rasmussen (ed.), *Smart Materials*,
https://doi.org/10.1007/978-3-030-70514-5

Dielectric elastomers (*cont.*)
high-resolution lightweight optical systems, 78
inflatable antenna experiment, 73
large-strain capability, 67
1-mm square flat actuator, 71
principle of operation, 64
resolution, 75
response times, 79
silicone and acrylic, 77
with lightweight space mirrors, 65–67
Direct inverse kinematics (IK)-based controller, 170
Discretized model verification
IK verification, 199–201, 203
model force verification, 197, 198
pressure verification, 198–200
scale factor analysis
constant offset, 209, 210
cross section analysis, 204–206
friction analysis, 206–209
frictionless analysis, 210–212
internal variation, 208, 209, 211
Discretized soft actuator bending model
global forces, 195, 196
inverse kinematics, 197
non-constant moments, 193
normal force calculation, 194
tangential forces calculation, 194, 195
Diversity, 58
DoF actuator application, 173
experiments, 183–186, 188, 189
fabrication, 174–177
feedback control, 181, 182
haptic wrist device worn, 175
path-following test, 177–179
physical capabilities, 177, 178
rPAM, 174
valve control, 183
virtual agent control, 178–181
5-DoF haptic arm exoskeleton, 133
1-DoF revolute kinematic model, 145
DoF soft-actuated modules
design and fabrication, 165
fabrication process, 165
module design, 165–167, 169
step-by-step fabrication process, 165
2-DoF module
experiments, 172, 173
kinematic model and control, 166–170, 172, 173
Dragon Skin 10 silicone rubber, 140

E
Eclipse Variable Emissivity Device (EclipseVED™), 237
Elastic modulus, 112
Electric motor techniques, 133
Electroactive polymers (EAPs)
aluminum screening, 14
contraction, 11
cylinders, 10
DEAs, 8
EDS comparison, 30
to electricity, 4
electric polarity, 11
electroactivity, 10
FingerTip™ soft sensors, 32
formulations, 14, 25
gamma radiation testing, 12
general and dielectric elastomers, 63
high-speed photography, 7
large 9-g EAP actuator, 5
light microscopy and SEM, 25
mechanical compression testing, 38
microclimate environments, 27, 29
pad systems, 36, 39
pads, 37
properties, Synthetic Muscle™ EAPs, 6
proprioception, 38
radiation testing, 12
Ras Labs, 4, 9
reactive actuation, 6
robotics era, 42
sensor, 31
sensor designs, 41
shape-morphing control, 39
smart materials, 4
space experiment, 16
Synthetic Muscle™, 5, 25, 31
2-gram EAP actuator system, 9
as variable resistors, 6
Electrochemically active species (EAS), 233–238, 248
Electrodes, 64
Electroemissive devices (EEDs)
active layers, 233–237
common design, 224
control of thermal performances, 246–248
at CY Cergy Paris Université, 244, 245
EAS, 233
ECP, 242
electrochemical cells, 223
emissivity, 223
IR-optical properties, 231
memory effect, 232

opaque materials with variable emissivity, 238–241
parameters for EED performances, 242
performance indicators, 232
potential difference, 231
principle, 231
redox process, 233
reflectivity variation, 232
for space application, 242, 243
Electromagnetic energy, 224
Electromechanical actuators, 103
Electronically conducting polymer (ECP), 233–235, 242–244, 248
Electrostatic model, 63
Electrostrictive polymers, 65
Emissivity, 223, 226–232, 234, 236–238, 246
Emitted radiation, 224
End effectors, 30, 32
Energy balance, 226
External sensors, 59

F
Feedback loop, 31, 42
Feedforward controller design, 149, 150
Fiber-reinforced elastomeric enclosure (FREE), 130, 131
Finite element analysis (FEA), 140, 143
Flexible magnetic curvature sensor, 155
Flow control valves, 84
Fluidic elastomer actuators (FEAs), 132
Force balance model, 115, 121–123
free-body diagram, 110
nonlinear elastic model, 113, 115, 117
non-optimized linear model, 115
push-PAM, 115
strains, 111
Free contraction, 106, 108, 115, 117, 118
Friction factor, 113
Friction force, 112, 113, 115, 120

G
Galvanic skin resistance (GRS), 59
Gaylord force, 109–112, 119, 120
Gravity compensation, 201, 203, 206, 208, 210–212

H
Hall effect sensor, 155
Haptic feedback, 133, 134, 178, 185–188, 213
Hot environment, 222, 228–230
Hubble space telescope, 62
Hubots, 10, 40, 41
Humanoids, 10, 40, 41
Hydraulic system, 91
Hydro Muscles
CFM/CMM Thermo-Anemometer, 89
characteristics, 84
CRFC Valve, 89, 91
efficiency, 84
elastic elements, 83
hip angle *vs.* time, 94
liquid-actuated Hydro Muscle, 85
McKibben artificial muscle, 84
McKibben elastomer, 84
McKibben muscle, 85
servo angle *vs.* average elongation rate, 91
and soft robotics, 90
strain, 92
strain *vs.* time, 93
valve's servo motor, 92
Hysteresis, 107, 108, 112, 113, 115, 120, 123

I
Inertial measurement units (IMUs), 177, 179, 181
Innovative biomimetic motion, 5
Intelligent materials, 4
Interpenetrating polymer network (IPN)-based matrixes, 245
Inverse dynamic controller, 158, 160, 161
Inverse kinematics (IK), 197
vs. Jacobian control, 172–174
verification, 199–204
IR-absorbing, 227, 233–235, 242, 244, 248
IR-electrochromic, 223
IR-electroemissive materials, 223
IR reflectivity, 239, 243
IR-thermochromic materials, 223, 230

J
6-m James Webb Space Telescope, 74
Joint control, 150–153

K
Kinesthetic haptic feedback, 133, 173–174

L
Large-aperture optics, 62
Lightweight space mirrors, 65–67

Linear elastic force balance model, 121
Liquid-actuated Hydro Muscle, 85
Luminance, 224, 225
Lunar, 56, 57

M
Machine learning (ML), 41, 42
Magnetostrictive materials, 103
Manganese perovskites, 230
Mars, 51, 54–56, 58
McKibben actuators, 130, 131
McKibben artificial muscle, 84, 85
Memory effect, 232
Metal oxides, 236
Mirrors
  actuator configurations, 69
  configurations, 66
  earth-facing, 67
  inflatable membrane, 73, 74
  laminated membrane, 69, 70, 72
  laminated thin mirror, 76–78
  lightweight space, 65–67
  membrane mirrors, 65
  positioning, rigid mirror, 74, 75
  3D mirror shapes, 70
Model reference adaptive controller (MRAC), 132, 157–159, 161–164
Moon, 50, 51, 56
Mooney-Rivlin model, 131

N
NASA, 58
Next-Generation Space Telescope (NGST), 66
No Gravity Compensation error, 204
Non-cylindrical tip shapes, 110, 121, 123
Nonlinear elastic force balance model, 116–123
Non-optimized linear force balance model, 115

O
Off-the-shelf sprue gripper, 36
Ogden hyperelastic solid model, 139
Optical solar reflector (OSR), 246, 247
Optical systems, 62
Optimized nonlinear elastic model, 123
OSR radiator, 247, 248

P
Passive coatings, 222
Passive means, 222
Passive thermal control, 229
Piezoelectric, 103
Piezoelectric PVDF materials, 65
Planar bending actuator, 206
Pneumatic artificial muscles (PAMs), 84
  actuation profiles, 108
  contractile, 104
  evolution of force, 107
  experimental characterization, 106
  feasibility, 109
  force balance (*see* Force balance model)
  Gaylord force, 109
  length correction term, 121
  manufacturing techniques, 105
  operating principle, 103
  physical actuators, 106
  pressurization, 105
  push-PAM, 104
  resting length, 107
  resulting motion, 104
  simple mechanical actuators, 103
Pneumatics, 8
PneuNet, 134
Poly(3,4-ethylenedioxythiophene) (PEDOT), 233, 235, 239, 240, 244–246, 249
Pressure-averaged blocked force, 113, 120
Pressure dead-band, 113, 114, 123
Pressure scale calibration, 199
Pressurization, 105, 110, 113, 123
Primary optics, 67
Proof-of-concept leg model, 91, 92
Proprioception, 38
Pulse width modulation (PWM) signals, 132, 183
Push-PAM, 104
  actuators, 106
  air inlet, 105
  with comparable accuracy, 120
  and contractile PAMs, 108, 119, 122, 123
  design goal, 105
  experimental characterization, 107, 123
  fabrication, 108
  friction factor, 113
  hysteresis, 108
  quasi-static testing, 104
  traditional contractile PAMs, 105

R
Radiation, 52, 60
Real bodies, 226
Redox process, 231, 233, 234
Reflectivity variation, 232
Rehabilitate, 53, 54
Reverse pneumatic artificial muscle (rPAM)

actuator deformation results, 142, 143
actuator fabrication process, 136
actuator static deformation response experimental setup, 141, 142
analytical model, 136, 137, 139, 140
antagonism, 153
design and fabrication, 153
fabrication process, 137
fiber-reinforced soft actuators, 131
geometric parameters, 138
linear extension, 136
linear rPAMs, 213
modeling and control, 135
numerical model, 140, 141
pneumatic actuators, 134
proposed joint, 154
soft actuators, 135, 138
soft linear actuator, 135
traditional PAM, 130
RMS errors (RMSE), 184, 185
Robotics, 40, 41, 51
CRFC Valve (*see* Compact Robotic Flow Control (CRFC) Valve)
grippers, 30, 31, 36, 41
insects, 4, 5
revolution, 31
sensing, 31
skills, 57
systems, 50
trade, 36
Robots, 9, 56, 59
rPAM-driven revolute joint
analytical model, 144–146
stroke length, 144

S

Sake Robotics® EOAT gripper, 33–34
Satellite
artificial, 228
geostationary satellite, 229
hot phase, 223
thermal control, 223
thermal exchanges, 229
thermal regulation, 233, 242, 245, 248, 249
visible solar absorptivity, 230
Sensors, 6
dome-shaped FingerTip-like sensor, 35
dome-shaped soft sensor, 35
FingerTip™, 30, 35
Synthetic Muscle™-based EAP pads, 31
Synthetic Muscle™-based EAPs, 32
Sliding mode controller design, 147–149, 151
Smart materials, 4, 8, 42, 55
Smart windows, 227
Society, 57
Soft actuator bending control, 154, 155
actuator design and experimental setup, 155, 156
experiments, 159
inverse dynamic controller, 158, 160, 161, 163, 164
MRAC, 157–161
system identification, 156, 157, 160
unstructured signal tracking, 163
Soft actuators, 130
control strategy, 146
rPAM (*see* Reverse pneumatic artificial muscle (rPAM))
Soft bending actuator static modeling
constant curvature bending, 190–192
constant curvature verification, 192
discretized bending model (*see* Discretized soft actuator bending model)
1-DoF bending module, 190
force experiment data, 193
Soft linear actuators, 131, 135, 143, 212
Soft pneumatic actuators, 130–131
Soft robotics, 6, 9
grippers, 9
model reference adaptive control, 135
models, 188
PneuNet, 134
robotic research, 132
rPAM, 212
snake-like, 131
system, 132
Solar absorptivity, 229, 230, 243, 245, 248
Solar arrays, 67
Solenoids, 103
Space applications, 85
Space-based sensing systems, 62
Spacesuit, 55, 59
Spectral luminance, 224, 225
Stefan-Boltzmann's law, 225
Summation upper limit, 111, 115
Survivability, 65, 80
Sympathetic nervous system, 60
Synthetic Muscle™-based actuation, 10
Synthetic Muscle™-based EAPs, 25, 29, 31, 36

T

Tactile haptic feedback, 133
Tangential forces, 194, 195
Technology-Engineering-Art-Math-Science (TEAMS), 57
Telescope, 62, 65, 74
Therapy, 53, 54

Thermal control, 222, 229, 230
Thermal energy, 228, 229
Thermal exchanges, 229
Thermal radiation, 223
Thermal regulation, 230, 232, 233, 240
  application, 228
  design EEDs, 242
  EclipseVED™, 237
  energy saving, 227
  extreme cases, 228
  "hot" case, 228
  IR emissivity control, 227
  satellite, 242, 245, 248, 249
Thermistors, 246
Three-dimensional (3D) printed mold, 136
Traditional actuation technologies, 62
Traditional gripper technology, 10
Traditional PAM, 130

V
Vacuum, 131

W
Wearable robotics, 84
Wrist device, 133–134

Z
Zero-Gravity aircraft, 54
Zygo interferometer, 76